MAPPING HUMAN AND NATURAL SYSTEMS

MAPPING HUMAN AND NATURAL SYSTEMS

PETE BETTINGER
Warnell School of Forestry and Natural Resources, University of Georgia, Athens, GA, United States

KRISTA MERRY
Warnell School of Forestry and Natural Resources, University of Georgia, Athens, GA, United States

KEVIN BOSTON
Department of Forestry and Wildland Resources, Humboldt State University, Arcata, CA, United States

ELSEVIER

ACADEMIC PRESS
An imprint of Elsevier

Academic Press is an imprint of Elsevier
125 London Wall, London EC2Y 5AS, United Kingdom
525 B Street, Suite 1650, San Diego, CA 92101, United States
50 Hampshire Street, 5th Floor, Cambridge, MA 02139, United States
The Boulevard, Langford Lane, Kidlington, Oxford OX5 1GB, United Kingdom

Notices

Knowledge and best practice in this field are constantly changing. As new research and experience broaden our understanding, changes in research methods, professional practices, or medical treatment may become necessary.

Practitioners and researchers must always rely on their own experience and knowledge in evaluating and using any information, methods, compounds, or experiments described herein. In using such information or methods they should be mindful of their own safety and the safety of others, including parties for whom they have a professional responsibility.

To the fullest extent of the law, neither the Publisher nor the authors, contributors, or editors, assume any liability for any injury and/or damage to persons or property as a matter of products liability, negligence or otherwise, or from any use or operation of any methods, products, instructions, or ideas contained in the material herein.

British Library Cataloguing-in-Publication Data
A catalogue record for this book is available from the British Library

Library of Congress Cataloging-in-Publication Data
A catalog record for this book is available from the Library of Congress

ISBN: 978-0-12-819229-0

For Information on all Academic Press publications
visit our website at https://www.elsevier.com/books-and-journals

Publisher: Joe Hayton
Acquisition Editor: Candice Janco
Editorial Project Manager: Sara Valentino
Production Project Manager: Surya Narayanan Jayachandran
Cover Designer: Christian Bilbow

Typeset by MPS Limited, Chennai, India

Working together
to grow libraries in
developing countries

www.elsevier.com • www.bookaid.org

Dedication

To our parents who always supported and encouraged us.

Contents

Contents

8. Map Errors

9. Maps in Popular Culture

Preface

Maps are a creative outcome of questions posed about land, air, water, and even outer space. Maps help answer questions such as *Where is the place of interest?* and *Where are we going?*, among many other curiosities we may have. Maps help us find our way, help support our decisions, and help illustrate the condition of human and natural systems. Arguably, popular interest in maps has never been greater than it is today (Hamerlinck, 2015). Maps can be found nearly everywhere, and they are widely available in paper and digital formats (Ooms et al., 2016). Although the technology for making maps has changed considerably over the past few decades, the purpose and principles regarding map components and the interpretation of maps have essentially remained the same.

Our fascination with maps and mapping began some time ago. With the exception of crude maps drawn in his youthful adventures, Pete Bettinger drew his first real map in 1985. The map had a purpose: to describe the forest resources within a *compartment* of land owned by a forestry company in Virginia. Compartments were a term used by his employer to represent the division of a district (the larger area managed) into relatively small contiguous areas, containing management units having different forest conditions (as described by tree age, tree species, etc.). The map was developed just prior to the widespread adoption of computerized mapping and drafting systems by forestry organizations. Drawn by hand to a specific scale, the map needed to be crafted in a careful, professional manner, as it would be used by other professionals for at least another 5 or 10 years. For him, the concepts of map development and interpretation were fostered by both academic coursework and these types of practical experiences.

Krista Merry's grandfather created his own topographic maps from stereo pairs of aerial photographs captured across the southwestern United States, a mapping skill he developed while he was in the Air Force. He noted on hand-drawn maps the latitude and longitude of the landscapes and the locations he photographed over the course of his professional photography career. His love of maps sparked Krista's early interest in geography and cartography, and his hobby instilled in her an appreciation of the effort required to make good maps. She frequently drew maps as a child, including mental maps of locations across her neighborhood, the location of her house in relation to the location of daily activities (school, neighborhood pool, friend's houses), and the perceived shortest route from one place to another. These maps included arrows identifying steep hills, landmarks, street names, cul-de-sacs, and general footprints of houses. Largely, scale was of no importance, with trees commonly bigger than houses, but attention was often paid to the accuracy of the spatial distribution of landscape features.

Kevin Boston's first map illustrated the location of a proposed timber sale on the Mammoth Ranger District on the Inyo National Forest in California. The purpose of the map was not only to indicate the location of the potential harvest units, but also the management prescriptions that might be employed. The map would become part of a contract and was drawn using a specific scale (1:15840). He drew this map on Mylar® film, using drafting pens and basic drafting skills he developed in his forestry surveying classes. The features on the map included the existing roads and boundaries of the potential harvest units. Kevin used a series of pen tip sizes to draw features on the map; large width tips were used for property boundaries, and narrow width tips were used for features such as roads or streams. Letter guides were also used to create parts of the title, north arrow, legend, and scale bar. He freely admits that developing maps such as these was the most enjoyable part of his job, as it allowed him to express his creativity.

Certainly, mapping technology has evolved, but perhaps not so much the purpose of maps. Our fascination with maps continues today, and in developing this book on mapping, we make an earnest attempt to present material in an interesting manner that may engage readers in the art, science, mathematics, and skill of mapping. Some concepts in our book focus on specific map components or principles of landscape description; however, we extend the conversation in several areas toward the broader field of geospatial analyses, or the *science of where*. Not only do we hope to build in our readers core mapping competencies, but we also hope to prompt readers to reflect upon and to synthesize mapping concepts, which may encourage personal growth through various explorations of mapping topics.

We have placed the vast array of topics associated with mapping into nine distinct chapters, in a manner we feel allows general thoughts regarding maps and mapping processes to flow logically. Intermixed within the core content of each chapter are *reflections*, *diversions*, *inspections*, and *translations* to encourage curiosity and to help develop creativity and critical thinking skills. The *reflections* serve as pauses from the topics at hand, and they encourage readers to think deeper about mapping ideas or concepts. Usually readers will be asked to formulate their personal perspectives and thoughts on a subject into a cohesive, short summary. These exercises are often not directly related to the development or viewing of a specific map. For example, when discussing map color schemes, we may ask readers to think about their preferred color combinations for illustrating differences in human or natural systems. Readers may then be encouraged to describe why they believe these colors would best communicate a message to an audience.

Reflection 1
Without using your computer, cellular phone, or other digital device, think about the last digital map that you used or viewed. In the background of the map, do you recall whether an aerial image, a street map, the topography and bathymetry, or a more basic landscape canvas was present? Perhaps some combination of these might have served as the background of the map. Of these options, which do you generally prefer to be contained in the background of a digital map?

The *diversions* ask readers to put the book aside and solve a problem. These may be as simple as short mathematical analyses, or may be as complex as the development of a map. For example, using a preferred geographic information system, readers may be asked to develop a map that focuses on the topic within which the diversion was introduced. In some of the diversions, readers will be encouraged to obtain and organize the necessary data, and ultimately develop a map to communicate a certain message.

Diversion 1
Using Google Earth, navigate to your hometown and focus on (zoom down to) a familiar area until you are positioned at an eye altitude of around 1000 m (3281 ft) above ground. Examine the landscape and map features that are visible, given the databases that have been selected in the *Layers* window. Now, begin the process of printing the map of this area. Aside from what was present across the landscape in the Google Earth window prior to printing the map, what components have been added to the map to enhance its usability? Finally, print the map.

The *inspections* encourage readers to analyze a map and determine the relative quality of that map with respect to the chapter topics within which each inspection is introduced. These activities may seem to overlap slightly with the diversions, yet the main difference is that the diversions encourage engagement in some part of the map development process or in some mathematical computation, whereas the inspections encourage engagement in the viewing and interpretation of a completed map. For example, with respect to discussions concerning map legends, we may direct readers to one or more specific maps available via the Internet. Readers will then inspect the legends of these maps along with the treatment of legend items within the maps and briefly summarize their thoughts on their technical quality. The inspections may also prompt readers to compare concepts presented in the book with real-world applications.

Inspection 1
Consider the Stonewall National Monument map (Fig. 1). Where exactly is the monument located (city and state)? Further, as you may see, the National Monument boundary is rather unique. How would you generally describe it, in a few words?

The *translations* provided throughout the book refer to the potential messages that the map developer may have been attempting to communicate through the use of words, symbols, or colors. While on the surface, the purpose of a map can be very distinct (a map of the road system of New York City), other underlying messages may be more subjectively argued (e.g., some roads are more important than others). For example, the developer of a map may highlight the outline of an area of perceived importance with the color red. At least two questions come to mind: why is the area highlighted of importance to the developer of the map, and why would they use the color red to indicate this fact? Other textual and symbolic content included in a map may further complement its overall message. The translation exercises may therefore help develop or expand the cognitive skills of readers.

Translation 1
Imagine in the far lower-left corner of a map, in very small print, resides the following statement:

This map has been produced using geospatial information compiled or developed by ___, which cannot warrant its reliability, accuracy, or suitability for any particular purpose. The data employed was compiled from a number of different sources, and may have been updated, adjusted, or otherwise modified. Any harm arising from the use of the map is solely the responsibility of the map user.

In a few simple words, what message is the map developer attempting to convey with this statement?

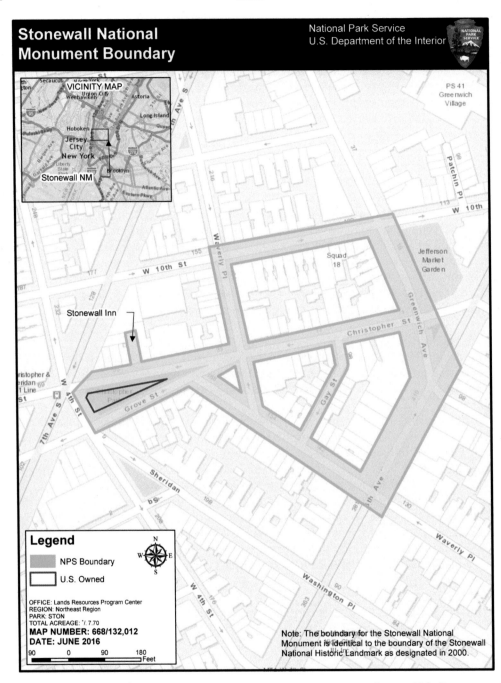

FIGURE 1   A map of the boundary of the Stonewall National Monument. Source: *U.S. Department of the Interior, National Park Service (2016)*.

A number of mapping and map interpretation concepts are important when developing the knowledge and skills one may need for lifelong appreciation of maps (Skryzhevska et al., 2013). Through discussions of these, we hope to engage readers in the process of *spatial thinking*, or the employment of cognitive skills when viewing a map. Spatial thinking helps one understand the location of the landscape, the units of measurement, the reasons why certain features are located where they reside, and the effects of map generalization and projection on realistic portrayal of an area (National Research Council of the National Academies, 2006). Further, we hope to improve readers' ability to communicate and receive messages through maps, and to ensure that they are better positioned to understand the purpose of maps and mapping processes.

References

Hamerlinck, J.D., 2015. Whither goes the "maps" course? Maintaining map-use concepts, skills, and appreciation in GIS&T curricula. Cartograph. Geograph. Inform. Sci. 42 (S1), S11–S17.

National Research Council of the National Academies, 2006. Learning to Think Spatially. The National Academies Press, Washington, D.C.

Ooms, K., De Maeyer, P., Dupont, L., Van Der Veken, N., Van de Weghe, N., Verplaetse, S., 2016. Education in cartography: what is the status of young people's map-reading skills? Cartograph. Geograph. Inform. Sci. 43 (2), 134–153.

Skryzhevska, L., Green, J., Abbitt, R., 2013. GIS textbook content as a basis for skill development in map interpretation. Cartographica 48 (1), 38–46.

U.S. Department of the Interior, National Park Service, 2016. Stonewall National Monument New York. U.S. Department of the Interior, National Park Service, Washington, D.C. <https://www.nps.gov/ston/planyour-visit/maps.htm> (accessed 14.01.19.).

Maps

A *map* is a small-scale representation of a much larger, usually real, system. Humans understand and make themselves comfortable by mapping the world (Pérez, 2013). A map is a form of cultural text, an inherently anthropogenic representation of a place or location, whether it includes land, water, or sky, or some combination of these (Harley, 1989). A map is drawn or rendered on flat media such as the screens of digital devices, paper, glass, plastic, or other types of surfaces. Ideally, it describes the position and condition of certain physical features such as roads and streams and others that seem to be of importance to humans. Through visual objects (symbols), colors, and messaging, maps present a story of an environment (Bestgen et al., 2017) at a scale suitable for the intended audience. We engage with maps as both a developer of them and a user of them (Skryzhevska et al., 2013). Maps often reference places on Earth from a bird's-eye or aerial perspective above a landscape or water body. Even maps of Earth's moon, Mars, or other astronomical entities may utilize this form of vertical perspective (Shevchenko et al., 2016). Maps of the solar system and other faraway constellations may use an alternative perspective that suggests the map user is standing on the Earth's surface, looking outward into space. For future reference, the map perspective we use exclusively in this book is the aerial perspective.

An essential purpose of a map is to communicate with geographic context information deemed by the map developer for the map user (Skupin and Skupin, 2009). While maps have been described simply as utilitarian devices created to convey messages (Robinson, 1974), they can be much more than that and creatively illustrate what we think we know

1

about landscapes or water bodies. Maps are often designed to initiate cognitive processes in map users that foster further understanding of the landscapes and water bodies that have captured our interest (von Ungern-Sternberg, 2009). Therefore, perhaps we should begin our discussion of maps and mapping by describing *what is not a map*. Of the various images or cultural texts we are exposed to on the screens of digital devices, on paper, or on other surfaces (the ground, Mylar®, a piece of lumber, the back of an envelope), maps typically do not include straightforward text, such as this paragraph, nor videos, histograms or pie charts, pictures, television or cinematic shows, or Internet sites. These communicative devices, though they may reference places or directions, are not technically maps because the messages they convey are not presented with an explicit geographic context. However, maps are often used in conjunction with these communicative devices to help deliver messages that have a distinct spatial or geographic context (Lee, 1995).

What would motivate someone to make a map? One reason may be that the process of making maps is a central task associated with a person's mode of employment. Many professionals employed in natural resource management fields (forestry, wildlife management, recreation management, hydrology, etc.) or land development fields (realtor, developer), for example, need to make maps that illustrate the condition of land and water resources. Another reason may be that a person has a keen interest in a specific topic, and the development of a map seems necessary to understand the spatial context of phenomena. Maps that are developed after political elections, to illustrate geographic differences in the preferences of society for politicians or political groups, might be an example of this motivation. Maps that complement research projects might be another example. A further reason may be more basic: a person wants to illustrate clearly what land is theirs, and how they plan to manage or develop this land. Real estate plats and maps developed in association with forest management plans are great examples of these, particularly maps developed for (or by) private individuals that illustrate the location of future management activities. A final example of the motivation to develop a map may involve a creative release of energy. Perhaps some people may view the map making process as a manner of creative expression. From an aesthetic perspective, some maps can also be considered works of art. As they integrate past, present, and future perspectives of a landscape or water body (Lechthaler, 2009; Metcalf, 2012), the artistry of the cartographer or map developer is exhibited through the use of the colors and symbols employed.

Reflection 2
You can draw a map by hand on a piece of paper, you can develop a map with geographic information system (GIS) or with computer-assisted drawing (CAD) software, and you can even develop a map as you engage with an online mapping program. The purpose can range from the need for very informal directional assistance to the need for a very formal illustration of your knowledge of a landscape. So, thinking ahead, what do you think would be the main motivation for you in making your next map?

The purpose of a map is often born from the ideas of *I want to show you something*, or *I want you to understand something*. The ideas and concepts that a map developer wants us to understand usually involve resources managed (or affected) by humans or that interest society. But, what would motivate someone to use a map? One reason may be that the use of a map is a necessary activity in the conduct of a job. For instance, a map may be needed in contracts that involve the purchase or sale of land, or in contracts associated with land development practices. People would then use the maps as guides for locating the places or resources involved in the associated contractual obligations. Another reason may be that the use of a map can assist us in acquiring pleasure through our personal activities. As an example, a national park hiking map will indicate the location of the trails people can hike, allowing a person to choose among the options and to feel confident about their progress along the route. Other reasons people may use a map may involve the search for a hidden treasure or the need to quell the fear of becoming lost in a big, unfamiliar city.

Throughout recent human history, maps have been developed to describe economies, societies, politics, ethics, and morals (Fiel, 2009). In his text *Complete Geography*, Frye (1895) presented maps describing various aspects of US commerce during the late 19th century. These maps illustrated the locations of major routes of transportation (railroads) and the approximate locations of areas where wheat, corn, oats, tobacco, timber, copper, and other commodities might be grown or supported. Maps may therefore represent a fundamental manner for communicating ideas through spatial cognition, employing symbols and colors to tell a story (Rambaldi, 2005). It may surprise some to know that maps are also often used as objects of persuasion. Maps may be developed to influence society's understanding of important issues or to convince society that the perspective of the map developer is correct (Wood, 1983). Maps associated with television commercials for cellular phone services are examples of these. These maps indicate the spatial coverage of service that a cellular phone company provides. They are meant to suggest that the services provided are quite broad, and that among the alternatives, the particular cellular phone company providing the commercial is the most suitable choice.

People can have different reactions to the design of maps based on individual tastes, educational backgrounds, and cultural sensitivities. For example, some people might find certain maps from the early modern period (about 1500–1800) to be quite beautiful, while others may conclude that those same maps are inherently distasteful, ugly, or busy due to the presence of extraneous information (Fig. 1.1). Many people expect maps to be accurate in describing features above, below, or on the surface of land and water. In fact, a number of descriptors can be associated with maps, such as *useful, familiar, boring,* or *surprising* (Oksanen et al., 2014). Further, some map users might invest a significant amount of trust in a map, while others may be skeptical of the information provided.

Inspection 2
Given the map provided in Fig. 1.2, illustrating the summer and winter ranges of mule deer (*Odocoileus hemionus*) on the Lassen National Forest in California, what five adjectives (descriptors denoting the quality of the map) immediately come to mind that describe your feelings about it? For instance, you might try to answer these questions: *the map is too ___* (the adjective), or *the map seems ___*. Once the five adjectives are noted, try to explain why you reacted to the map in this manner.

One purpose of a map is to provide the map user with orientation and understanding, perhaps navigational or cultural. Along these lines, fundamental questions that come to

FIGURE 1.1   Map originating from the Early Modern Period in Europe (1562) illustrating North and South America and the adjacent seas. Source: *Gutiérrez et al. (1562).*

mind include: *Where are things located? Why are things located there?*, and *What are the connections between places?* (Sack, 1972). Perhaps most often the question answered using a map is *Where am I?* But of course, maps can be of interest to people who are not actually situated within the landscape represented by the map. For example, a person may be interested in understanding *How far is it from point X to point Y?* before actually traveling to that place. If one were contemplating travel to Rome, for example, they may be interested in

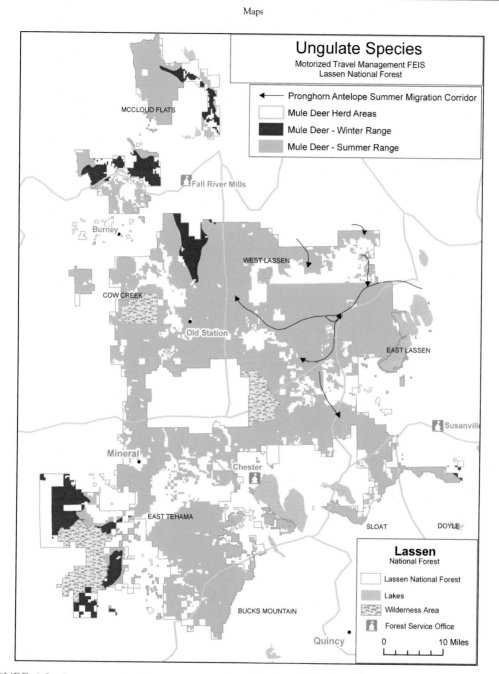

FIGURE 1.2 Summer and winter range areas for mule deer (*Odocoileus hemionus*) on the Lassen National Forest, California. *Source: U.S. Department of Agriculture, Forest Service (2009).*

understanding the walking distance from their preferred hotel to the entrance of the
Colosseum, the Vatican museums, or the Trevi fountain.

Maps, while created in present times, can have the purpose of describing events or conditions of the past (Heffernan, 2014), or desired conditions of the future. With Internet-
based technologies available today and the ability of organizations to create near real-time
maps, the length of time between the present and the past can be as short as a few minutes. Digital weather maps are great examples of near real-time maps, as they may show
us the places where it has rained recently. Digital fire maps are another example, as they
may show us the progress of the fire line, the leading outer edge of a fire as it progresses.
Maps can act to stabilize people's sense of identity, to demonstrate economic and political
power (Caquard and Wright, 2009; Petto, 2016), or to assist in visualizing both reality and
fantasy (Lechthaler, 2009). Through maps, one can illustrate political power through political values and territorial assertions (Harley, 1989). In the former case, a map may illustrate
the positive values held by one political group and the negative values held by another
political group simply through the cunning use of color or symbols. In the latter case, a
good example might be to consider whether a disputed area should be illustrated on
a map in a way that suggests it may be an integral part of a nearby larger country, or in a
way that suggests it is indeed independent of the larger country. These issues are not to
be taken lightly, as in some cases the consequences can be severe. In India, for example,
Section 2 of The Criminal Law Amendment Act, 1961 (Act No. 23 of 1961) notes:

> Whoever by words either spoken or written, or by signs, or visible representation or otherwise, ques-
> tions the territorial integrity or frontiers of India in a manner which is, or is likely to be prejudicial to the
> interests of the safety or security of India, shall be punishable with imprisonment for a term which may
> extended to three years, or with fine, or with both. (Government of India, 2015)

Aside from these serious matters, map users should be able to easily interpret a map;
this is as essential as the accuracy of the content of a map (Imhof, 1974). While all maps
are imprecise due to measurement error, drafting or digitizing errors, omissions, or falsifi-
cation of information, it is through the simplification of a real-world system that maps can
convey a measure of knowledge covering space and time that people may find informative
(Plath, 2009).

Translation 2

We make judgments every day about people, places, and events. Our interpretation
of the importance of these can in part be related to our up-bringing, education, life
experiences. With some thoughtful consideration, a person can interpret a message
from a map that does not explicitly and clearly offer one. If you were to make one
statement about the message that might be conveyed in the map provided in Fig. 1.3,
what would that be?

We emphasize the sound development of maps in several chapters of this book. To do
this involves the use of *cartography*, a term that arises from the Greek words *chartès* and
graphein (to describe), forming the concept we now widely use to generally describe the
art and process of making a map (van der Krogt, 2015). Practicing cartographers may even

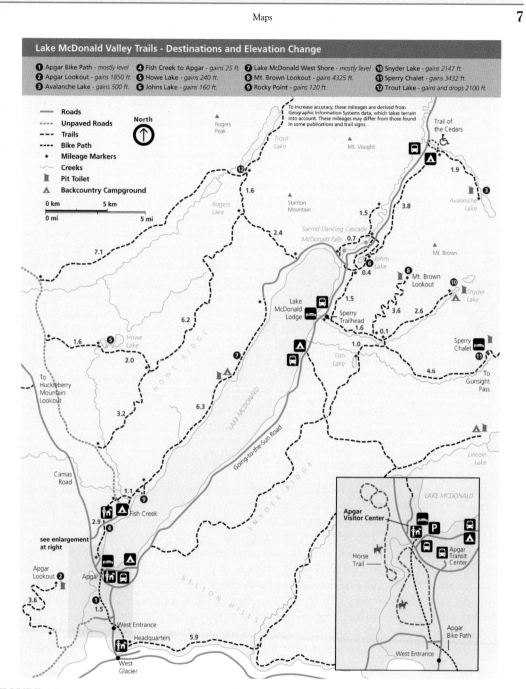

Lake McDonald Valley Trails - Destinations and Elevation Change

1. Apgar Bike Path - *mostly level*
2. Apgar Lookout - *gains 1850 ft.*
3. Avalanche Lake - *gains 500 ft.*
4. Fish Creek to Apgar - *gains 25 ft.*
5. Howe Lake - *gains 240 ft.*
6. Johns Lake - *gains 160 ft.*
7. Lake McDonald West Shore - *mostly level*
8. Mt. Brown Lookout - *gains 4325 ft.*
9. Rocky Point - *gains 120 ft.*
10. Snyder Lake - *gains 2147 ft.*
11. Sperry Chalet - *gains 3432 ft.*
12. Trout Lake - *gains and drops 2100 ft.*

FIGURE 1.3 A map of the Lake McDonald area in Montana. Source: *U.S. Department of the Interior, National Park Service (2016).*

go further to suggest that cartography represents the technology and science associated with interpreting and analyzing spatial relationships (Harley, 1989). Through cartography, the making of a map requires one to describe a landscape (or water body) with clarity, order, balance, and contrast sufficient to communicate a message (Tyner, 2010). The sum of the cartographic effort can convey the story of the map, provide an aesthetically pleasing product, express intellectual merit on a subject, contribute to social goals, or describe technical or scientific matters (Wolodtschenko, 2007). In a general sense, a *cartographer* or (in a much less formal sense) a *map developer* is the person that attempts to do these things and thus explain the importance of a place and its descriptive features (Metcalf, 2012). Cartographers and map developers endeavor to join the characteristics of information and location, and although the level of formal training may be the only measure that separates the two, either would benefit from having a decent amount of imagination and creativity. Philosophically, one might view the existence of a map as the outcome of the desires of the cartographer or map developer, as they create new conceptions of a place, reject proposed conceptions of a place, or reinforce older conceptions of a place (Morrison, 1974). Some suggest that the traditional mapping discipline of *cartography* is being transformed and reinvented as advances in technology (e.g., new sources of data, new mapping processes, new platforms and devices for displaying maps) facilitate the evolution of different perspectives on the visualization of landscapes (Bevington-Attardi and Rice, 2015). We agree that computer-assisted drawings and renderings have mostly replaced the manual drafting of maps, but the development and use of maps remain the same. One purpose of this text is to help bridge the connection between evolving mapping technologies and the timeless principles of cartography which are grounded in geometry, measurement, and design (Edney, 2015; Howarth, 2015).

Maps as Models

Maps are *models* of the surface of the Earth and its water bodies, or even the phenomena below (mines, oceanic trenches) or above (atmosphere) the surface of our planet. Some of the oldest maps describe not only society's geographic understanding of the world at that time, but also religious and historical perspectives concerning life on Earth. These maps may have been developed as models or *compendiums of knowledge* for others to enjoy (Griffiths and Buttery, 2019). The process of converting what one sees and experiences to a drawn or rendered model is challenging. Creating a model entails selecting a set of landscape features, environmental variables, and mappable elements of a landscape, and then mirroring them on analog media (e.g., paper) or rendering them in a digital environment at a portable scale (Downton, 2009). Whether a map describes managed or natural phenomena, or characteristics of human civilization and associated demographics, they can be viewed as a pictorial-symbolic representation of an area of interest (Lisitskii, 2016).

Drawn or rendered to a specific scale, maps provide people with a model of the spatial arrangement of natural or artificial landscape features found within a defined area (Fig. 1.4). Maps may be quite utilitarian, simply describing a parcel of land that someone or some organization owns, with boundaries delineated using distances and directions

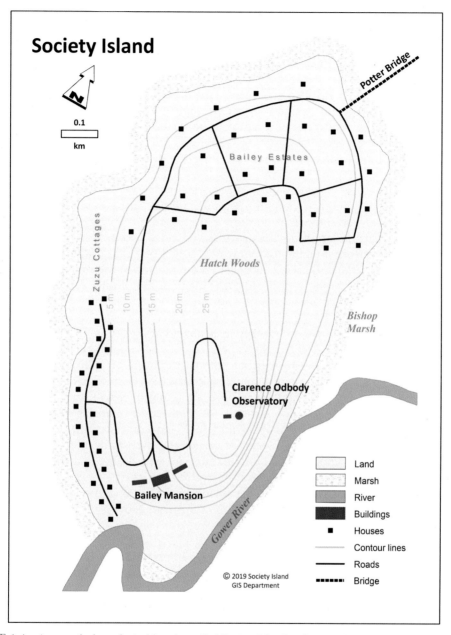

FIGURE 1.4 A map of a hypothetical location called Society Island, indicating the roads, buildings, water bodies, and other important landscape features that, in sum, represent a model of the island.

measured in the field (*in situ*) to allow transformation and reduction to a smaller two-dimensional model, as if looking down upon the land from quite a distance above. For example, a map (Fig. 1.5) may indicate to a forest worker the environment in which they should be working, which may include the boundaries of the tract of land that they manage, the topography within the tract, and any other potential features of interest that may affect the course of their work (e.g., streams, roads, old wells). Although maps could be used to illustrate the actions or conditions that might be possible in the future, for example, the desired future conditions of a national forest, the underlying structure of the landscape often includes representations of activities conducted in the past (previously developed roads, buildings, waterways).

Diversion 2

This exercise may require some improvision and creativity on your part. Certainly, you may use any resource that you can access to inform your response. Imagine you wanted to create a map of the major roads and cities within the country of Togo in West Africa. Further imagine that your model will be 1/3,000,000th of the real size of the country (i.e., the scale is 1:3,000,000). At a minimum, how much space would you need on a piece of paper in order to draw the boundary (outer edge) of the mapped model of this country?

As models, maps can be viewed simply as abstract surveys of landscapes influenced by the various cognitive filters of the cartographer or map developer, and their interpretation is subsequently influenced by the various filters employed by map users (Müller-Funk and Holter, 2009). In order to create a model with the best representation of a real system and to communicate the important ideas of a landscape to map users, a map developer usually includes only a limited number of real-world features on the landscape (Cartwright, 2009). Otherwise, a map would look too complex and its communicative value may be reduced. The map developer selects the aspects of a landscape to include and to exclude in the model, and the manner in which the features are illustrated (Kinman, 2009); therefore, objectivity and subjectivity might be questioned. In whatever manner these models are created, maps provide insight into human nature and culture, and how these characteristics change over time (Thrower, 2008).

Reflection 3

Maps, as models of a landscape, may contain information and traces of human and animal presence through an illustration of developments (buildings, roads, bridges, etc.) and trails (hiking, horse, all-terrain vehicle, etc.). However, maps nearly universally exclude the specific depiction of human beings and other animals that move around the landscape. The exclusion of humans and animals on maps is not necessarily due to the fact that they move, there may simply be too many of them to depict on a map, and their locations may be too transient. Yet there are some instances where maps, as models of a landscape, contain representations of other features that move or inherently are in motion. In some of the recent maps you have used or viewed, what mapped features actually might arguably be considered to be moving in real life?

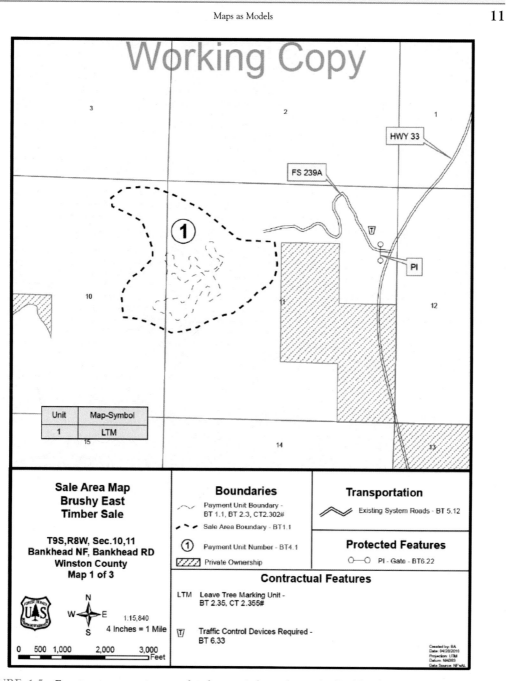

FIGURE 1.5 Forest management map related to a timber sale on the Bankhead National Forest in 2017. Source: *U.S. Department of Agriculture, Forest Service (2016).*

As we suggested, as models, maps can formally be represented as fixed features drawn to scale on flat (paper, wood, composites such as Mylar®) or round (globes) analog media, or represented or rendered as dynamically colored features displayed on the screens of digital devices (computers, cellular phones, tablets). Certainly, the preference for a map product is situational, as paper maps can tear and digital maps can become unavailable as electricity or batteries wane. When one finds themselves in the dark, with a hazy memory of places and routes, a cellular phone equipped with global positioning system (GPS) capabilities might be preferable for viewing and navigating the model landscape. When one finds themselves without electrical power, one might prefer a paper map model to provide navigational guidance. Admittedly, with their increased prevalence, maps presented in a digital environment can enhance a map user's experience through direct interaction with a model of the landscape.

Historical maps can provide us with models of the condition of Earth over long periods of time (Heffernan, 2014). These types of maps can illustrate long-term use or development of a landscape; they can indicate the works of both humans and nature (Greenwood, 1964). Depicted on historical maps might be the boundaries of land governed by the rulers of various countries, and these maps may have acted as expressions of power. The use of maps to subtly express power may be reflected in the ethnocentricity and social order under which the map developer operated (Harley, 1987), however limited this power may have been (Rose-Redwood, 2015). Maps often assist debates that occur during war and peace negotiations to divide conquered territories into new political units (Culcasi, 2014). Old world maps often placed landscapes best understood, such as Europe or Africa, in the center, perhaps to describe routes of trade within the scope of knowledge of the map developer (Metcalf, 2017). The color pink was used in historical maps of the British Empire to convey power, and it typically spanned from edge to edge of a map to indicate how the Sun never set on the Empire. Although today we use maps printed on paper or delivered through various digital devices, prehistoric maps have even been found on rocks (petroglyphs), ceramics, and parchments underscoring humans' long use of models to communicate spatial relationships (Wolodtschenko, 2007).

Inspection 3
Located on this book's website (mapping-book.uga.edu) is a map of New York City that was published in 1777. While this may be a rather old model of the human use and development of Manhattan Island, what landscape features, natural or developed by humans, are present on this map? In terms of buildings developed by humans, which types are present on the map and which types seem to be absent from the map?

As previously noted, maps are also often developed as models to illustrate routes across both land and sea. For example, if we were to locate a map of the Puget Sound in the State of Washington, we would likely find the current routes for the Bainbridge Island or Bremerton ferries that initiate (or terminate) in Seattle. Prior to the incorporation of GPS services into cellular phones and prior to the development of GPS units for cars, people often used printed maps (Fig. 1.6) to navigate road systems. Printed

FIGURE 1.6 A road map covering a portion of the Chattahoochee National Forest in northern Georgia. Source: *U.S. Department of Agriculture, Forest Service (2012).*

road maps, or road atlases, provide society with a model of the transportation infrastructure of a country, state, city, or other administrative unit depending on the scale employed and the area of interest. Until recently, printed road maps were an indispensable tool for navigating one's car or truck through unfamiliar places. These map models illustrate routes between places and are often accompanied by thematic illustrations representing the quality of the roads. These were also often made to attract potential customers to the business patrons who sponsored their development (Petto, 2016). The famous *Michelin* maps were the definitive models of the European road system, as were the *Thomas Brother* road maps for the United States. One can also likely easily locate maps illustrating routes of bird migration, rail or subway systems, and waterways with a little online investigation. These models provide humans guidance in understanding how things might effectively move around the world, however large or small the world is envisioned.

As one might recall from movies or television programs, maps may also be developed as informal models (perhaps hand-drawn on used paper) that illustrate such things as escape routes or directions to a hidden treasure. These map uses can be very important (the consequences may be life or death) and can be developed with great haste (after burying a treasure and quickly leaving the area). Issues of scale are often ignored in models made in haste, as are many less important landscape details. Maps of this nature can be considered expendable (discarded after use), secretive models, using uncommon symbols and text to represent in a disguised manner those landscape features others might easily recognize.

Maps as Memories

There may be times when we remember explicitly the features contained in certain maps. Maps as diverse as those that describe broad expanses of our solar system, or those that illustrate small, local places within which a person frequently navigates, such as a 100 foot by 200 foot parcel of land (Fig. 1.7) can be burned into our minds. Maps are often portrayed as static models of real systems, but in our memories they may dynamically change as society changes around us. A printed map generally represents conditions at a particular point in time in the history of a landscape. Each geographic feature seems definitive and fixed in the mind of the map user. Since a map is a model of a place, it allows both the map developer and the map user to interpret the original place in their own unique manner. The matters of concern on a map are often portrayed as factual evidence of the landscape, yet these representations can unfairly portray experience and objectivity of the world, the map, and the map-maker themselves (Fiel, 2009) since it is our observations and memories of a place that influence the development of a map. However, maps need not be viewed as stoic presentations of land, sky, or sea; they can be developed to disrupt our mental impression of spaces and prompt new forms of inquiry into spatial relationships (Earhart, 2014). Certainly, even though one may assume maps are complete and precise models of a landscape, faults may be noticed (Edney, 2015). Trade-offs and deep examinations such as these that people face in modern society should empower the curious to explore the various technologies and abilities of mapping.

Diversion 3
From memory and without assistance from other resources, using a single piece of paper and a writing utensil, draw a map of a place you remember. This place can be as large as a country, state, province, or county, or as small as a city or a local neighborhood. Include the most important anthropogenic features with the map, such as roads and other developments. Use your own judgment with respect to the importance of features. Also, and where appropriate, include natural features such as water bodies and mountains. Drawing this map to a specific scale will be difficult; therefore, worry less about the scale of the map and concentrate more on the spatial relationships among the mapped features. Label certain features that are important for the map reader (perhaps only yourself) to understand. Once completed, compare this map to another representation of the same place, such as another printed map or a digital map available through the Internet. What aspects of the place in question did you remember well and what aspects had you forgotten?

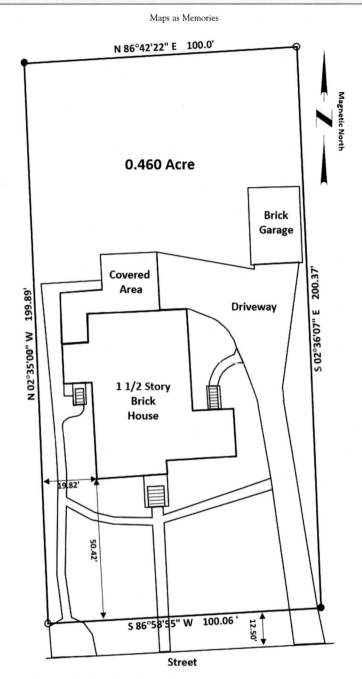

FIGURE 1.7   A map of a parcel of land in Athens, Georgia.

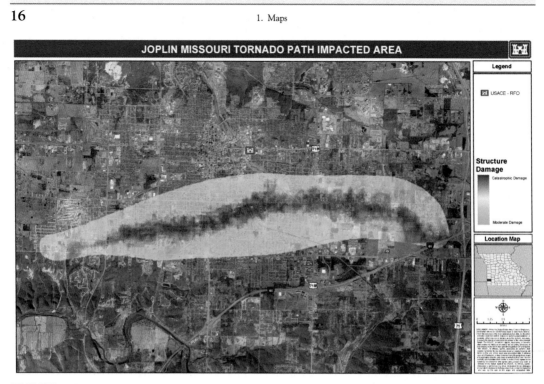

FIGURE 1.8   Path and intensity of the impact of a tornado, Joplin, Missouri, May 22, 2011. Source: *U.S. Department of Defense, Army Corps of Engineers (2011).*

Often in the management and reporting of events affecting natural resources or human development, maps have been created to help society remember what occurred within a given range of time. For example, disaster management professionals use maps to track the impacts of hurricanes, tornados (Fig. 1.8), fires, and other events that affect both the developed and natural world. Analysts may then use these maps to understand how and why significant impacts occurred. Land managers also make maps to remember where tree harvests, plantings, road development activities, and other uses of land and manipulations of resources have occurred. These records serve to refresh society's memory of past events or conditions, and they may be referenced periodically in the planning of future management activities.

Translation 3
Maps can be used to document a portion of a story describing what happened within a landscape or water body during some specific period of time. For the map illustrated in Fig. 1.9 explain in one or two sentences what might have occurred in this location over the period of time that is noted. What one or two map features can help remind us about the severity of events?

Government agencies also use maps to remember the size and dimension of properties owned legally within their jurisdiction by local or absentee landowners. Registered land

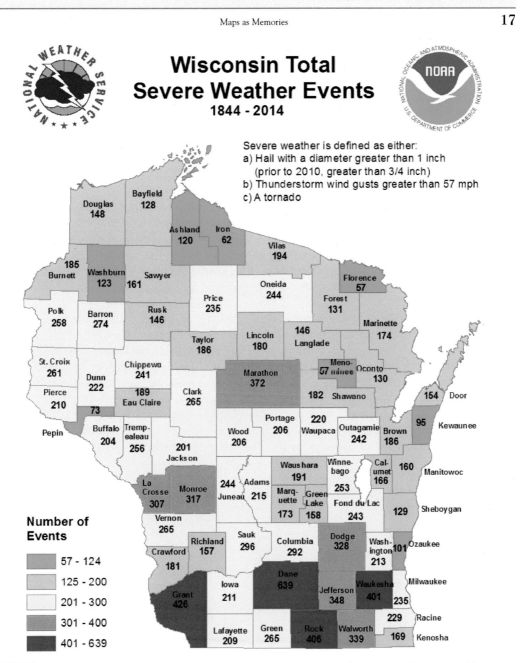

FIGURE 1.9 Severe weather events in Wisconsin, by county, 1844–2014. Source: *U.S. Department of Commerce, National Oceanic and Atmospheric Administration (2014).*

surveyors often develop these *plat maps*, which describe land with an accuracy and pre-scription beyond the capabilities of normal people who may only have simple measure-ment equipment (compass and measuring tapes). These maps, along with their accompanying warranty deeds, are official memory of land tenure and ownership. Changes in land uses can also be estimated through the comparison of maps made at different times. U.S. Geological Survey 7.5-minute series topographic maps, when revised, often contain information (often displayed as purple symbology or text) that describe what has changed since the previous map of each area was developed.

> **Reflection 4**
> Printed land ownership and land use maps are often very useful for guiding community planning discussions. For the area where you or your family currently live, about how often would you feel it would be necessary to update the maps that are used to document and remember the conditions of the human settlements (homes) and associated developments (e.g., roads)? Why would you suggest this time interval to document a society's memory of its landscape?

Of course, there are challenges associated with the ability of different map developers to portray landscapes at a reasonable scale, with similar land classification schemes. But in general, the use of maps as a memory of resources and conditions can contribute to our understanding of how landscapes may have changed, given the social and economic evolution of humankind, or given the occurrence of natural disasters. As an example, maps of Latvia spanning over 200 years were compared to help understand changes in woodland location and composition (Fescenko et al., 2016). Similarly, maps of Poland spanning about 215 years have been used to estimate forest cover changes (Ciupa et al., 2016). An historical series of maps such as these, memories of the past as portrayed by several different cartographers, can offer insight into how and why a landscape has changed. Fig. 1.10 provides an example of mapped historical features, and if compared to Fig. 1.4, one may be able to generally describe how this landscape may have changed over time.

Maps as Inspiration

Maps may act to inspire people to create stories or to express visually the metaphoric possibilities that landscape features represent (Mourãu, 2009). For example, a writer may create a short piece of fiction based on what is present within a favorite map of theirs. Or, an artist may develop an embellished map based on their romantic association with a real or fictional place (Fig. 1.11). Creative map renderings are prolific through popular culture. A few examples of pop-culture mapping include the animated map used in the introduction sequence for *Game of Thrones* highlighting the kingdoms of Westeros, navigation maps incorporated into video games such as the *Legend of Zelda*, or maps placed in books, such as Tolkien's map of *Middle-Earth* in the *Lord of the Rings* and Robert Louis Stevenson's map of *Treasure Island*.

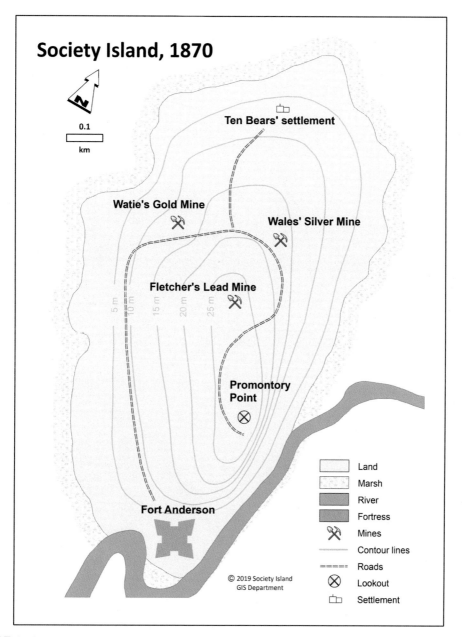

FIGURE 1.10 An historical map of the hypothetical location called Society Island, indicating the locations of significant landscape features where archeological surveys might be conducted.

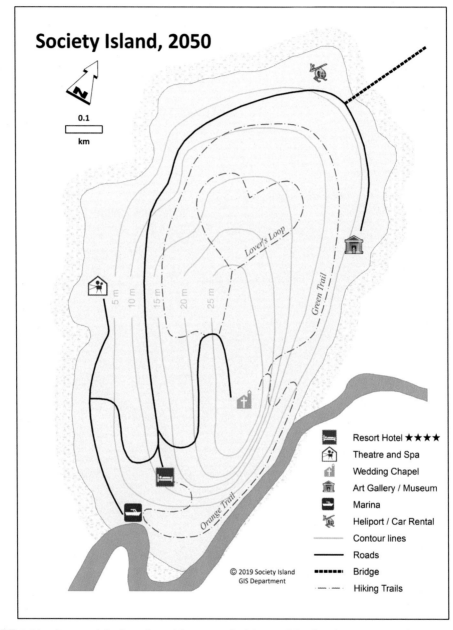

FIGURE 1.11   A map of the hypothetical location called Society Island, indicating a set of idealized features, assuming one had the resources to create them, which would act to promote the location as a destination for people seeking a romantic holiday.

Inspection 4

Using the Internet, search for maps depicting some aspect of popular culture. Select a map and document as well as possible the date of the map, the developer, and the original source (publication or Internet site). Then, describe what you think may have inspired the map maker to create the divisions contained in the legend (if one exists) and what you think may have inspired them to select the color scheme that has been employed throughout the map.

Occasionally, maps have been developed to brag or express the perceived dominance of organizations within certain fields of sport or commerce. For example, while not explicitly comparing control of resources against a competitor, land management organizations develop maps illustrating size and distribution of resources they control. Restaurant chains often produce maps illustrating the widespread distribution of their facilities. Further, sporting conferences produce maps to illustrate the presence and distribution of their teams across a geographic region. As we mentioned earlier, in centuries past, cartographers for governments often created maps to demonstrate their control over land or sea. These types of maps arguably were designed to inspire people with similar dispositions (or thought) to devote attention (or resources) to an organization.

Maps as Products

On a basic level, the maps we produce need to be of value to a map user. Maps are essentially products of a map development process and are ideally devised to communicate a message. The legitimacy of a map may depend on whether the map user believes in the accuracy and legitimacy of the message. For example, maps can facilitate comparisons of natural resources among or within different spatially defined regions (Fig. 1.12). Maps can depict serious issues facing humanity, or less critical, and perhaps humorous, aspects of our experiences. Maps are also integral pieces of business contracts and proposals. Further, maps are used by academics and others in presentations and papers describing their research. On the more serious side, maps have been developed to describe demographic characteristics, disease susceptibility, natural disaster occurrences, and general economic condition of a place. These map products often tell a story of human condition within a geographic area. On a less serious note, maps have also been developed by individuals simply for fun, to illustrate such things as liquor preferences among various populaces, or their most-favored restaurants (Fig. 1.13). These map products also tell a story of human preference and tastes.

Diversion 4

Using the data on GPS accuracy that can be accessed through this book's website (mapping-book.uga.edu) create a map using your preferred mapping system that very clearly indicates the locations of these places with respect to the major surrounding geographic features. While we have not yet discussed the use of color or symbols to represent landscape features, use your best judgment to graphically emphasize those aspects of the accuracy of GPS data that you feel are important.

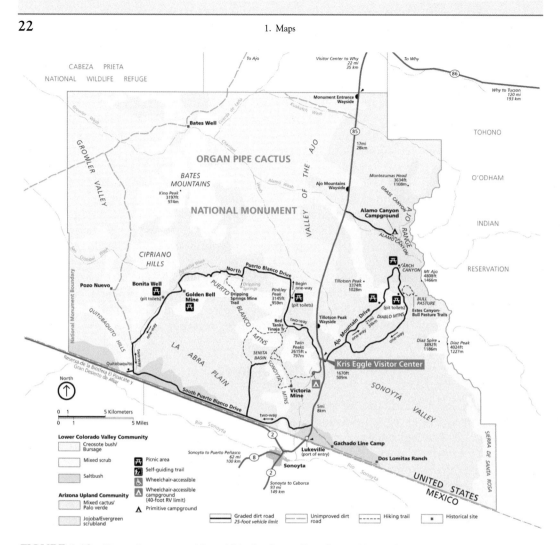

FIGURE 1.12   Vegetation communities within the Organ Pipe Cactus National Monument in Arizona. *Source:* *U.S. Department of the Interior, National Park Service (2015).*

In terms of alternative forms of directional guidance, understanding the location of a place may be better communicated using a map (Fig. 1.14) as opposed to a textual description (Drive south on Milledge Avenue in Athens, Georgia, to where it ends, then drive 0.6 miles down Phoenix Road) or a set of common geographic coordinates (33.8917°N latitude, 83.3586°W longitude using the Greenwich Standard Meridian). Maps, therefore, allow the presenter of information to establish geographic context in reader's or listener's minds. These products then allow people to better relate the purpose and outcomes of work conducted to other experiences that they may know or remember about these places. As a result, map products can provide essential context to a communication process.

Maps are now offered as products on paper, personal computers, cellular phones, tablets, GPS data recorders, and other digital devices (Fig. 1.15). Maps are ubiquitous in

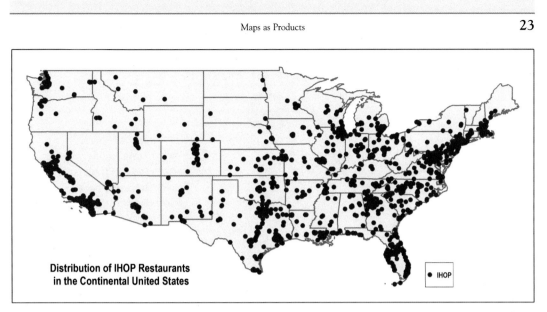

**Distribution of IHOP Restaurants
in the Continental United States**

● IHOP

FIGURE 1.13 Locations of *International House of Pancakes* (*IHOP*) restaurants in the United States.

digital and social media, and are offered as products to inform people as well as elicit their reactions to patterns and behaviors (Roth, 2015). Understanding each media in which a map can be presented is an important aspect of personal success in modern society. A person may frequently be confronted with a need to understand and interpret a variety of maps to inform their activities (*Is it raining in the city I am traveling to? Are there scenic highways I can travel during my trip? Where is the closest gas station to my current location?*). This information might be acquired through interpretation of maps as diverse as those provided on cellular phones or those posted on the walls of subway (metro) stations. Further enhancing human interaction with maps, multiscale and pan-able maps are now facilitated by digital devices (personal computers, cellular phones, tablets) and *the cloud*. In this case, *the cloud* refers not to a mass of frozen crystal and liquid droplets suspended in the troposphere, but to a collection of computers, communication technology, and storage capacity housed within distributed data centers that represent a shared pool of computing resources, and a virtual workspace for map developers (Peterson, 2015).

Translation 4
Access the image of the digital map of Congaree National Park (South Carolina) that is provided through this book's website (mapping-book.uga.edu). This image was created through a service, *topoView*, which is administered by the U.S. Geological Survey. Aside from the fact that the interactive website illustrates the location and extent of 1:24,000 scale topographic maps (e.g., Gadsden and Wateree), name four other things that this digital map can also assist the map user in understanding.

Digital maps are frequently used today to help map users locate places such as restaurants or businesses. They also provide map users with directions, perhaps shortest

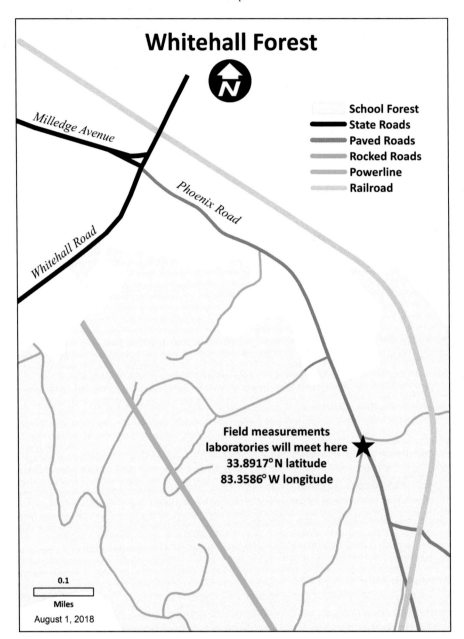

FIGURE 1.14   A specific place on University of Georgia's Whitehall Forest where field laboratories will meet.

paths based on distance or driving time, from where they currently are situated to some desired destination. However, even though society has experienced significant changes in how maps are produced in the last three decades, mapping and carto-graphic principles remain important. Common symbols and representations of features,

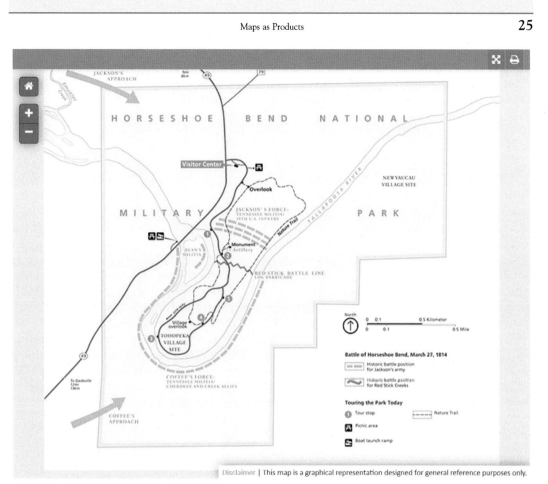

FIGURE 1.15 A digital map of Horseshoe Bend National Military Park in Alabama. Source: *U.S. Department of the Interior, National Park Service (2019)*.

accepted projections of the Earth onto flat surfaces (paper or digital screens), accurate measurements of distances and areas, and appropriate generalization of map components with respect to mapping scale (or to changes in map scale) are all important aspects of map products regardless of the manner in which a map is delivered (Lisitskii, 2016).

Maps, in a variety of settings and for many different purposes, are developed to reduce or eliminate uncertainty about locations or landscapes. *The place is here* is a fairly direct statement of locational uncertainty. *These places contain old trees*, perhaps illustrated in color, is a fairly straightforward statement that eliminates uncertainty regarding what might be found across a landscape. Many other examples (*I own this place*, or *The road from one place to another follows this route*) can be depicted on maps to add clarity to previously uncertain ideas that have a geographic concern. Thus there are many situations where map products can be of great value to society.

Inspection 5
Using a digital device (television, computer, cell phone, tablet), locate a map of the current weather conditions for your home area (city, county, state, country). What weather events or conditions are your family or home town friends experiencing right now? In what manner is the digital device providing information through this map product that informs your understanding of the current weather conditions in your home area?

Often, maps are the result of *geospatial analyses* conducted by people who are interested in understanding where features are located, and why they are located there. These map products provide society with insight into a problem or issue that would not otherwise be obvious in the display of common landscape features. Maps are often being used in collaborative natural resource management efforts to allow the public to highlight places of interest. Many forms of geospatial analyses can be facilitated by GIS or computerized mapping programs. GIS includes the software, hardware, data, and distribution technology that allow a person to capture, store, manipulate, analyze, and display geographic phenomena. Geospatial analyses can involve a mathematical interaction or integration of two disparate maps (e.g., combining a forest database with a soil database) or one of a vast number of processes that examine landscapes or water bodies and use their characteristics to understand our physical environment. With advances in computer technology over the past three decades, geospatial analyses are now very efficiently conducted in a digital environment. In many respects, map products derived from these processes can help people reduce the risks of making bad decisions.

Concluding Remarks

As a form of cultural text, a map is subject to the objectivities and subjectivities of the map developer. While we may assume that maps might represent a neutral, objective measurement and display of real-world phenomena, we should also bear in mind that they may also reflect aspects of the social dimension within which they were created, perhaps defining or reinforcing social relationships, rules, and values (Harley, 1989). Map development involves a set of synergistic concepts related to the design, compilation, description, and perhaps generalization of spatial information; the map development process blends classic cartographic techniques with contemporary practice and social demand (Harley, 1987). Maps can now be developed to inform common activities, such as decisions about where to go to have dinner, or to inform communities on issues of concern, such as proposed public policies (Griffin and Fabrikant, 2012). Although only about 30 years mature, our current digital environment allows both expert and naïve cartographers to develop maps and deliver them quickly to a large audience through the Internet (Griffin and Fabrikant, 2012). In today's postindustrial information era, we often find it necessary for maps to be produced quickly and to be displayed easily on computers, tablets, and cellular phones. People now naturally access cartographic information on digital devices, and while the traditional role of a map as a depository and source of information still remains, its role has shifted in many respects to a tool of interaction with the map user (Lisitskii,

2016). However, it is important to understand that maps may never be neutral in their role as authoritarian images, and may act to persuade changes in beliefs or to reinforce the status quo (Harley, 1989).

For better or worse, ordinary people with little training can now create their own maps, rather than rely upon the skill and knowledge of trained cartographers (Perkins, 2008). The citizen cartographer may want to make a map that most effectively expresses their feelings about human or natural phenomena. Yet there is still a need to describe how cartographic practices and principles guide the development of, or are represented in, various forms of maps through texts such as this (Griffin and Fabrikant, 2012). One rubric that has been offered to evaluate the design and effectiveness of maps (Robinson and Nelson, 2015) poses four interrelated questions of a map:

1. Does a map communicate a complete story?
2. Does a map communicate a compelling story?
3. Has a map been developed with attention to cartographic design and geospatial analysis?
4. Does a map have aesthetic characteristics that complement the story?

Questions concerning whether we should develop a map as a model of a landscape, to inspire people, to record experiences or activities, or simply to complement commerce or science should be kept in mind. The design of a map can influence the spatial cognition people have for natural or human-influenced phenomena. While our conceptions of the world can be influenced by maps, they often are not associated with direct experience with the places described. However, learning about maps, especially at a younger age, can also influence spatial cognition capacity in people and influence how we might interpret reality (Uttal, 2000).

Reflection 5
Using memory alone, without accessing your digital devices, consider the last map you viewed on a digital device (cellular phone, tablet, personal computer). Perhaps it was a weather map, a map of bus or train routes, or a road map. Briefly describe the map and try to answer the four questions posed above.

With respect to the study of mapping, a concentration on common cartographic practices (the procedures and technologies for making maps) seems essential. A focus on the sociocultural aspects of maps, the deeper meaning of maps and mapping, also seems essential (Edney, 2014a). While this book may be more heavily weighted to the former, we allocate some discussion, and many of the *reflections* and *inspections* contained herein, to understanding the nature of maps. Aspects of both the *empiricist* and *critical* understanding of maps (Edney, 2014b) are therefore pursued in this book. The empiricist view bases the development of knowledge on experience, observation, and experimentation, while the critical view bases the development of knowledge on investigations into the spatial relations among resources, human, and animal behavior and the testing of geographic theory. Although we feel that there is certainly a need for map developers to understand the technology behind maps, such as the concepts of vector (point, line, polygon) and raster (pixel, voxel) data structures in conjunction with similar concerns regarding geographic literacy

(Saarinen and MacCabe, 1995), we are not suggesting that the ability to develop a map is the end point of an education in geography. Knowledge and understanding of mapping processes are foundations upon which geographic thought can be further developed through applied experiences and creative efforts.

References

Bestgen, A.-K., Edler, D., Miller, C., Schulze, P., Dickman, F., Kuchinke, L., 2017. Where is it (in the map)?: recall and recognition of spatial information. Cartographica 52 (1), 80—97.

Bevington-Attardi, D., Rice, M., 2015. On the map: American cartography in 2015. Cartography Geogr. Inform. Sci. 42 (Suppl. 1), 1—5.

Caquard, S., Wright, B., 2009. Challenging the digital cartographic continuity system: lessons from cinema. In: Cartwright, W., Gartner, G., Lehn, A. (Eds.), Cartography and Art. Lecture Notes in Geoinformation and Geography. Springer-Verlag, Berlin, pp. 193—206.

Cartwright, W., 2009. Art and geographic communication. In: Cartwright, W., Gartner, G., Lehn, A. (Eds.), Cartography and Art. Lecture Notes in Geoinformation and Geography. Springer-Verlag, Berlin, pp. 9—22.

Ciupa, T., Suligowski, R., Wałek, G., 2016. Use of GIS-supported comparative cartography and historical maps in long-term forest cover changes analysis in the Holy Cross Mountains (Poland). Baltic Forestry 22 (1), 63—73.

Culcasi, K., 2014. Disordered ordering: mapping the divisions of the Ottoman Empire. Cartographica 49 (1), 2—17.

Downton, P., 2009. Maps of what might be: a dozen works on ideas and possibilities. In: Cartwright, W., Gartner, G., Lehn, A. (Eds.), Cartography and Art. Lecture Notes in Geoinformation and Geography. Springer-Verlag, Berlin, pp. 327—344.

Earhart, A.E., 2014. "After a hundred years / Nobody knows the Place": notes toward spatial visualizations of Emily Dickinson. Emily Dickinson J. 23 (1), 98—105.

Edney, M.H., 2014a. A content analysis of Imago Mundi, 1935-2010. Imago Mundi 66 (Suppl. 1), 107—131.

Edney, M.H., 2014b. Academic cartography, internal map history, and the critical study of mapping processes. Imago Mundi 66 (Suppl. 1), 83—106.

Edney, M.H., 2015. Cartography and its discontents. Cartographica 50 (1), 9—13.

Fescenko, A., Lukins, M., Fescenko, I., 2016. Validation of medium-scale historical maps of southern Latvia for evaluation of impact of continuous forest cover on the present-day mean stand area and tree species richness. Baltic Forestry 22 (1), 51—62.

Fiel, W., 2009. Dream but conflict. In: Cartwright, W., Gartner, G., Lehn, A. (Eds.), Cartography and Art. Lecture Notes in Geoinformation and Geography. Springer-Verlag, Berlin, pp. 369—375.

Frye, A.E., 1895. Complete Geography. Ginn & Company, Publishers, Boston, MA.

Government of India, 2015. Instructions for publication of maps by central/state government departments/offices and private publishers. Survey of India, Dehra Dun, India.

Greenwood, D., 1964. Mapping. The University of Chicago Press, Chicago, IL.

Griffin, A.L., Fabrikant, S.I., 2012. More maps, more users, more devices means more cartographic challenges. Cartogr. J. 49 (4), 298—301.

Griffiths, C., Buttery, T., 2019. The world's oldest medieval map. BBC Travel, 25 March 2019. <http://www.bbc.com/travel/gallery/20190324-the-worlds-oldest-medieval-map> (accessed 30.03.19).

Gutiérrez, D., Cock, H., L.J. Rosenwald Collection. 1562. Americae sive qvartae orbis partis nova et exactissima descriptio. [Antwerp: s.n] [Map] Retrieved from the Library of Congress, <https://www.loc.gov/item/map49000970/> (accessed 02.04.19).

Harley, J.B., 1987. Innovation, social context and the history of cartography. Cartographica 24 (4), 59—68.

Harley, J.B., 1989. Deconstructing the map. Cartographica 26 (2), 1—20.

Heffernan, M., 2014. A paper city: on history, maps, and map collections in 18th and 19th century Paris. Imago Mundi 66 (Suppl. 1), 5—20.

Howarth, J.T., 2015. A framework for teaching the timeless way of mapmaking. Cartogr. Geogr. Inform. Sci. 42 (Suppl. 1), 6—10.

Imhof, E., 1974. Preface of the founder of the yearbook on the occasion of change of redaction and editorship. In: Kirschbaum, G.A., Meine, K.-H. (Eds.), International Yearbook of Cartography, 14. Kirschbaum Verlag, Bonn-Bad Godeberg, Germany, pp. 9–10.

Kinman, E.L., 2009. Sculpting place through ceramic maps. In: Cartwright, W., Gartner, G., Lehn, A. (Eds.), Cartography and Art. Lecture Notes in Geoinformation and Geography. Springer-Verlag, Berlin, pp. 309–316.

Lechthaler, M., 2009. The world image in maps — from the old ages to Mercator. In: Cartwright, W., Gartner, G., Lehn, A. (Eds.), Cartography and Art. Lecture Notes in Geoinformation and Geography. Springer-Verlag, Berlin, pp. 155–174.

Lee, J., 1995. Map design and GIS — a survey of map usage amongst GIS users. Cartogr. J. 32 (1), 33–39.

Lisitskii, D.V., 2016. Cartography in the era of informatization: new problems and possibilities. Geograp. Natural Resour. 37 (4), 296–301.

Metcalf, A.C., 2012. Amerigo Vespucci and the Four Finger (Kuntsmann II) world map. e-Perimetron 7 (1), 36–44.

Metcalf, A.C., 2017. Who cares who made the map? La Carta del Cantino and its anonymous maker. e-Perimetron 12 (1), 1–23.

Morrison, J.L., 1974. A theoretical framework for cartographic generalization with emphasis on the process of symbolization. In: Kirschbaum, G.A., Meine, K.-H. (Eds.), International Yearbook of Cartography, 14. Kirschbaum Verlag, Bonn-Bad Godeberg, Germany, pp. 115–127.

Mourãu, M., 2009. Early-modern Portuguese maps: metaphors of space and spaces of metaphor. In: Cartwright, W., Gartner, G., Lehn, A. (Eds.), Cartography and Art. Lecture Notes in Geoinformation and Geography. Springer-Verlag, Berlin, pp. 379–384.

Müller-Funk, S., Holter, M.C., 2009. This is not a map. In: Cartwright, W., Gartner, G., Lehn, A. (Eds.), Cartography and Art. Lecture Notes in Geoinformation and Geography. Springer-Verlag, Berlin, pp. 385–391.

Oksanen, J., Halkosaari, H.-M., Sarjakoski, T., Sarjakoski, L.T., 2014. A user study of experimental maps for outdoor activities. Cartographica 49 (3), 188–201.

Pérez, M.A., 2013. Lines underground: exploring and mapping Venezuela's cave environment. Cartographica 48 (4), 293–308.

Perkins, C., 2008. Cultures of map use. Cartogr. J. 45 (2), 150–158.

Peterson, M.P., 2015. Maps and the meaning of the cloud. Cartographica 50 (4), 238–247.

Petto, C.M., 2016. To know the distance: wayfinding and roadmaps of early modern England and France. Cartographica 51 (4), 240–262.

Plath, N., 2009. Robert Smithson's pursuit: mapping some traces of remnants. In: Cartwright, W., Gartner, G., Lehn, A. (Eds.), Cartography and Art. Lecture Notes in Geoinformation and Geography. Springer-Verlag, Berlin, pp. 253–265.

Rambaldi, G., 2005. Who owns the map legend? URISA J. 17 (1), 5–13.

Robinson, A.C., Nelson, J.K., 2015. Evaluating maps in a massive open online course. Cartographic Perspectives 80, 6–17.

Robinson, A.H., 1974. A new map projection: its development and characteristics. In: Kirschbaum, G.A., Meine, K.-H. (Eds.), International Yearbook of Cartography, 14. Kirschbaum Verlag, Bonn-Bad Godeberg, Germany, pp. 145–155.

Rose-Redwood, R., 2015. Introduction: the limits to deconstructing the map. Cartographica 50 (1), 1–8.

Roth, R.E., 2015. Interactivity and cartography: a contemporary perspective on user interface and user experience design from geospatial professionals. Cartographica 50 (2), 94–115.

Saarinen, T.F., MacCabe, C.L., 1995. World patterns of geographic literacy based on sketch map quality. Prof. Geogr. 47 (2), 196–204.

Sack, R.D., 1972. Geography, geometry, and explanation. Ann. Assoc. Am. Geogr. 62 (1), 61–78.

Shevchenko, V., Rodionova, Z., Michael, G., 2016. Lunar and planetary cartography in Russia, Astrophysics and Space Science Library, vol. 425. Springer, Cham, Switzerland.

Skryzhevska, L., Green, J., Abbitt, R., 2013. GIS textbook content as a basis for skill development in map interpretation. Cartographica 48 (1), 38–46.

Skupin, A., Skupin, M., 2009. On written language in works of art and cartography. In: Cartwright, W., Gartner, G., Lehn, A. (Eds.), Cartography and Art. Lecture Notes in Geoinformation and Geography. Springer-Verlag, Berlin, pp. 207–222.

Thrower, N.J.W., 2008. Maps and Civilization. The University of Chicago Press, Chicago, IL.

Tyner, J.A., 2010. Principles of Map Design. The Guilford Press, New York.

U.S. Department of Agriculture, Forest Service, 2009. Final Environmental Impact Statement, Motorized Travel Management, Lassen National Forest. U.S. Department of Agriculture, Forest Service, Pacific Southwest Region, Vallejo, CA. R5-MB-207.

U.S. Department of Agriculture, Forest Service, 2012. Chattahoochee National Forest map. U.S. Department of Agriculture, Forest Service, Southern Region, Atlanta, GA.

U.S. Department of Agriculture, Forest Service, 2016. Sale Area Map, Brushy East Timber Sale. U.S. Department of Agriculture, Forest Service, Bankhead National Forest, Montgomery, AL, <https://www.fs.usda.gov/Internet/FSE_DOCUMENTS/fseprd541894.pdf> (accessed 22.01.19).

U.S. Department of Commerce, National Oceanic and Atmospheric Administration, 2014. Wisconsin Tornado and Severe Weather Statistics. U.S. Department of Commerce, National Oceanic and Atmospheric Administration, Weather Forecast Office, Green Bay, WI. <https://www.weather.gov/grb/WI_tornado_stats> (accessed 23.01.19).

U.S. Department of Defense, Army Corps of Engineers, 2011. Joplin Missouri Tornado Path Impacted Area. U.S. Department of Defense, Army Corps of Engineers, Kansas City, MO.

U.S. Department of the Interior, National Park Service, 2015. Organ Pipe Cactus National Monument Arizona. U.S. Department of the Interior, National Park Service, Washington, D.C., <https://www.nps.gov/orpi/planyourvisit/upload/Park-Map-2015-2.pdf> (accessed 22.02.19).

U.S. Department of the Interior, National Park Service, 2016. Lake McDonald Valley trails – destinations and elevation change. U.S. Department of the Interior, National Park Service, Washington, D.C.

U.S. Department of the Interior, National Park Service, 2019. Horseshoe Bend National Military Park Alabama. U.S. Department of the Interior, National Park Service, Washington, D.C., <https://www.nps.gov/hobe/planyourvisit/maps.htm> (accessed 22.02.19).

Uttal, D.H., 2000. Seeing the big picture: map use and the development of spatial cognition. Dev. Sci. 3 (3), 247–264.

van der Krogt, P., 2015. The origin of the word 'Cartography'. e-Perimetron 10 (3), 124–142.

von Ungern-Sternberg, A., 2009. Dots, lines, areas and words: mapping literature and narration (with some remarks on Kate Chopin's "The Awakening"). In: Cartwright, W., Gartner, G., Lehn, A. (Eds.), Cartography and Art. Lecture Notes in Geoinformation and Geography. Springer-Verlag, Berlin, pp. 229–252.

Wolodtschenko, A., 2007. Some aspects of the prehistoric maps as cultural heritage. e-Perimetron 2 (1), 48–51.

Wood, D., 1983. Early thematic mapping in the history of cartography/Arthur H. Robinson [Book review]. Cartographica 20 (3), 109–112.

Map Types

Maps are distillations of reality; they are models that serve many purposes (Barbour, 2001). Maps attempt to communicate a message, or tell a story, using geography as a base. Some maps are very similar, for instance, a choropleth map (as we will soon describe) is very similar in nature to a thematic map. Yet some types of maps are very different, for instance three-dimensional tactile surface maps of a city (with perhaps no color nor attribution) are much different than two-dimensional soil maps of a rural landscape that illustrate soil depth, chemical composition, and pH levels. Maps tell a story, through symbols and color or other constructs, and through an emphasis on one type of resource or phenomena over another, this may prove to be the basis for which we call them one type or another.

Reflection 6

Before we get too far along with this discussion of map types, think about when, where, and how you have recently used maps. Briefly, describe the last two unique maps that you used. In general, what types of maps would you consider them to be?

In the not so distant past, only two broad classes of maps, topographic and thematic, were recognized by geographers. However, additional consideration of map style, use, geographic scope, and the projection system applied suggested to map taxonomists many variants and subvariants of these exist (Benová and Pravda, 2009). The theme, purpose, and accuracy of a map, along with the subjective proclivities of the map developer for individual styles and colors, combine to reflect the type of map produced (Benová and Pravda, 2009). Interestingly, one can observe overlapping styles and purposes within a single map, making a distinct classification of it a cause for debate. For example, one can imagine a map (Fig. 2.1) that may contain both thematic and topographic qualities.

In one approach to map classification, Casti (2015) placed maps into three categories: general, special subject, and special purpose. In this system, general maps could include road maps or maps highlighting cultural places of interest, such as the locations of libraries, museums, and theaters within a specific city. Special subject maps within this system could include maps that portray climate information, political or administrative units, and commercial interests across a landscape or water body. Then, special-purpose maps could consist of those that describe unique geological features, locations of matters of interest to the military, or areas of forest and agricultural activities. In classifying maps with this system, one might imagine the debates that could arise, particularly involving similarities between the latter two classes. In developing a map of forest ownership within a county, for example, one might argue that it represents both a special subject (noting commercial interests and administrative units) and a special purpose (identifying locations of forests).

From a technological perspective, maps can be characterized by how they are presented (i.e., analog or digital). Analog maps are traditionally displayed on paper or other similar flat media, such as Mylar®, plastic, or even wood. They are often considered static (they have no moving elements), and are often archived in printed collections, books, and reports. Analog maps may be more appropriate for situations where an energy source (other than the sun or a fire) is unavailable and where people need to compare different themes of a landscape (Mendonça and Delazari, 2011). Digital maps are often considered dynamic (some have moving and scalable elements), and are presented on the screens of televisions, computers, cellular phones, and other electronic devices. They need energy to be viewed. In some cases, through complex mathematical operations, personal computers can facilitate comparisons of multiple maps. Digital maps delivered on cellular phones or tablets may likely be preferable for navigational purposes in today's world.

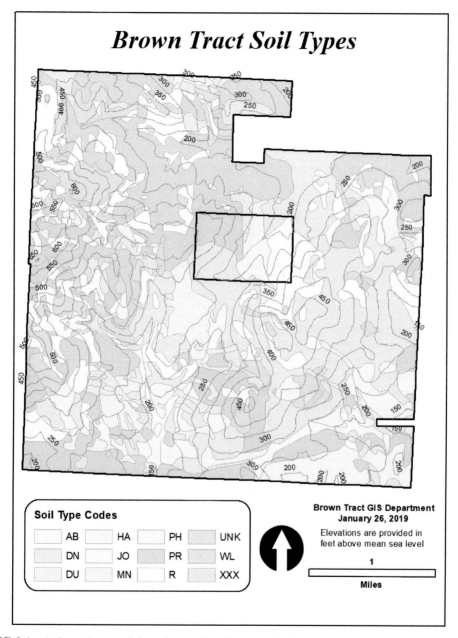

FIGURE 2.1   A thematic map of the soil types found across a landscape, along with contour lines that reflect the topographic relief.

Traditional Types of Maps

Thematic Map

In a rather straightforward sense, thematic maps illustrate the relative occurrence of conditions and phenomena that may be of interest to society. The term *thematic* is often used to describe how a map, through color or symbology, illustrates the various intensities or classes of these conditions and phenomena. For example, a map of the concentration of a disease based on records of affected people, might be considered a thematic map (Fig. 2.2). Thematic maps such as these can be of great importance to society. Areas of concern can be analyzed in conjunction with other mapped information (housing, water systems, etc.) to help society understand the causes and vectors of disease transmission. Through thematic maps, outbreaks of contemporary infectious diseases, using data representing hospital admissions or records of medical treatment, may provide both spatial and temporal perspectives on human health and condition. In one such study, the distribution of norovirus cases in humans was mapped to help understand trends (rise and fall in cases) associated with an epidemic (Inaida et al., 2013). Some have even attempted to reconstruct the spatial dimension of disease through historical tax records and their associated demographic information, such as a plague epidemic that affected Dijon (France) in the 15th century (Galanaud et al., 2015).

For more basic purposes, thematic maps can be of value in describing general land uses or land characteristics across a landscape. For example, a thematic map might represent the type of property tax designation that individual land parcels may have been assigned by a tax department (Fig. 2.3). As might be evident from cursory inspection of the map, parcels further from urban or developed areas tend to fall into forested or agricultural property tax classes.

> **Diversion 5**
> Using the databases provided on this book's website (mapping-book.uga.edu) and your preferred mapping program, develop a thematic map that represents a range of different forested conditions of the area described. Any of the attributes contained in the databases can be used for this purpose. However, consider carefully the use of color and the number of classes to display within the map. At this point, do the best you can in developing the map; we will discuss issues of cartography later in this book.

The activity of humans, both good and bad, can be communicated through thematic maps. In one example, thematic maps related to the location and injuries sustained from farm tractor accidents in Kentucky were developed to help understand the risks farmers face. The maps revealed an association between accidents, landscape conditions, and demographic characteristics of the farmers (Saman et al., 2012). Increasingly, thematic maps can also provide perspectives on human social networks and may illustrate potential patterns of human activity among communities without the use of complex statistical analyses. Advances in data collection technology, particularly through mobile digital devices, can allow human activity and behavior to be monitored. One particular analysis

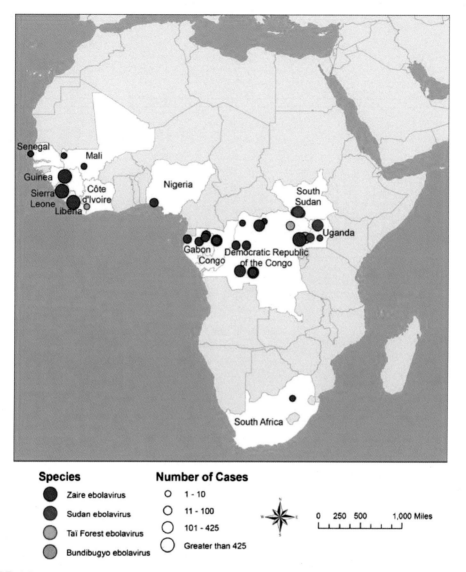

FIGURE 2.2   Ebola virus distribution map in Africa. Source: *U.S. Department of Health and Human Services, Centers for Disease Control and Prevention (2018).*

of human activity in Milan, Italy, facilitated the development of a thematic map that illustrated how and when people use telephones (Botta and del Genio, 2017). For example, a general change in people's behavior was noted during a holiday season, as people communicated less with acquaintances outside of their family and their close circle of friends.

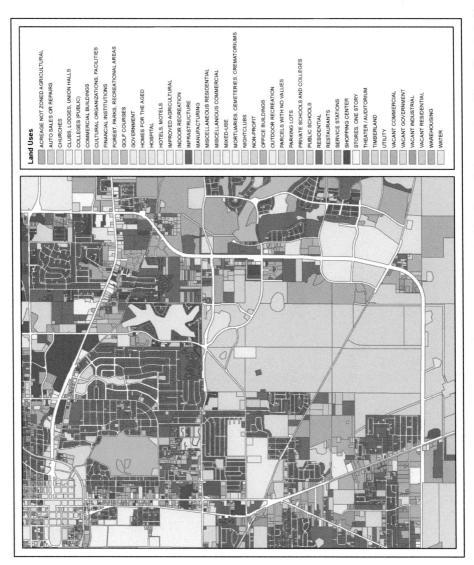

Land Uses

ACREAGE NOT ZONED AGRICULTURAL
AUTO SALES OR REPAIRS
CHURCHES
CLUBS, LODGES, UNION HALLS
COLLEGES (PUBLIC)
COMMERCIAL BUILDINGS
CULTURAL ORGANIZATIONS, FACILITIES
FINANCIAL INSTITUTIONS
FOREST, PARKS, RECREATIONAL AREAS
GOLF COURSES
GOVERNMENT
HOMES FOR THE AGED
HOSPITAL
HOTELS, MOTELS
IMPROVED AGRICULTURAL
INDOOR RECREATION
INFRASTRUCTURE
MANUFACTURING
MISCELLANEOUS RESIDENTIAL
MISCELLANEOUS COMMERCIAL
MIXED-USE
MORTUARIES, CEMETERIES, CREMATORIUMS
NIGHTCLUBS
NON-PROFIT
OFFICE BUILDINGS
OUTDOOR RECREATION
PARCELS WITH NO VALUES
PARKING LOTS
PRIVATE SCHOOLS AND COLLEGES
PUBLIC SCHOOLS
RESIDENTIAL
RESTAURANTS
SERVICE STATIONS
SHOPPING CENTER
STORES, ONE STORY
THEATER / AUDITORIUM
TIMBERLAND
UTILITY
VACANT COMMERCIAL
VACANT GOVERNMENT
VACANT INDUSTRIAL
VACANT RESIDENTIAL
WAREHOUSING
WATER

FIGURE 2.3   A thematic map indicating the property tax designation for individual land parcels in a portion of Leon County in Florida.

Inspection 6

Using the Internet, locate the paper "A spatial cluster analysis of tractor overturns in Kentucky from 1960 to 2002" written by Saman et al. (2012). This paper was published in 2012 in the journal *PLoS ONE*, which is freely available through the Internet. Figure 1 of this paper provides an estimate of the number of tractor overturns per 100,000 tractors in each county, and Figure 4 provides a set of thematic maps to illustrate county-level landscape and demographic characteristics. Given this information, in what part of the state did there seem to be a higher incidence of farm tractor accidents over the period that was analyzed? Through visual inspection of the maps, how does the risk of farm tractor accidents seem to correlate with landscape and demographic characteristics?

One type of thematic map associated with the management of natural resources involves the scope of destruction imposed by wildfires or the threat of potential destruction given current landscape characteristics (vegetation, topography, etc.). Particularly in the western United States and Canada, and across Australia, wildfires in the last decade have burned considerable forest and rangeland areas. Perhaps out of fear and fascination thematic maps portraying burned areas or those of high fire risk are of interest to society (Fig. 2.4). Enhancements to these types of maps might include the portrayal of risks at a larger scale (smaller area), and the presence of human development, especially homes and businesses. In a similar vein, maps portraying insect or disease outbreaks in forests can be of great interest to society. Once assessed, the risk of damage to vegetation can be presented thematically to communicate potential problems to stakeholders interested in the management of these resources.

As we began this section, we suggested that thematic maps can be developed to illustrate the geographic occurrence of the natural resource phenomena or other resources of concern to humans, of which there is a broad array that interests society today. As will be illustrated later in this chapter, thematic maps can portray pollution or air quality issues, caused by humans or occurring naturally, and their potential distribution can spread across a country or continent. Also, tree and plant distribution maps can be designed as simple thematic maps that inform society of past, current, or future areas where these plants might naturally be found. Of further interest to society may be variations in lower atmospheric temperature levels and temperatures of water bodies, which can be displayed as thematic maps.

Translation 5

View the November 2017 US monthly drought outlook map stored on the book's website (mapping-book.uga.edu). For the time period that the map was valid, describe in general terms the areas of the United States that were estimated to likely improve with respect to current droughts and the areas that were estimated likely to face new pressure from potential droughts.

Choropleth Map

A choropleth map is a general type of map that presents ranges of the quantitative characteristics of spatial features through changes in colors or symbols. With this in mind, one

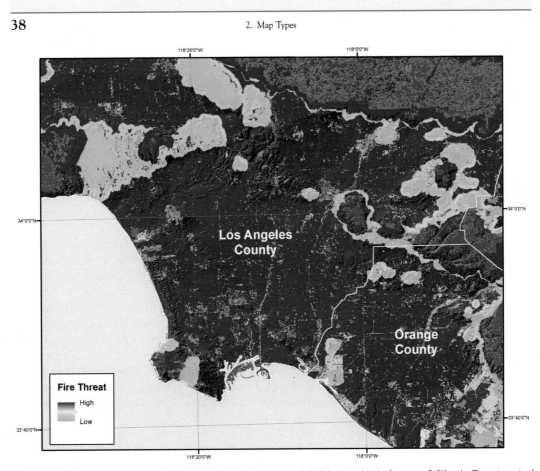

FIGURE 2.4   Fire threat levels in the Los Angeles area of California, 2018. Source: *California Department of Forestry and Fire Protection (2018).*

might consider choropleth maps to technically be a subset of thematic maps. Areas or features with the same symbol or color are meant to represent similar conditions. For example, Fig. 2.5 illustrates the average age of trees within various stands (management units) of a forest. In developing a choropleth map, one would classify their underlying geographic data into interval ranges, then select the appropriate graphical symbols or colors to represent those values (Cromley and Mrozinski, 1999). Within Fig. 2.5 map, the average ages of the trees within specific areas of land (often called *stands of trees*) are assigned to intervals of 20-year age classes; then each age class is depicted or drawn using a different color. Of the many other applications of choropleth mapping, some examples include the mapping of crime rates (Poulsen and Kennedy, 2004) and the incidence of disease (Kronenfeld and Wong, 2017) among and between politically defined land areas. The class intervals chosen for a choropleth map can be designed to favor the intended message; however, classless choropleth maps (Fig. 2.6) which present quantifiable data associated

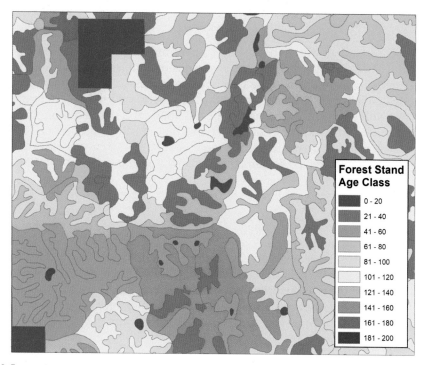

FIGURE 2.5 A choropleth map of forest stand age classes for a portion of the Talladega National Forest in Alabama (2013).

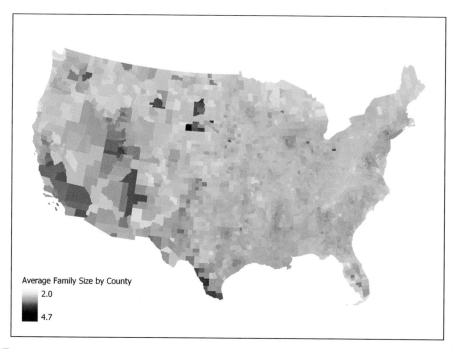

FIGURE 2.6 A choropleth map illustrating the average family size of each county in the conterminous United States, using a continuous scale color range.

with divisions of land or water bodies as a spectrum of continuous color combinations may also be of value (Tobler, 1973). One of the disadvantages of a choropleth map is that when two adjacent areas are assigned different symbols (colors or otherwise), the map may suggest a rather abrupt change in condition, whereas a gradual change in color may better reflect subtle changes in the nature of the landscape. For example, with respect to the soil map presented as Fig. 2.1, the soil types that are illustrated likely do not change abruptly from one to another, but rather more gradually change as elevation, aspect, or underlying geology change. To be fair, this perceived weakness is not unique to a choropleth map.

> **Translation 6**
> Examine the choropleth map illustrated in Fig. 2.6. Imagine that you have a family member, an international friend, or a colleague who is interested in demographic aspects of the United States. Develop a short story for them about the range of average family sizes across the conterminous United States, using this map as a basis. What are the two or three main messages you might convey to them from this graphical representation of a single aspect of the US demographics?

Isopleth Map

Within an *isopleth map*, various ranges in the values of landscape or atmospheric conditions are presented as line symbols (Fig. 2.7). An isopleth map requires information that is distributed (or modeled) across a landscape in at least two dimensions, as well as information that is described with continuous numeric values (rather than discrete or binary values). Along each line within an isopleth map, it is assumed that conditions are constant and have the same measurable value or quantity with respect to some phenomena. Examples include elevations of land, precipitation amounts, and atmospheric pressure conditions. The lines within an isopleth map should never cross, and the distance or interval between each line within the map would ideally be consistent. One common isopleth map used in meteorology contains *isobars*, mapped lines representing locations of equal atmospheric pressure conditions within the tolerances of their measurement and models. Similarly, *isotherms* are lines representing locations of equal air temperature conditions. Contour intervals found on topographic maps, those lines of equal land elevation, are technically called *isoheight* lines. Certainly, it is possible to produce a map that represents differences in elevation, air temperature, atmospheric pressure, and other phenomena simply with ranges of color, yet without the distinct labeling of classes, these maps may be unacceptable as true isopleth maps.

As we suggested, a topographic map (Fig. 2.8) is technically an isopleth map. A topographic map presents and describes elevations of land with respect to average sea level and may include many other thematic features. In cases involving features below water, *isobaths* are used to illustrate changes in depth to a solid surface under water. Each line on a topographic map represents a distinct elevation or depth, and the lines are drawn to represent a constant change in elevation, such as every 20 feet. These *class intervals* of values are convenient groups designed to divide the data presented for informational purposes (Mackay, 1955). In this case, the class intervals, or contour intervals, help us understand

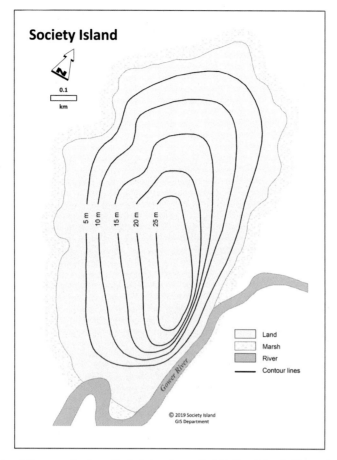

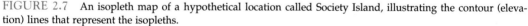

FIGURE 2.7 An isopleth map of a hypothetical location called Society Island, illustrating the contour (elevation) lines that represent the isopleths.

relative changes in values. When the lines are placed closely together on a map, the values that they represent change quickly across the landscape. In the case of land contours, this implies that elevation changes quickly (steep inclines). When the lines are placed far apart on a map, the values that they represent change slowly across the landscape. Again, in the case of land contours, this implies that elevation does not change much (relatively flat land). The scale of a map can influence the number of classes presented, as the density of the lines drawn needs to be considered to avoid the pitfalls of omission (too few lines) and confusion (too many lines), and to deliver a meaningful message to the user of the map (Lin and Cromley, 2017).

Inspection 7
Using the Bucksport (Maine) U.S. Geological Survey 7.5-minute topographic map that is provided on the book's website (mapping-book.uga.edu), what is the contour

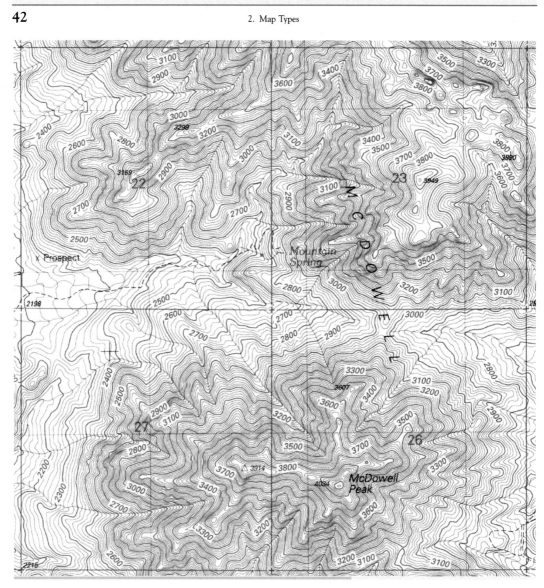

FIGURE 2.8   A portion of the U.S. Geological Survey McDowell Peak 7.5-minute quadrangle in the State of Arizona (33111-F7-TF-024).

interval associated with the isopleth portion of the map? What would be your assessment of the elevation of Treat Hill (approximately 4,933,000 m north, 517,000 m east)? What would be your assessment of the elevation of McCloud Mountain (approximately 4,939,000 m north, 511,000 m east)? About how deep is the deepest part of the Penobscot River between the town of Bucksport and Fort Knox/Prospect Ferry?

Isarithmic Map

An *isarithmic map* is a form of a thematic map that uses *isolines* (lines drawn on maps to represent points that have similar values) and areas representing similar values to illustrate similar conditions of phenomena across a continuous surface. A precipitation or an air temperature map (Fig. 2.9) that illustrates these phenomena across a landscape are examples. We may only have observations of these at specific points within the landscape, perhaps from official weather stations. An interpolation process would be used to create

FIGURE 2.9 Average annual minimum temperature for countries of North America. Source: *U.S. Department of Agriculture, Agricultural Research Service (1990).*

realistic, continuous surface maps representing their estimated distribution over the entire mapped area. Isarithmic maps can therefore be used to illustrate and estimate the precipitation or air temperature everywhere, in places where these phenomena were not measured. An *isometric map* is similar, although emphasis is placed on the use of two-dimensional isolines in such a way as to make a map appear to represent the underlying data in three dimensions (Dooley and Lavin, 2007). A set of *isohyet* lines would be used on isarithmic maps in a manner such as contour lines, to represent those places where the precipitation experienced during a rainfall event would have been equal. When an isarithmic map represents the density of phenomena, it may be referred to as a *heat map*.

Characteristics of natural features that vary continuously across an area may be well suited for this purpose. For example, across a vast forested area, one might sample forest fuel loads at specific sample locations, then model or interpolate the spatial distribution of fuel loads across the landscape using an isarithmic process called *kriging* (Kalabokidis and Omi, 1995). Another method for spatial interpolation in this sense is *inverse distance weighting*. Both processes are often available in modern geographic information system (GIS) programs. For over 150 years human populations have also been modeled using isarithmic maps, even though critical arguments can be made that the spatial interpolation of these phenomena (*the density of humans across a landscape*) may not accurately reflect the presence of humans in certain places of the landscape (Nordbeck and Rystedt, 1970).

Hypsometric Map

A hypsometric map is one that explicitly presents and describes elevations continuously across a landscape, using changes in colors or the shades of colors. Very similar to isarithmic maps, with the exception perhaps of the absence of drawn lines, the term *hypsometric* relates to the measurement of heights. A tool for measuring tree heights (e.g., clinometer, laser) would be considered a *hypsometer*. Rather than using lines, as in the case of isopleth maps, hypsometric tinting (i.e., coloring) is commonly used to represent differences in elevation in conjunction with raster data (grid cells, pixels). Although color schemes may be misleading in some cases, in temperate zones on Earth the green color might represent low-lying valley bottoms and colors from orange to red might represent higher elevations (Fig. 2.10). If a landscape included very high elevation geographic features, such as mountains, the white color might represent the mountain tops (Darbyshire and Jenny, 2017). Divergent color schemes have been developed for other places, such as Earth's moon. Here, warm orange and red color tints have been used to represent highlands (Fig. 2.11), and cold blue colors have been used to represent broad plains, or lunar mares (Zeng et al., 2015).

Reflection 7

We have presented several general map types that are only slightly different in how features or colors are to be displayed, or slightly different with respect to their purpose. Thematic maps seem to be the broader class, and choropleth (areas), isopleth (lines), isarithmic (areas and lines), and hypsometric (pixels) seem to be the subsets of this. Of those presented thus far, which general type of map appeals to

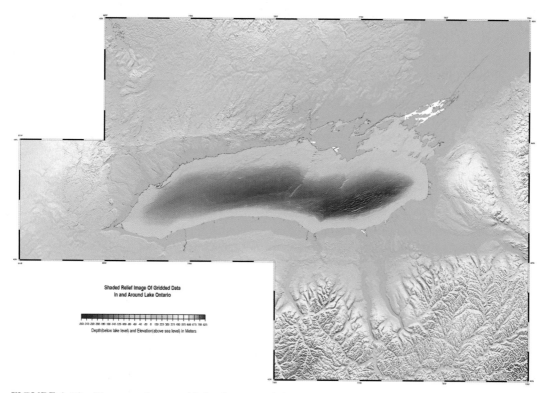

FIGURE 2.10 Hypsometric map of Lake Ontario and the surrounding Canadian and US countryside. Source: *Courtesy U.S. National Oceanic and Atmospheric Administration through Wikimedia Commons.*

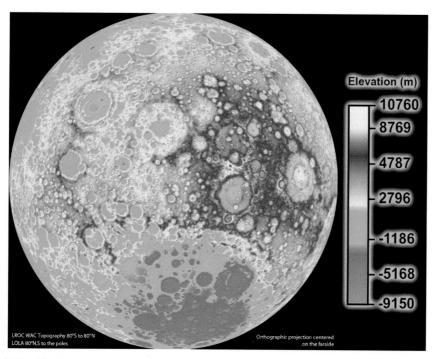

FIGURE 2.11 Hypsometric map of the Earth's moon. Source: *NASA/Goddard Space Flight Center/DLR/ASU.*

you the most? Did you base this selection on the potential use of the map, on philosophical preference, or for some other reason?

While the color models selected for a map are essential within the context of the landscape (see Chapter 6: Map colors), hypsometric maps are useful for communicating topographic detail in a manner different from isopleth maps, using divergent color schemes applied to grid cells (raster data) rather than lines (vector data) to represent places of similar elevation. However, the choice of hypsometric classes, and the associated color scheme, should be carefully considered so that essential landscape features that have slightly different elevations are not grouped together and obscured (Denil, 2016). *Hypsometric analysis,* or *hypsometry*, has become an area of geographic research that concentrates on the analysis of landscapes and the bottom surfaces of water bodies to describe natural and anthropogenic processes that have resulted in their formation. For example, digital elevation models might be used to describe the influence of glaciers on the characteristics of mountains (Sternai et al., 2011).

Dasymetric Map

Dasymetric mapping is a concept that formally began as a method for describing human population densities across a landscape. Although alluded to in works earlier than 1900, the term *dasymetric* was coined by Russian geographer Benjamin Semenov-Tian-Shansky as a Greek translation of the term for *measuring density* (Petrov, 2012). A dasymetric map is a form of thematic map, perhaps best viewed as a compromise between choropleth and isopleth maps. This type of map is used to display volumetric information (e.g., human population density) using changes in map colors (Fig. 2.12), thus enhancing the visualization of differences in values over a landscape. A dasymetric map uses statistical processes and ancillary geographic data to divide a landscape into homogeneous zones, so that one can visualize spatial distributions (McKenzie, 2008; Petrov, 2012). This type of map would be similar in purpose to the modeled isarithmic maps of human density described earlier.

> **Translation 7**
> Using Google Earth or some other online imagery program, find the city of Augusta in Georgia (United States). Compare what you see in the imagery to the dasymetric map described in Fig. 2.12. Do you agree with the representation of areas where there are high human population densities (places where people live)? Other than the water bodies illustrated in the map, do you agree with the representation of areas of low human population densities?

The ancillary geographic information on a dasymetric map would recognize that certain areas of the landscape would not contain the feature being mapped. For example, in Fig. 2.12 you can see that one class ("population 0") represented by a light cyan color would not contain people; these specific areas are ponds, wetlands, and industrial areas. In contrast to a choropleth map, which may display statistical characteristics of a landscape, a dasymetric map interpolates statistical characteristics across the landscape,

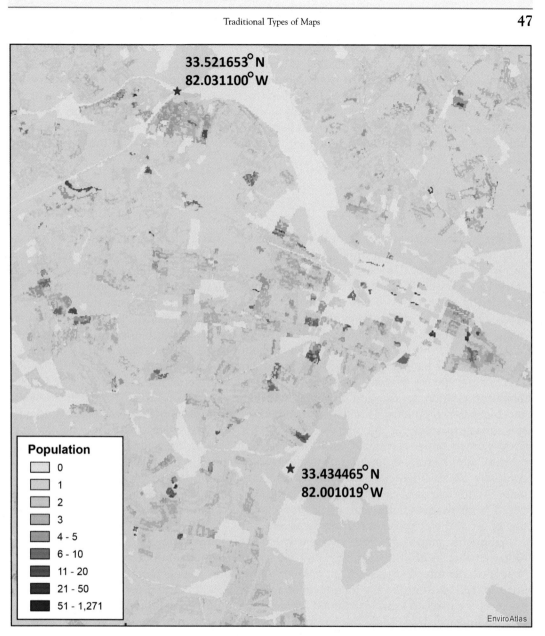

FIGURE 2.12   A dasymetric map reallocating the 2010 population of Augusta, Georgia (United States) to 30 m grid cells based on land cover and slope of the landscape. Source: *U.S. Environmental Protection Agency (2017)*.

creating a population density surface that looks more natural than arbitrary lines defining the boundaries of census blocks, cities, counties, or other political units (Petrov, 2012). The example presented in Fig. 2.12 distributes census block population to raster cells based on land cover and ground slope.

The advent of high-resolution imagery and advances in computer systems (and software) have facilitated a great interest in dasymetric mapping over the last two decades (Petrov, 2012). While often used to describe the spatial distribution of human populations, a dasymetric map can likely be developed to describe the spatial distribution of any resource (human or otherwise). For example, spatial data regarding the location and intensity of lobster fishing off the shores of Maine has been used to develop a dasymetric map indicating patterns of fishing activity within about 50 nautical miles of the coast (Brehme et al., 2015).

Planimetric Map

Planimetric maps are those developed through direct surveys of land areas and are often used to indicate prominent landscape features, land uses, or ownership of land within settled areas. They are said to depict only natural and cultural features of a landscape (Brinker and Wolf, 1984), and may lack indications of topographic relief that may be included in other types of maps, such as those described earlier. Often, they are referred to as *line maps*, yet older surveying texts (e.g., Breed and Hosmer, 1928) suggest that line maps can indeed include contour lines that represent topographic relief. In any event, a planimetric map illustrates the horizontal position of features across a landscape and is characterized as having a uniform scale throughout the map (Bouchard and Moffitt, 1965). Therefore, distances between features on a map should be highly accurate. These types of maps can include the location of roads, building footprints, railroads, bridges, sidewalks, and other developments, as well as water bodies and forests, if these features are identifiable through surveys or the interpretation of areal images. Other structures, such as grave sites, storage tanks, terminals, and airports may also be depicted in planimetric maps. Further, administrative boundaries (state, county, and city boundaries), administrative or corporate limits, district lines, and parcel boundaries may be included in planimetric maps. While indications of topography are often absent in these maps, positions of surveyed benchmarks (along with associated elevations) can provide a map user an indication of ground elevations at various locations.

Planimetric maps have been of great value to urban planning efforts, but also add significant value to utilities planning processes, land acquisition and administration efforts, recreation planning purposes (e.g., park or golf course management), and general navigational purposes. With a very high level of accuracy, planimetric maps (in this sense, called *plats*) can be used as legal instruments to describe land ownership or other features of cultural significance (Fig. 2.13). One example of a planimetric map used for archeological purposes is an earthlodge village map created by Wood (1993) that describes a settlement of Native Americans along the Missouri River in North Dakota. For recreational purposes, planimetric maps have also been noted to be of value in the development of Nordic ski trail maps (Tait, 2010).

Inspection 8
Using Google Earth or some other online imagery program, find Fort Clark State Historic Site in North Dakota. Compare what you see in the imagery provided to the

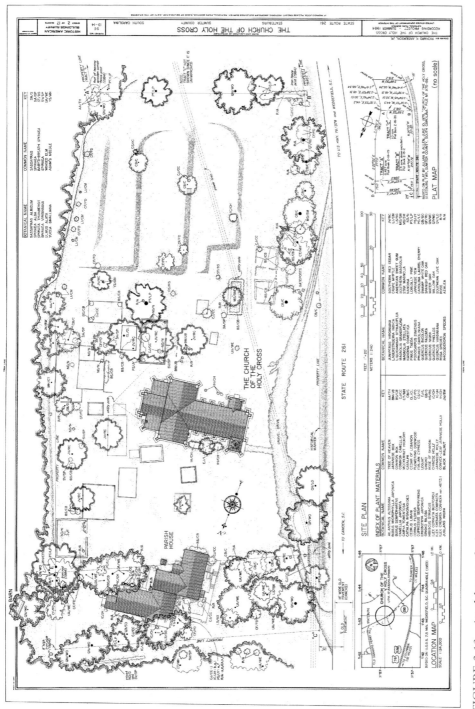

FIGURE 2.13   An elaborate planimetric map of the land containing the Church of the Holy Cross in Stateburg, South Carolina. Source: *Anderson (1984)*.

planimetric map (noted as "Figure 3") in *Integrating ethnohistory and archaeology at Fort Clark State Historic Site, North Dakota* by Wood (1993). This paper was published in the journal *American Antiquity*, which can be accessed through Cambridge University Press or JSTOR.org, depending on access restrictions. Describe, with specific examples, your impression of the planimetric map as it compares to the actual resources found at this site.

Base Map

Base maps are maps and images often used in the background of a broader map product. Base maps can provide detailed information on aspects of a landscape, such as topography, waterways, roadways, administrative boundaries, land uses, and aerial or satellite imagery. Base maps are used as both as a reference point for a map product (*What else is around my area of interest?*) and as an analytical tool (*How does my area of interest relate to the broader land-scape?*). A base map should be visually and thematically acceptable, and used in a manner that will not overshadow other thematic information provided through the map product. For example, a base map should not be so elaborate that it draws attention away from the features that are more important to the message of the map (Raposo et al., 2016).

Diversion 6

Using the Internet, search for the U.S. Department of Agriculture's Web Soil Survey. Once this link has been located, press the green button labeled *Start WSS*, and then zoom in and define (create) a specific area of interest. Select the soil map tab near the top left side of the Internet page, then press the printable version button off to the right side of the page to create a map. What type(s) of base map(s) is(are) used as default, background reference information for this digital soil mapping service?

In recent years, GIS software packages like ESRI's ArcMap and ArcGIS Online have provided base map services to users of their mapping technologies through both software and Internet-based computer mapping platforms. Similarly, as was suggested in the diversion noted above, government entities and private companies utilize base maps as a common informational backdrop in association with the services that they offer. Some government agencies, such as U.S. Geological Survey's National Map Viewer and the United States Department of Agriculture's Natural Resources Conservation Service also directly provide publicly available base map products through online map portals, downloadable files, and GIS data servers.

Specific Types of Maps

World Maps

A world map describes the physical features, political divisions, or perhaps topography of the entire planet Earth (Fig. 2.14), or most of it, as the map developer understood it at

Political Map of the World, January 2015

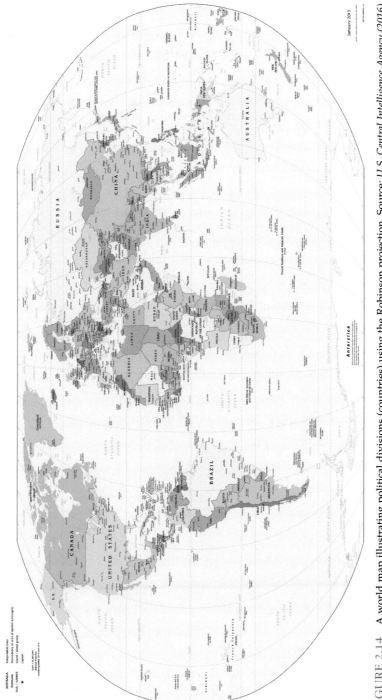

FIGURE 2.14   A world map illustrating political divisions (countries) using the Robinson projection. Source: *U.S. Central Intelligence Agency (2016)*.

the time the map was developed. As only about 29% of Earth is land, all world maps show considerable areas of water and therefore some land areas appear so small that they perhaps fail to provide the map user any useful detail (Jefferson, 1930). World maps underwent a continuous evolution from about the 13th century through the 20th century as society's understanding of the existence and shape of land areas evolved. Famous examples of world maps began with areas society understood well, Europe, North Africa, Japan, and China, and later expanded to the world as we now know it through land and sea expeditions and more recently remote sensing technologies (satellite imagery, aerial photography). The Hereford World Map (c.1300) is among the earliest still preserved images of Europe and Africa as it was known, and embellished with iconic depictions of events occurring during this period. An elaborately drawn and painted image of land masses, water bodies, cities expressing the political aspirations of King Edward I of England, the Hereford World Map is perhaps the oldest, fully preserved medieval map (Strickland, 2018). Some of the early maps of the world included imaginary lands that were added due to incomplete knowledge, and may have been copied or emulated from older maps that presumed those lands to exist (Heawood, 1923).

As a modern society, we now have a solid understanding of the physical location of land masses, due to the advent of satellite imagery and global positioning systems (GPS). However, political and other physical descriptions of the world will continue to change or are in dispute, and thus thematic world maps will continue to evolve. Because of the shape of the Earth, a significant challenge in the development of a world map is to represent areas or distances appropriately. Imagine peeling an orange (or dissecting a soccer ball) and then attempting to flatten the surfaces of the peel (or ball). Therefore, one major challenge of developing a world map is to represent a three-dimensional oblate spheroid on a two-dimensional surface (Fig. 2.15), whether this surface is a piece of paper or the screen of a digital device. The projections employed distort distance and direction, and perhaps areas; we discuss these in more detail in Chapter 4, Map reference systems.

Navigation Maps

Navigation maps can include ground-based transportation maps and maps related to the movement of ships or other vehicles across water bodies or air routes. At its most basic level, a navigation map either (1) represents the path one might take to get from one place to another using the mode of travel selected or (2) represents all potential paths throughout a transportation system. Many people are accustomed to printed navigation maps that are commonly available as foldable travel guides (Fig. 2.16). In addition to displaying traditional or approved routes across land or water bodies, a navigation map might describe the flow of airline traffic throughout the sky above a country (Fig. 2.17).

Reflection 8
Of course, navigation maps are very often viewed on portable digital devices such as cell phones and tablets. They are also available through Internet services such as *MapQuest* or *Google Maps*. Think about the last navigation map you used. What was the form of the map (digital, printed) and what was your purpose for using it?

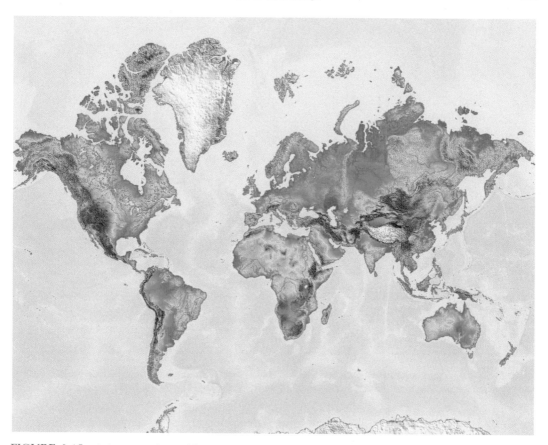

FIGURE 2.15 A hypsometric world map using the Mercator projection. *Source:* Physical world Map *image courtesy PublicDomainPictures.net.*

An *atlas* may be envisioned as a collection or sequence of regional navigation maps depicting various parts of the world that, when considered as a whole, represent the geography, political division, and perhaps other natural resources or features of human interest around the planet. In contrast to single-use thematic maps, an *atlas* is a collection of information, containing maps with associated photographs, graphics, illustrations, and supporting content. Examples of the type of navigation atlas we reference here are the very popular road atlases developed by *Rand McNally, Delorme, Michelin Guide,* and other publishers which focus on smaller geographic areas (cities, counties, states, single countries). The common navigation maps found in road atlases and on walls of classrooms have been called *prototypical maps,* illustrating typical qualities of maps on which others are based (Uttal, 2000). Atlases can also help describe the natural and human systems of countries and regions around the world. For example, atlases of traditional place names and undocumented routes on landscapes inhabited by aboriginal communities have been developed to preserve traditional histories

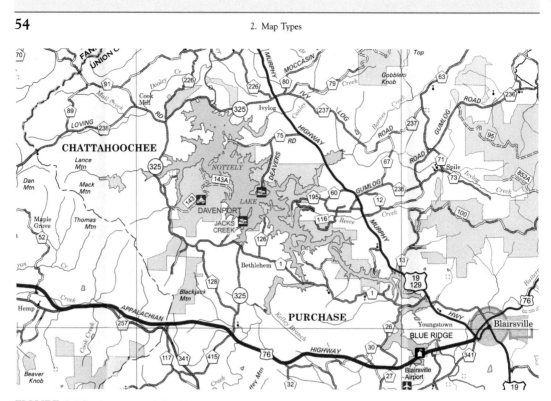

FIGURE 2.16   A portion of the Chattahoochee National Forest (Georgia) indicating the local road transportation system. Source: *U.S. Department of Agriculture, Forest Service (2012).*

regarding the use of land, to illustrate resource ownership and sovereignty of a region, and to engage these communities in the dissemination of their knowledge (Engler et al., 2013).

Nautical Charts

The purpose of a nautical chart is to describe coastal areas and adjacent water bodies so that map users can navigate safely (Yan et al., 2015). The existence of nautical charts can be traced back to the 2nd century and the Ptolemaic Atlas. These original nautical charts were essentially basic representations of coastlines, and with an increased interest in geography and advances in technology, the accuracy of nautical charts improved over time. In the 13th century, there was an emergence of highly accurate nautical charts, called portolan or harbor-finding charts, that depicted the Mediterranean Sea and other parts of Europe (Andrews, 1926). Contemporary nautical charts show not only coastline characteristics and place names, but also shipping channels, seafloors, waterway obstructions, water depths, and submarine descriptions (Fig. 2.18). The technology for developing modern nautical charts has evolved from the use of aerial photography to depict coastlines and hydrographic features (Jones, 1957; Theurer, 1959), to the use of lasers (specifically LiDAR technology) and augmented reality to develop highly accurate spatial datasets (Burdziakowski et al., 2015) and the use of echo sounding to develop submarine terrain models (Yan et al., 2015)

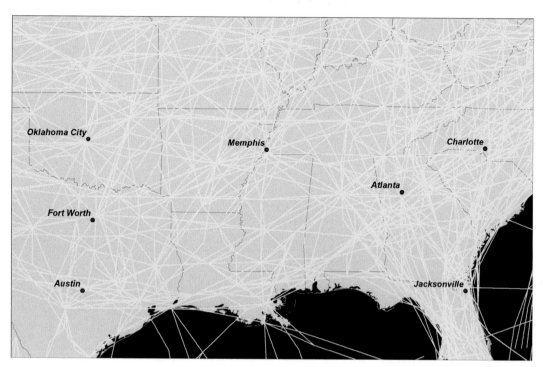

FIGURE 2.17   Routes for channeling the flow of airline traffic for the provision of air traffic services across the southern United States. Source: *US Department of Transportation, Federal Aviation Administration (2018)*.

Location Maps

Location maps are a broad set of map products that provide a reference for the placement of people, businesses, or resources. Through digital devices such as cellular phones, tablets, or personal computers, for example, one might be able to view a location map such as *Snap Map*, a service provided by Snapchat, to view the locations of friends who have shared this information (Fig. 2.19). Another digital application, *Locus map*, allows users to navigate and record points of interest, primarily related to outdoor recreation activities. For marketing and commerce purposes, location maps may have been developed to provide a graphic that allows a potential customer to understand where a place (business, park, etc.) is located. These can be delivered to a digital device or they can be included as a colorful inset of pamphlets or advertising materials (Fig. 2.20). Within the WhatsApp application, people can also share their current location (or rather, the location of their cellular phones) with others. Location maps in the vein of our discussion here could also consist of small-scale maps inserted within larger maps, illustrating the general location of the landscape and thus providing geographic context to map users. However, these types of location maps might be more appropriately considered as *locational insets*, which we will describe in Chapter 4, Map reference systems, and they would not normally stand alone as a separate map, but be incorporated as a small element of a larger map.

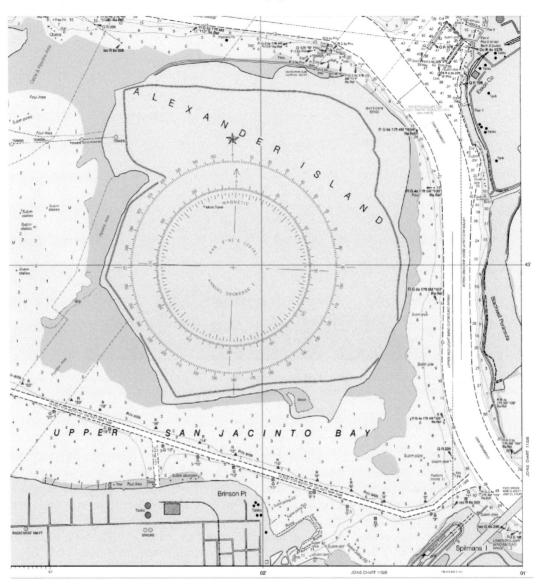

FIGURE 2.18 A portion of the Houston (Texas) ship channel nautical chart. Source: *US Department of Transportation, Federal Aviation Administration (2018). Full map is available on the book's Internet site (mapping-book. uga.edu).*

Inspection 9
Visit the website associated with this book (mapping-book.uga.edu) and locate the Central California District, Mother Lode Field Office map. The map includes a locational inset in the upper right corner that illustrates the general location of the field office administrative area within the State of California. This map could be

FIGURE 2.19   A digital map of the location of people of interest to the map user. Source: *Stephens (2017)*.

considered a navigation map, since major roads are noted throughout the landscape, yet aside from human developments such as cities, counties, and roads, does the map indicate the location of natural or human resources? If so, what resources are illustrated?

Geologic Maps

Geology involves the science and practices for understanding the history, physical structure, and composition of the Earth. Geologic maps, for the most part, illustrate aspects of Earth that will likely never be seen by humans, perhaps developed from extensive subsurface sampling efforts (drilled wells or seismic interpretation). Subsurface contour maps have been developed by geologists to illustrate their knowledge of underground reservoirs of resources. *Isopach maps* are contour maps that indicate the thickness of rock layers and layering of subsurface materials (Fig. 2.21). Although perhaps used interchangeably with the term *isopach map*, an *isochore map* is essentially an underground contour map that illustrates locations of equal thickness values of subsurface materials. Similar to an isopach map, an *isolith map* would display lines that represent different physical characteristics of rocks (the lithology), and therefore presents other information about the rock formations than simply their thickness. These types of maps may all be

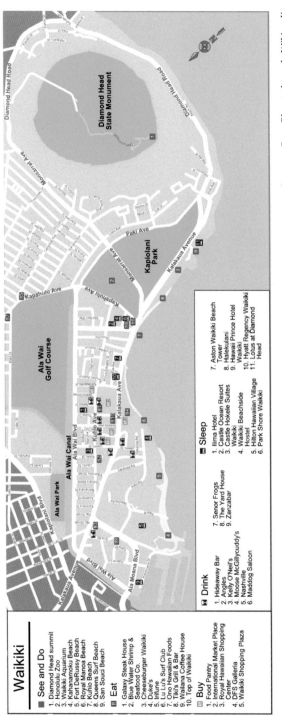

FIGURE 2.20 Places of interest, restaurants, hotels, and businesses in the Waikiki (Hawaii) area. Source: *Courtesy PerryPlanet through Wikimedia Commons.*

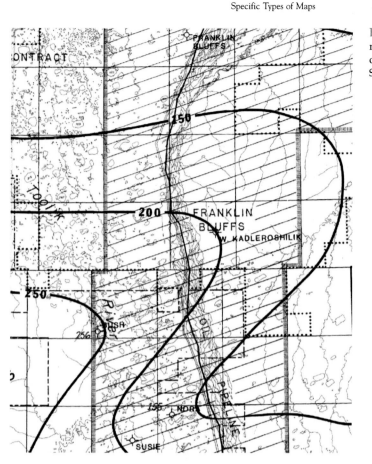

FIGURE 2.21 Outcrop thickness noted in a portion of an isopach map of the eastern north slope of Alaska. Source: *Pessel et al. (1978).*

considered forms of *facies maps* that divide an underground mapped area into subareas based on different expressions of the character of rock, through changes in material composition and formation (Krumbein, 1955). Combinations of these and other map types can enhance the representation and communication of the geologic structure to a broader audience. For example, the general geologic setting of an area in southern California was embedded within a topographic map (Fig. 2.22), to illustrate parts of the San Andreas Fault and other faults (e.g., Bannin, Chicken Hill) across the landscape (Matti et al., 2003). *Land use hazard maps,* or *geologic hazard maps,* are often developed for areas where society needs to improve emergency response and disaster management procedures. For example, areas of potential seismic activity (Fig. 2.23), landslides, or debris flows from natural or human-caused events may be displayed in a land use hazard map.

While many geologic maps describe what can be found on Earth today, the Earth is currently estimated to be 4.5 billion years old (Dalrymple, 1991). We might therefore use other terms to describe the character of Earth at previous points in time. For example, the word *paleo* refers to old or ancient aspects of the Earth (the geology) or life on Earth (the diet). *Paleogeologic maps* therefore illustrate the state of the geologic nature of Earth as it was

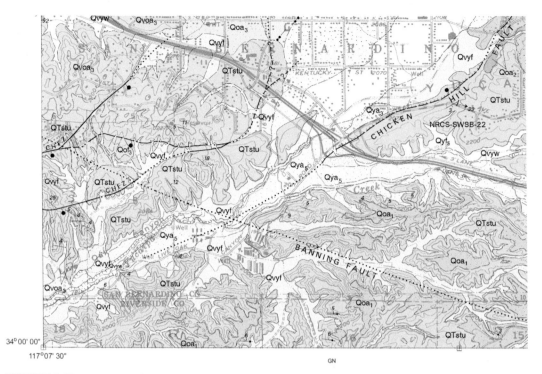

FIGURE 2.22   A portion of the Yucaipa 7.5-minute quadrangle geologic map. Source: *Matti et al. (2003).*

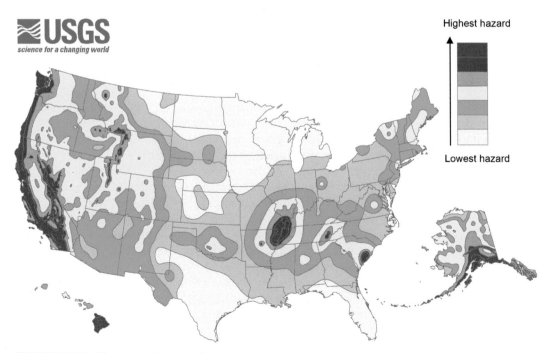

FIGURE 2.23   The potential seismic hazard in the United States. Source: *Image courtesy of the U.S. Geological Survey, through Wikimedia Commons.*

understood to be some time in the past. *Paleogeographic maps* more generally illustrate the distribution of land and water bodies as they were understood during some point in the past (Loomis, 1961). And finally, *paleotectonic maps* similarly illustrate the structural patterns of the tectonic framework of the Earth (Loomis, 1961).

Topographic Maps

A topographic map is an isopleth map, a scaled representation of the surface of a landscape that indicates relief and aspect. With these maps, one can infer some characteristics of hydrography (rivers, streams). A few pages back, we provided a portion of a common topographic map in Fig. 2.8. Topographic maps are developed to provide an indication of the general characteristics of the surface of a landscape through the placement of contour intervals; by association, they then display some information that may be useful in the development of roads, reservoirs, and communication towers, among other useful land management purposes (Brinker and Wolf, 1984). In one study of map preferences, it was noted that the more opportunities people had to use maps in their lives, the more they preferred this type of map (Oksanen et al., 2014).

Translation 8
Access the Pikes Peak (Colorado) 7.5-minute topographic map from the book's website (mapping-book.uga.edu). In the marginalia (notes in the margin) of the map there are a few references to the landscape's topography. Imagine that you had to describe what the irregularly shaped brown lines on the map meant to a friend unfamiliar with these types of maps. After reading the information contained in the marginalia, in a sentence or two describe how one would use contour lines. As an interesting side-story, discuss the time period that the contour lines were developed for this landscape, and whether you feel these represent current conditions.

Bathymetric Maps

Isopleth maps describing the floor of water bodies are considered *bathymetric maps*. These types of topographic maps (Fig. 2.24) are developed to describe underwater features that may be of interest to recreationists, fishers, people in the shipping business, and people interested in the development of marine resources. In one mapping effort, the mountains under the Pacific Ocean near Hawaii were described in a bathymetric map, providing a physical reference for the seafloor topography that most humans will never see (Patterson, 2013).

Inspection 10
Locate the Patterson (2013) paper noted above through a search of the Internet. The journal that published this paper is *Cartographic Perspectives*; the paper can be found in volume 76. Locate Figure 10 in this paper. How were colors used to represent the bathymetry of this area? How deep is the water, in terms of miles or kilometers, in the deepest regions of the Pacific Ocean around the islands depicted in the map?

U.S. Fish & Wildlife Service

Kenai Fish & Wildlife Field Office
43655 Kalifornsky Beach Road
Soldotna, AK 99669   907-262-9863

Skilak Lake

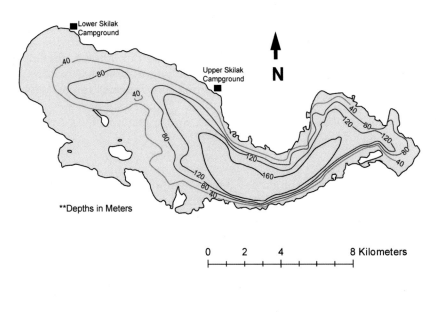

0   2   4   8 Kilometers

Surface Acres: 24,512

Maximum depth: Unknown

Shoreline Length: 26,480 meters (86,882 feet)

Most recent fish survey: 2006

Sport fish species observed: lake trout, rainbow trout, Dolly Varden

Latitude: 60°26.30'N

Longitude: -150°24.88'W

Access: Road

<u>Lake Public Access:</u> Mile 75.3 Sterling Hwy, 5.3 miles (mile 14.3) east on Skilak Lake Loop Road, 1mile southwest on Lower Skilak Campground Road, or Mile 75.3 Sterling Hwy, 10.6 miles (mile 8.6) east on Skilak Lake Loop Road, 2 miles south to the upper Skilak Campground. Facilities at Lower Skilak Campground include toilets, 14 campsites, tables, firepits, water, and boat launch. Facilities at Upper Skilak Campground include 26 campsites, tables, firepits, water, toilets, and boat launch.

FIGURE 2.24.   A bathymetric map of Skilak Lake, located approximately 60.42° north and 150.35° west on the Kenai Peninsula in Alaska. Source: *US Department of the Interior, Fish and Wildlife Service (2012).*

Water depth is obviously of interest when using bathymetric maps, as are the locations of underwater ridges, plains, canyons, and mountains. As dams in the western United States are removed, bathymetry maps are also used to estimate the sediment produced when the dam is breached. A recent bathymetric map developed for an area of Naples Bay on the western coast of Italy provides a digital underwater elevation model that illustrates the underwater relief of an active volcanic complex (Passaro et al., 2016). Shaded relief, aspect, and slope maps (all for underwater surfaces) can all be created to illustrate the complex features that are influenced by sedimentary and tectonic processes. Remote sensing technology, such as photographic images or devices that collect acoustic backscatter, has been used to assist in the development of bathymetric maps (Kostylev et al., 2001).

Property Maps

Planimetric maps that formally describe the dimensions of real estate, or real property, are a form of property map. Although one can develop a map that uses the features associated with a property boundary (the lines or polygons that describe a property's shape), what we are referring to here are those maps that may often be called *plat maps*. Plat maps are cadastral maps that display a thorough description of the boundary of a property through the use of scale, direction, and distance. In Fig. 1.7, we offered an example of a plat map. These property maps often include the location of surveyed corners, from which the property measurements begin or end, and a statement on the reliability of the land measurements. On the ground, the corners would be monumented by a metal pipe or some other type of permanent feature that can be repeatedly located, sometimes etchings on nearby trees may serve as reference marks to these. These property maps also include orientation (north arrow) and certification (surveyors mark or seal), and are often considered to be the highest level of knowledge concerning the location and description of a piece of land. Variations on this theme include platted subdivisions or townships that provide direction and distance information for the boundaries of an extensive collection of individual land parcels. Further, *tax maps*, while generally lacking boundary descriptors, may provide a name or tax identification code for each land parcel.

Soil Maps

Soils are the unconsolidated materials lying on top of the Earth, above the parent material (often solid rock). Living organisms and climate alter soils. Soils also act as a medium for plant growth. A *soil map* indicates the distribution of different soil map units, boundaries. Distributed across the landscape, these polygons represent the varying soil types or soil complexes (or nonsoil areas) based on a soil classification system (Soil Science Division Staff, 2017). Often these types of maps (Fig. 2.25) are created using aerial photography, digital terrain models, and field samples of soil characteristics. For example, the development of detailed soil maps for an area of northern Spain utilized measurements of the depth and texture of soils, judgments on the drainage and water retention capacity of soils, and other soil and geomorphic variables (Castañeda et al., 2017).

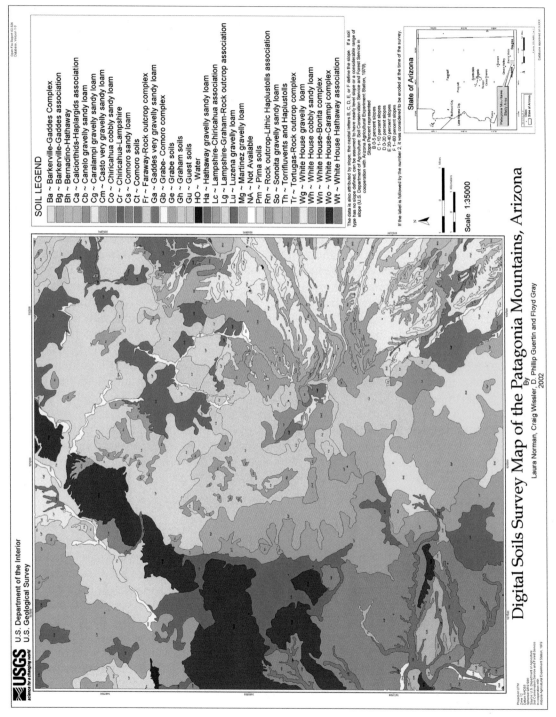

FIGURE 2.25   A map of the soils found in the Patagonia Mountains in Arizona. Source: *Norman et al. (2002).*

Diversion 7

Using the data provided through this book's website (mapping-book.uga.edu) concerning a small area of farmland in Nebraska, develop a thematic map of the soil types of this area. Although we discuss the use of color and the inclusion of map elements in subsequent chapters of this book, make a concerted attempt to create the most aesthetically pleasing map product that illustrates the different types of soils in this area.

As suggested earlier in this chapter, the delineation of one soil map unit from another seems very abrupt due to the lines that are drawn. In reality, the transition from one soil map type to another may be rather gradual. Local models of soil characteristics can be developed in this manner, as well as those that address much broader landscapes. Digital soil maps can be created using spatial and statistical inference processes applied to aerial or satellite imagery. The result would be a raster database of predicted soil patterns or properties in discrete or continuous (values representing different soil characteristics) forms (Soil Science Division Staff, 2017). For example, using a sample of 4857 soil sample points distributed across the United States, concentrations of the chemical composition of soils were interpolated using an inverse distance weighting process (samples closer to an unsampled area have greater weight than those further away), broad-scale soil maps (Fig. 2.26) for the entire country were created (Smith et al., 2014). Global models of soil properties have also been estimated in this manner using soil databases, topography, air temperature, precipitation, land cover, and land use information (Stoorvogel et al., 2017).

Vegetation/Forest Maps

Vegetation and forest maps are thematic maps that illustrate the different vegetative conditions across a landscape or within a specific forest. These conditions are often defined through divisions in tree species (or species groups), tree ages, and management emphases. They may represent current conditions within a small property or across a broad landscape, or potential natural vegetation ranges (Fig. 2.27). For some time, foresters and natural resource managers have been interested in these types of maps since they inform both planning and policy development processes. These maps differ from geologic and soil maps in that they likely need to be updated at regular intervals as vegetation and forests change (Weir, 1997).

Diversion 8

Using the data provided on the book's website (mapping-book.uga.edu) for a portion of the Talladega National Forest in Alabama, and using a GIS program or other similar mapping program such as Google Earth, create a thematic map that illustrates the different vegetation or forest types. Again, although we discuss the use of color and the inclusion of map elements in subsequent chapters of this book, make a concerted attempt to create an aesthetically pleasing map product that illustrates the different vegetation or forest types in this relatively small forested area.

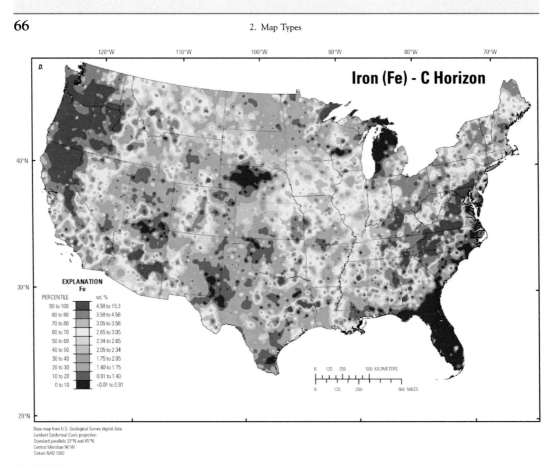

FIGURE 2.26　Distribution of iron in the soil C horizon across the conterminous United States. Source: *Smith et al. (2014)*.

Within the last decade or so, interactive digital maps have been devised to allow one to use a computing device to select and focus on (zoom into) an area of interest, then view current or historical vegetation types. One interactive map service of this type was developed for eastern Africa (van Breugel et al., 2015); another was developed for North American tree species (plantmaps.com). Further, with advances in satellite remote sensing technology, society is now able to understand quite well both the distribution of forest resources across broad areas and the relative quantity or density of these resources (Fig. 2.28).

Zoogeographic Maps

Zoogeographic maps are designed to illustrate the association between animals and the environment in which they nest, roost, forage, and reproduce. Zoogeographic maps have been developed to delineate population presence or absence, and regional distributions of species of animals (Fig. 2.29), across landscapes. For example, one could use geographic and climatic conditions to develop zoogeographic maps indicating the potential presence

FIGURE 2.27 Natural distribution or range of western juniper (*Juniperus occidentalis* Hook.). *Source: Little (1971).*

or absence of Griselda's striped mouse (*Lemniscomys griselda*) and an antelope, the lechwe (*Kobus leche*), across a country such as Angola. Given historical and potential changes in land use, it is not unreasonable to assume that changes in these conditions over time can also be analyzed and mapped. Zoogeographic maps can also be developed to illustrate specific animal migration patterns and other behavior and phenomena associated with

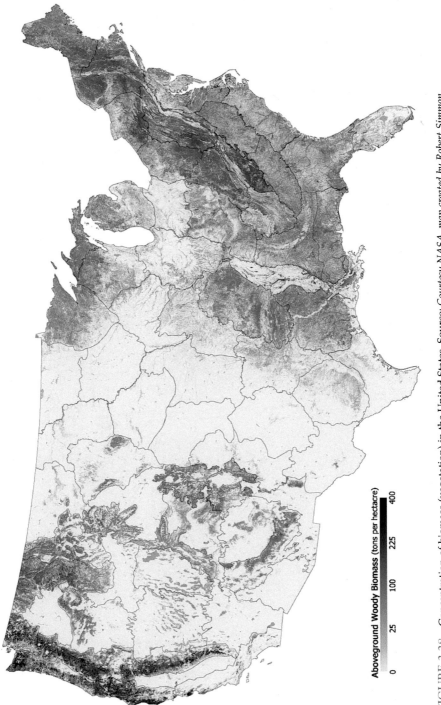

Aboveground Woody Biomass (tons per hectare)

0   25   100   225   400

FIGURE 2.28   Concentration of biomass (vegetation) in the United States. Source: *Courtesy NASA, map created by Robert Simmon.*

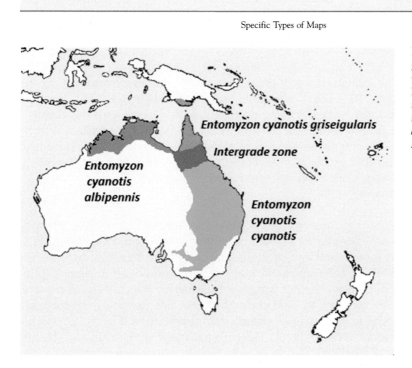

FIGURE 2.29 A zoogeographic range map of the three subspecies of blue-faced honeyeater, *Entomyzon cyanotis*. Source: *Courtesy Cashliber through Wikimedia Commons and adapted from Schodde and Mason (1999).*

landscapes of water bodies. More generally, zoogeographic maps can be developed to illustrate how groups of animals might be distributed across landscapes or how groups of animals utilize natural resources. General zoogeographic maps can also be developed, for example, to illustrate the presence or spread of disease, such as the Ebola virus, by groups of animals across broad landscapes (Olivero et al., 2016).

Environmental Maps

Environmental maps describe the association between human health and well-being and environmental conditions such as water quality and infectious diseases. Maps of this type might not only include forests, wetlands, transportation routes, and other urban or developed themes regarding a landscape, but also include information on the specific locations of potential risks to humans or other animals (e.g., potential sources of toxins), the broader spatial distribution of these risks, or perhaps the sites where these issues of concern are monitored. These types of maps often focus on societal vulnerabilities (Opach and Rød, 2013) and societal risks (Kostelnick et al., 2013), but more simply, some may consider environmental maps to include those that portray stocks of resources of interest to society, such as forest carbon. However, in this short section we are referring more specifically to those maps that directly affect human health and well-being. In addition to infectious disease and water quality issues, these types of maps may also consider urban energy consumption and renewable energy potential, or other types of maps that help society make informed decisions about the environment from engineering and economic perspectives (Fish and Calvert, 2016).

Translation 9
Using the map provided in Fig. 2.30, describe in general terms how it communicates a message to society. The location is a plot of land in west Texas (approximately 33.594481° north latitude, 102.022209° west longitude). The 2004 image (US Department of Defense, Air Force Civil Engineer Center, 2014) represents the extent and severity of underground contamination (the *Tower Plume,* as it was called) which was caused by the release into the environment of cleaning solvent that was once used to repair airplanes. The 2012 image represents the same area, after 10 years of *pump and treat* and *in situ biodegradation* technologies have been applied. The small black points represent monitoring wells.

In addition to static maps that illustrate environmental concerns, interactive mapping applications have also been developed to dynamically explore the spatial location of environmental health data and the relationship of these data to human demographic information. For example, TOXMAP, an Internet-based mapping application, allows one to view the location of Superfund (a US government program funding for the cleanup of contaminated lands) sites and other facilities tracked through the Toxics Release Inventory, and overlay these with demographic data such as census data (population density, population by race or age) and personal income levels (U.S. Department of Health and Human Services, National Library of Medicine, 2018). In response to a severe weather event that affected the northeastern seaboard of the United States (Hurricane Sandy, 2012), the Sediment-Bound Contaminant Resiliency and Response Strategy (SCoRR) application (Fig. 2.31) was developed. This interactive mapping application allows one to view the locations of potential sediment-bound environmental health stressor sampling locations and their association with soils, protected areas, wetlands, federal lands, and demographic data such as census data (population density) and personal income levels (U.S. Department of the Interior, Geological Survey, 2018).

Weather Maps

Weather maps (Fig. 2.32) are ubiquitous in society today. Maps of continuously changing weather conditions are offered by government agencies and news organizations through computers, television, and portable digital devices. They are arguably one of the most frequently used types of maps by the general public (Greenland, 2000). The more technical forms of these types of maps show cold and warm fronts of air (Fig. 2.33), the direction winds were headed, and other details of interest to climatologists, pilots, and others. With the advances provided by Doppler radar, weather maps often show the movement of precipitation events (storms), as radar reflectivity is proportional to atmospheric water content (Grappel, 1999). General weather maps often display air temperature conditions along with the location and intensity of precipitation events. Given the manner in which they are delivered today, they are not usually immortalized as static maps, unless they refer to some significant weather event such as a hurricane (tropical cyclone).

As opposed to climate maps, which present general average conditions for an area from data collected over a longer period, weather maps present short-term or immediate,

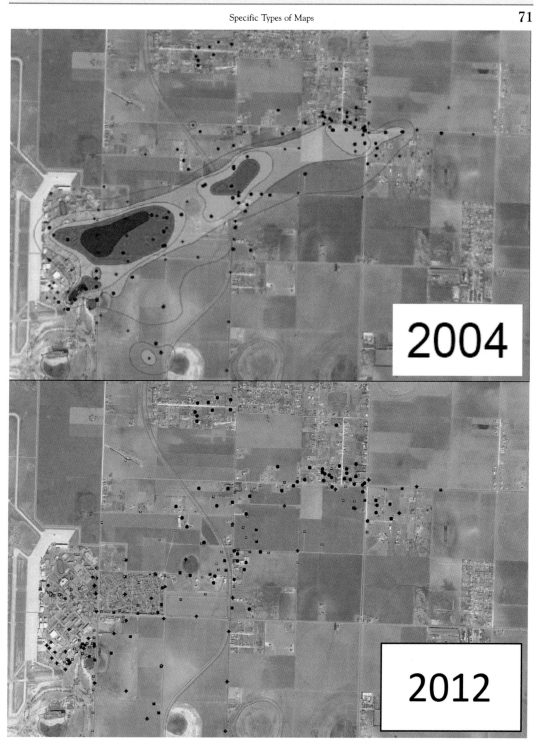

FIGURE 2.30 An environmental map illustrating two points in time and the extent of underground contamination. Source: *US Department of Defense, Air Force Civil Engineer Center (2014).*

Mapping Human and Natural Systems

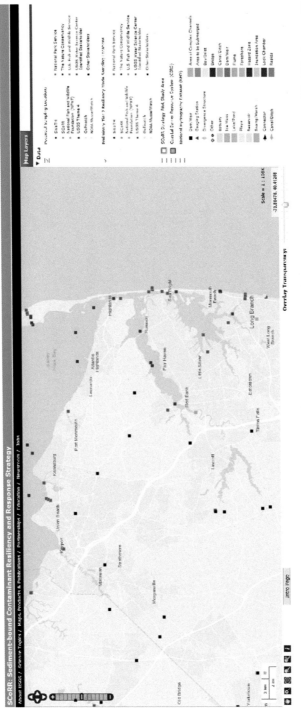
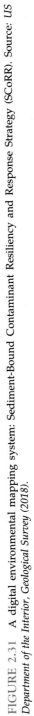

FIGURE 2.31 A digital environmental mapping system: Sediment-Bound Contaminant Resiliency and Response Strategy (SCoRR). Source: *US Department of the Interior, Geological Survey (2018).*

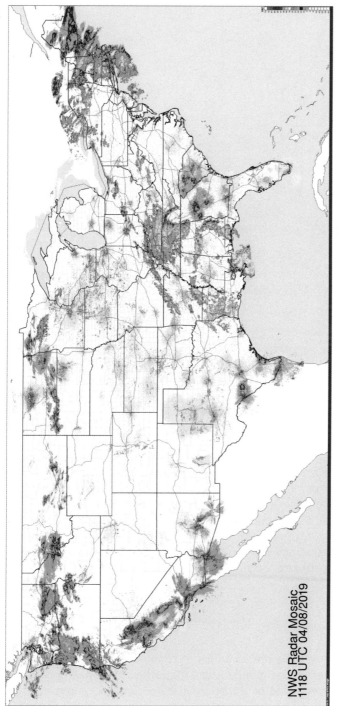

NWS Radar Mosaic
1118 UTC 04/08/2019

FIGURE 2.32 A weather map indicating areas of precipitation events in the conterminous United States on April 8, 2019, at around 07:30 a.m. in the eastern time zone. Source: *US Department of Commerce, National Oceanic and Atmospheric Administration (2019).*

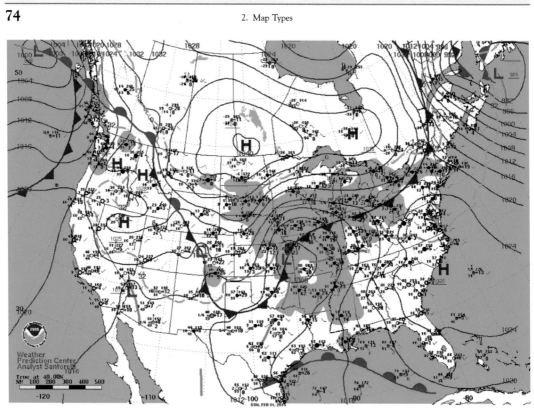

Surface Weather Map and Station Weather at 7:00 A.M. E.S.T.

FIGURE 2.33 A weather map indicating surface conditions, cold and warm fronts, and other detail. Source: *Courtesy National Centers for Environmental Prediction through Wikimedia Commons.*

real-time conditions. Most often, people want to see immediate weather conditions, recent trends, and near-term predictions of weather events. Roads, county boundaries, and other landmarks placed on maps help people understand the locations of storms. To facilitate this, location maps (or base maps) are used as spatial reference for weather conditions. Weather maps provide a synoptic view of the landscape, from observations collected over broad areas such as formal weather stations, to describe earth surface conditions. However, some organizations are developing crowd-sourced weather maps, utilizing temperature, barometric pressure, humidity, and magnetic fields sensed by cell phones of registered users (Marks, 2013).

Planning Maps

In the process of planning the activities of cities, towns, or other government jurisdictions, geographic zones or designated areas might be developed and mapped to illustrate general categories of proposed or approved uses of land. A *comprehensive plan map* represents the proposed long-term vision of an administrative body (e.g., the city planning office). This type of map is mostly thematic, illustrating through color and other annotation the thoughts that a planning office would have about potential changes to the various zones over some fixed planning horizon (10 years, 20 years, etc.). A *zoning map* illustrates approved uses of land, guided by current zoning restrictions and codes. These maps thus describe the types of land management or land development actions that would be acceptable in each zone. A zoning map is complementary to a comprehensive plan map, yet provides more detail concerning the activities that are acceptable today. Land use plans, transportation plans, parks and recreation plans, and water resources plans would all be accompanied by planning maps that illustrate the priorities and proposed activities of an administrative body.

Inspection 11

Using the Internet, search for *City of Saint Paul 2030 Comprehensive plan maps* (City of Saint Paul, 2018). From the Land Use Plan section, access one of the generalized 2030 land use maps. Of the land use descriptors noted in the legend of the map, which were you most surprised that the city would want to designate separately from the others? Of these land use designations, which were you unaware of before to viewing the map?

A *photo plan* uses aerial imagery to express the outcomes of proposed planning efforts (Gomis and Turón 2017). For example, Fig. 2.34 was developed to illustrate a program meant to preserve the design integrity of a culturally significant area of Washington, D.C. and to encourage the placement of new museums and memorials in places that complement existing cultural sites. While a photo plan may be related to other planning maps, it appears to be more specific and specialized in communicating a message to map users.

Noise Maps

Human voices, environmental action (e.g., animals, wind, and water), vehicular traffic, and the various noises associated with shops, restaurants, schools, and other places where people congregate accumulate to produce noise. This mixture of sound may be pleasant to some, yet annoying to others. Within urban areas in particular, people have collected spatially referenced information regarding the location and level of noise, assessed society's perceptions of these, and subsequently mapped sound quality (Fig. 2.35) through indices they developed to represent noise exposure or noise pleasantness (Lavandier et al., 2016). These *noise maps* display predictions of the noise levels that may be encountered within a landscape (Borelli et al., 2014).

Reflection 9

In your mind, select a place where you would like to sit and read a book, or to engage in a conversation with your friends through social media. Perhaps this place is a coffee shop, your living room, or an overlook along a mountain trail. Think about the noises that might be observed at the place you have selected. Make a list of the noises, and categorize them as natural (e.g., running stream) or as originating

Figure 1-2 Adopted Commemorative Zones

Memorial Zones

■ The Reserve

□ Area I

■ Area II

▨ Architect of the Capitol

FIGURE 2.34 A photo plan illustration of a museum and memorial zoning policy in Washington, DC. Source: *National Capital Planning Commission (2001).*

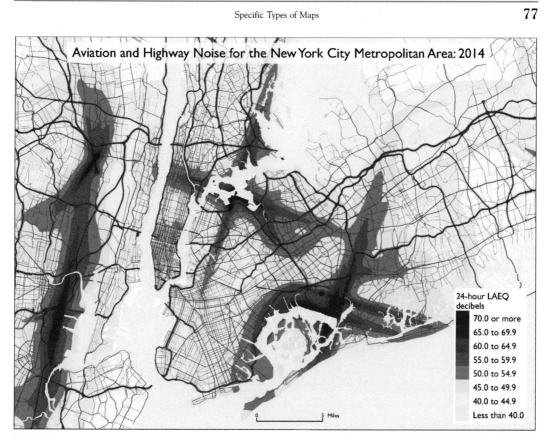

FIGURE 2.35   Aviation and highway noise map of New York, 2014. Source: *US Department of Transportation, Bureau of Transportation Statistics (2014)*.

through human activity (e.g., motorized vehicles). If you were to make a map of noise sources around this place, could the map be used to help you select the place where you would like to sit? How might you develop the map to best illustrate noise direction, intensity, and form?

Flow Maps

Thematic maps that allow one to visualize the movement of phenomena from one place to another are considered *flow maps* (Verbeek et al., 2011). These types of maps can illustrate volume or intensity of flow (Fig. 2.36) and perhaps even the direction of flow. For example, through advances in digital mapping and associated services, people now may be able to understand current vehicular flow volume along roads in their cities. These types of maps can illustrate one-way movement of people, animals, or commercial resources across a landscape. One-way flow maps may illustrate the migration of people or animals across continents. Two-way flow maps illustrate the movement of phenomena

Average Daily Long-Haul Truck Traffic on the National Highway System: 2015

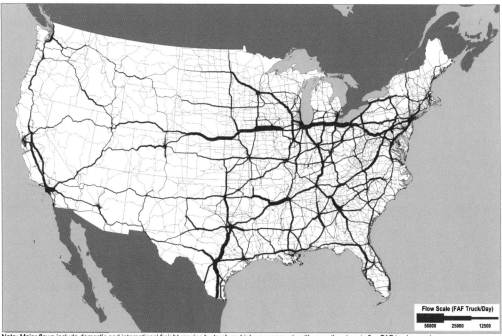

Note: Major flows include domestic and international freight moving by truck on highway segments with more than twenty five FAF trucks per day and between places typically more than fifty miles apart.
Source: U.S. Department of Transportation, Federal Highway Administration, Office of Freight Management and Operations, Freight Analysis Framework, version 4.3, 2017.

FIGURE 2.36   Average daily long-haul truck traffic on the US national highway system in 2015. Source: *US Department of Transportation, Federal Highway Administration (2017).*

to, and from, various locations. For example, the transport of electrical signals across data networks might be illustrated through a two-way flow map. In addition to the direction of flow, flow maps may indicate the quantity or quality of flow using *flow trees*. These flow trees consist of branches that ordinarily do not cross. The thickness of the branches represents the amount or magnitude of items flowing from a source to a destination (Verbeek et al., 2011). Curves through the tree are ideally as smooth as possible, and are ideally designed to avoid covering other important mapped features (Buchin et al., 2015). Other design considerations involve the color of the branches, the tapering of the edges, and the characteristics of the ends of branches (arrowheads). These considerations can help a map communicate importance, dominance, magnitude, origin, destination, and clustering of resources (or people, animals, etc.) within a geographic context (Koylu and Guo, 2017).

Historical Maps

As might be inferred through the previous discussions, historical maps can provide a record of the events that have led to the present state of a landscape (Weir, 1997). Historical maps, perhaps in conjunction with written works and local knowledge, have been used to examine long-term transitions in land cover, settlement patterns, and boundaries of political entities as they were known at the time of map development (Buterez et al., 2016). A map of New York City, as it was understood in 1829, is presented in Fig. 2.37 as an example of a historical map. These types of maps may have illustrated trade routes, and may have elaborated on the size and location of landscapes that were not well known at the time of map development. Sometimes referred to simply as *old maps*, they represent the thematic and topographic knowledge of land and water bodies at some time in human history. The University of Texas Libraries' Perry-Castañeda Library Map Collection is one example of a collection of historical maps, another is The Lionel Pincus & Princess Firyal Map Division of The New York Public Library. While often very fascinating, the accuracy of historical maps has been an area of concern for geographers. The spatial accuracy of historical maps can affect their suitability to assist society in evaluating changes to anthropogenic and natural systems (Lukas, 2014).

Three-dimensional and Tactile Maps

Facilitating the ability of map interpretation through the third dimension (depth) in some cases can enhance the transfer of information to the map user (Edler et al., 2014). As opposed to two-dimensional maps, three-dimensional (3D) maps allow maps users to not only view a landscape of interest but also understand the character and composition of a landscape through alternative modes of sensory perception: stereographic imaging or touching the surfaces of the land, trees, buildings, and other features presented (Kete, 2016). In one study of recreationists, preference for three-dimensional trail maps was evident in people aged 26–40 and in women (Schobesberger and Patterson, 2008). One way to experience a three-dimensional landscape is through a map that was developed using a lenticular (conical) imaging process, which involves two or more map images that are dissected into thin strips and merged into one. This combined image is then presented under a transparent (half-cylinder) foil (Fig. 2.38) that refracts the light interacting with the map, and prompts the human brain to perceive the illusion of a third dimension (Edler et al., 2014). Interestingly, the foils are so small that lenticular images can consist of relatively flat two-dimensional maps which can be presented on the screens of modern digital devices. These types of maps provide one the ability to imagine the third dimension through the illusion of depth, without the need for special optical lenses; therefore, they can assist in the visualization of three-dimensional features (Edler et al., 2014). In essence, this is a stereographic imaging process that may similarly be employed in viewing aerial photographs, yet without the need for special eye glasses or stereoscopes. Many readers may connect more closely with this concept by remembering the postcards, posters, cups, pictures, and other products that produce optical illusions and the so-called holographic effects with slight changes in orientation.

Translation 10
View the Uniform California Earthquake Rupture Forecast map which is available on this book's website (mapping-book.uga.edu). Briefly, what is the message that this map intends to communicate? Describe in brief, simple terms how this map utilizes three dimensions to communicate an idea to the general public.

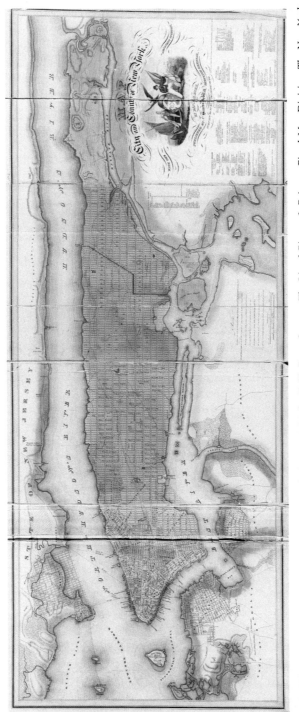

FIGURE 2.37 An historical map of New York City in 1829, by D.H. Burr. Source: *The Lionel Pincus & Princess Firyal Map Division, The New York Public Library.*

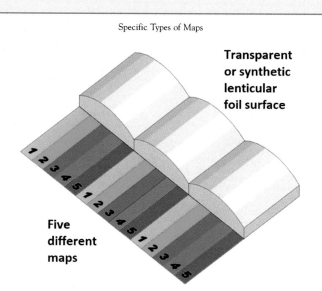

Transparent
or synthetic
lenticular
foil surface

Five
different
maps

FIGURE 2.38 Conceptual model of a lenticular foil process. Source: *Courtesy Koperczak through Wikimedia Commons.*

In an effort to overcome vision-centric map cartography, and provide access to maps for people whose vision is limited, research is being conducted on the development of tactile maps (Perdue and Lobben, 2016). These types of maps provide a person the ability to cognitively imagine a landscape, through haptic perception (the sense of touch), based on stimuli experienced when the map is touched (Kete, 2016). *Tactile cartography* is therefore an area of cartography aimed at translating geographic information onto media (Fig. 2.39), which ultimately may facilitate the comprehension of spatial features of a landscape in a clear and manageable way for people who have vision limitations (Perdue and Lobben, 2016). Tactile maps can be developed using various plastic media and a thermoform process, where the media is rendered pliable. In this type of process, raised three-dimensional features (lines, symbols) are created to represent different aspects of the landscape (Ungar et al., 1997). Heat, pressure, and perhaps moisture are needed to change a flat media into a tactile map. These types of maps can also be created through 3D printing processes, some of which use special paper coated with a chemical surface that reacts to heat applied during a printing process. These types of maps can also be created through a *collagraphy* printing process, where various materials are attached to a rigid, flat base, creating a *collage* that is in essence a map.

Informal and Creative Maps

Virtual world maps, fantasy maps, and other places described in analog or digital fashion with no formal reference to real places on Earth are called maps of *informal geographies* (Kinberger, 2009). Virtual world maps include informal maps of word usage. *Topic bursts* or *word clouds* might fall into this class (Fig. 2.40). Topic bursts can be found within the publications of scholarly journals and other publications (Mane and Börner, 2004). *Cartograms* are another example of informal, creative maps that emphasize geographic distribution of various phenomena (Tobler, 2004). For example, within a cartogram map the

FIGURE 2.39   A tactile map. Source: *Courtesy Samuli Kärkkäinen.*

states within the United States might be presented as land areas proportional to their human population, drawn somewhat in the same location using somewhat of their original shape (Fig. 2.41). While the cartogram itself might resemble geographic features or reasonable depictions of land areas, it may also be quite abstract in design and contain colorful or thematic content (Fig. 2.42). Further, distinct land areas such as states or countries can be displayed as rectangles or other regular shapes, juxtaposed appropriately, yet of a size proportionate to some common statistical measure or proportionate to some other value of interest. For example, a map such as this might illustrate the geographic popularity of such things as preferences for social network applications (Jhong et al., 2011).

Inspection 12
Through a search of the Internet, locate a rectangular cartogram developed by Raisz (1934), that describes the population of the world, or the population of a specific country such as the United States. Describe how this informal, creative type of map is composed, and how it either attracts or repels your interest.

Game maps, whether considered to be *open world* (where video game players can freely navigate through their virtual world) or whether they are more limited in scope, are examples of informal and creative maps. Recently, procedural content generation algorithms

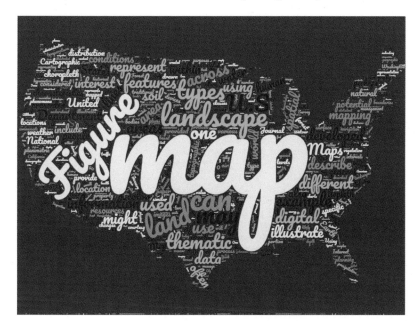

FIGURE 2.40 A word cloud based on the text of this chapter and the boundary of the contiguous United States. Source: *Courtesy WordClouds.com.*

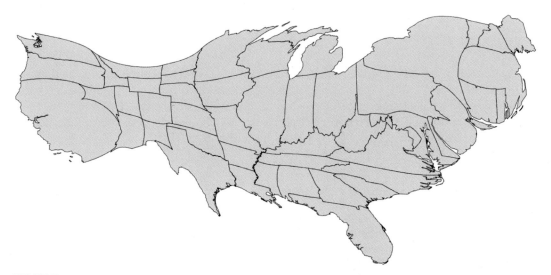

FIGURE 2.41 Cartogram of the contiguous United States indicating the relative number of geotagged Wikipedia entries as of 2011. Source: *Courtesy Dispenser through Wikipedia.*

have been developed to take into account the history of a game user's experience, and then create fresh, interactive landscapes in which a person navigates (Korn et al., 2017). This use of knowledge allows gamers to interact with fresh, dynamic maps of virtual landscapes, rather than forcing them to interact with a set of static maps. Computer

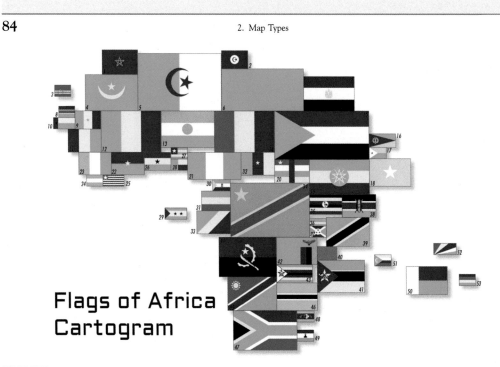

FIGURE 2.42   Flags of Africa cartogram. Source: *U.S. Central Intelligence Agency, Cartography Center, Washington, D.C.*

programmers and video game designers are therefore actively creating processes that generate real-time maps for gamers who demand fluid navigability and appealing aesthetics. Contemporary video game maps may also contain large-scale local map elements (vehicles, plant, buildings) that represent typical natural challenges encountered in real life (Anand and Wei, 2016).

Digital Maps

Digital maps displayed on the screens of cellular phones, tablets, or personal computers are mainly used for locational and navigational services, and they may provide thematic information of a given environment (Nuñez, 2013). Internet-based maps (Fig. 2.43) are ubiquitous today, and we have mentioned the use of digital maps frequently in various parts of this chapter. Given the apparent flexibility available in delivery and presentation, these maps are often used to present thematic content of interest to society (navigation, land use, etc.) at a variety of scales relatively quickly (Battersby et al., 2014). Digital mapping is a relatively new method for communicating knowledge of a landscape, and both Internet-based mapping activities and navigation activities can be conducted on electronic devices through *planimetric browsers* (e.g., Google Maps) or *geobrowsers* (e.g., Google Earth). Geobrowsers are also considered *virtual globes*, given their

FIGURE 2.43   Air quality index map of the United States on December 20, 2018. Source: *Courtesy Airnow.gov.*

model (the Earth) for initiating operations. While we include these types of maps in this book, it should be acknowledged that some people may not necessarily consider the information provided by digital navigation tools to technically be considered maps (Speake and Axon, 2012).

Diversion 9
Using your favorite geobrowser, develop a map of your home neighborhood, and if possible, add annotation that expresses features that may be of interest to visitors. How might you deliver this map to your potential visitors?

Digital maps may be needed to attract the attention of *digital natives* (Oksanen et al., 2014), people born during the digital age (1980 to today), who can interact well with various digital devices, and whose social interactions are often mediated by digital technologies (Palfrey and Gasser, 2008). Interactivity with maps has become a common part of society's use of digital devices, and the effectiveness and efficiency of representation on the devices that deliver these map products may impact how a person views the potential of the map to provide information of value (Mendonça and Delazari, 2014). Digital maps allow a map user to access a map's database information over the Internet or from the *Cloud* (a data center connected to the Internet), to view aerial images of a place that can further be panned and zoomed, to toggle on or off specific map layers (i.e., roads, town labels, administrative boundaries), and to search for specific locations of interest. New digital formats may increase a user's interaction with mapped content. Real-time spatial data can also be incorporated directly into a digital map through applications that accommodate these services. For example, Google Maps' users can view traffic data, and those using weather applications can view the locations of current precipitation events (Kraak, 2001). In theory, digital maps would never go out of date, as they can be regularly updated (van Elzakker, 2001).

Concluding Remarks

Cartographers, geographers, and others trained in classical mapping systems categorize maps in the manner we discussed in the early stages of this chapter. Although many maps may technically be considered thematic in nature, the differences between subsets of these, the choropleth, isopleth, dasymetric, and others, can be important, even though the lines of demarcation among traditional map types may seem to overlap. For example, a single map may include specific characteristics that define several of these types of maps. We also mentioned many examples of specific maps in this chapter, but obviously others were omitted due to space limitations.

Reflection 10
Think of a specific type of map that we omitted from this chapter. How would you describe its typical format and use?

Some of the specific maps we described in this chapter are only available today in printed, scanned (i.e., .jpg, .png format), or other (e.g., .pdf) formats. In this sense, they are static and reflect past conditions or desired (or projected) future conditions of landscapes or water bodies. However, with advances in computing technology, certain types of maps that require constant updates can now be delivered in a dynamic manner to the screens of digital devices such as cellular phones, tablets, and personal computers. Current weather or traffic maps are prime examples of these, where a map user may be interested in conditions or events occurring now, rather than last or 100 years ago. Debate has occurred concerning whether these should technically be considered maps; one may imagine a map purist discounting the value of some of these products due to their lack of essential map components. However, digital maps are ubiquitous today, and where possible we make an attempt to incorporate them into our discussions of mapping.

References

Anand, B., Wei, W.H., 2016. TIŢAL — Asynchronous multiplayer shooter with procedurally generated maps. Entertainment Computing 16, 81—93.

Anderson Jr., R.K., 1984. The Church of the Holy Cross Recording Project — Summer 1984. U.S. Department of the Interior, National Park Service, Historic American Buildings Survey, Washington, D.C.

Andrews, M.C., 1926. The British Isles in the nautical charts of the XIVth and XVth centuries. Geographic J. 68 (6), 474—481.

Barbour, D., 2001. Maps for fun and profit: one small company's experience. Cartogr. J. 38 (1), 81—85.

Battersby, S.E., Finn, M.P., Usery, E.L., Yamamoto, K.H., 2014. Implications of Web Mercator and its use in online mapping. Cartographica 49 (2), 85—101.

Benová, A., Pravda, J., 2009. Map style. In: Cartwright, W., Gartner, G., Lehn, A. (Eds.), Cartography and Art. Lecture Notes in Geoinformation and Geography. Springer-Verlag, Berlin, pp. 144—154.

Borelli, D., Repetto, S., Schenone, C., 2014. Noise mapping of the flyover highway in Genoa: comparison of different methods. Noise Mapping 1, 59—73.

Botta, F., del Genio, C.I., 2017. Analysis of the communities of an urban mobile phone network. PLoS ONE 12 (3), e0174198.

Bouchard, H., Moffitt, F.H. (Eds.), 1965. Surveying. International Textbook Company, Scranton, PA.

Breed, C.B., Hosmer, G.L., 1928. The Principles and Practice of Surveying, Volume II, Higher Surveying. John Wiley & Sons, Inc, New York.

Brehme, C.E., McCarron, P., Tetreault, H., 2015. A dasymetric map of Maine lobster trap distribution using local knowledge. Prof. Geogr. 67 (1), 98—109.

Brinker, R.C., Wolf, P.R., 1984. Elementary Surveying, seventh ed. Harper & Row, Publishers, New York.

Buchin, K., Speckmann, B., Verbeek, K., 2015. Angle-restricted Steiner arborescences for flow map layout. Algorithmica 72 (2), 656—685.

Burdziakowski, P., Janowski, A., Kholodkov, A., Matysik, K., Matysik, M., Przyborski, M., et al., 2015. Maritime laser scanning as the source for spatial data. Polish Maritime Res. 22 (4), 9—14.

Buterez, C.-I., Popa, A., Gava, R., Dumitru, R., Gruia, A.R., 2016. On the trail of a legend. The legacy of Lady Neaga seen through historical maps. e-Perimetron 11 (2), 77—89.

California Department of Forestry and Fire Protection, 2018. FRAP Mapping, GIS Data. California Department of Forestry and Fire Protection, Sacramento, CA. <http://frap.fire.ca.gov/data/frapgisdata-sw-fire-perimeters_download> (accessed 26.01.19).

Castañeda, C., Herrero, J., Nogués, J., 2017. Soils of Barbués and Torres de Barbués, Ebro Basin, NE Spain. J. Maps 13 (2), 47—54.

Casti, E., 2015. Reflexive Cartography: A New Perspective on Mapping. Elsevier, Amsterdam.

City of Saint Paul, 2018. 2030 Comprehensive Plan Maps. Saint Paul, MN. <https://www.stpaul.gov/departments/planning-economic-development/planning/comprehensive-plan-maps> (accessed 30.01.19.).

Cromley, R.G., Mrozinski, R.D., 1999. The classification or ordinal data for choropleth mapping. Cartogr. J. 36 (2), 101—109.

Dalrymple, G.B., 1991. The Age of the Earth. Stanford University Press, Stanford, CA.

Darbyshire, J.E., Jenny, B., 2017. Natural-color maps via coloring of bivariate grid data. Comput. Geosci. 106, 130—138.

Denil, M., 2016. Review of *a historical atlas of Tibet*, by K.E. Ryavec. Cartogr. Perspect. 85, 44—47.

Dooley, M.A., Lavin, S.J., 2007. Visualizing method-produced uncertainty in isometric mapping. Cartogr. Perspect. 56, 17—36.

Edler, D., Huber, O., Knust, C., Buchroithner, M.F., Dickmann, F., 2014. Spreading map information over different depth layers—an improvement for map reading efficiency? Cartographica 49 (3), 153−163.

Engler, N.J., Scassa, T., Taylor, D.R.F., 2013. Mapping traditional knowledge: digital cartography in the Canadian North. Cartographica 48 (3), 189−199.

Fish, C.S., Calvert, K., 2016. Analysis of interactive solar energy web maps for urban energy sustainability. Cartogr. Perspect. 85, 5−22.

Galanaud, P., Galanaud, A., Giraudoux, P., 2015. Historical epidemics cartography generated by spatial analysis: mapping the heterogeneity of three medieval "plagues" in Dijon. PLoS ONE 10 (12), e0143866.

Gomis, J., Turón, C., 2017. From layout to photoplan: reflections on the "rePresentation" of urban planning. Geographia Technica 12 (1), 57−63.

Grappel, R.D., 1999. Rotating a weather map. Dr. Dobb's J. 24 (6), 80−83.

Greenland, D., 2000. Air apparent: how meteorologists learned to map, predict, and dramatize the weather by Mark Monmonier. Ann. Assoc. Am. Geogr. 90 (1), 192−195.

Heawood, E., 1923. A hitherto unknown world map of A.D. 1506. Geogr. J. 62 (4), 279−293.

Inaida, S., Shobugawa, Y., Matsuno, S., Saito, R., Suzuki, H., 2013. The south to north variation of norovirus epidemics from 2006-07 to 2008-09 in Japan. PLoS ONE 8 (8), e71696.

Jefferson, M., 1930. The six-six world map: giving larger, better continents. Ann. Assoc. Am. Geogr. 20 (1), 1−7.

Jhong, S.-Y., Lin, C.-C., Liu, W.-Y., Huang, W., 2011. Rectangular cartogram visualization interface for social networks. In: Chang, R.-S., Kim, T.-h, Peng, S.-L. (Eds.), Security-Enriched Urban Computing and Smart Grid Communications in Computer and Information Science, 223. Springer-Verlag, Berlin, pp. 336−347.

Jones, B.G., 1957. Photogrammetric surveys for nautical charts. Int. Hydrogr. Review 32 (2), 111−128.

Kalabokidis, K.D., Omi, P.N., 1995. Isarithmic analysis of forest fire fuelbed arrays. Ecol. Modell. 80 (1), 47−55.

Kete, P., 2016. Physical 3D map of the Planica Nordic Center, Slovenia: cartographic principles and techniques used with 3D printing. Cartographica 51 (1), 1−11.

Kinberger, M., 2009. Mapping informal geographies. In: Cartwright, W., Gartner, G., Lehn, A. (Eds.), Cartography and Art. Lecture Notes in Geoinformation and Geography. Springer-Verlag, Berlin, pp. 281−291.

Korn, O., Blatz, M., Rees, A., Schaal, J., Schwind, V., Gorlich, D., 2017. Procedural content generation for game props? A study on the effects on user experience. Comput. Entertainment 15 (2), Article 1.

Kostelnick, J.C., McDermott, D., Rowley, R.J., Bunnyfield, N., 2013. A cartographic framework for visualizing risk. Cartographica 48 (3), 200−224.

Kostylev, V.E., Todd, B.J., Fader, G.B.J., Courtney, R.C., Cameron, C.D.M., Pickrill, R.A., 2001. Benthic habitat mapping on the Scotian Shelf based on multibeam bathymetry, surficial geology and sea floor photographs. Mar. Ecol. Prog. Ser. 219, 121−137.

Koylu, C., Guo, D., 2017. Design and evaluation of line symbolizations for origin-destination flow maps. Information Vis. 16 (4), 309−331.

Kraak, M.-J., 2001. Settings and needs for web cartography. In: Kraak, M.-J., Brown, A. (Eds.), Web Cartography: Developments and Prospects. Taylor and Francis, London, pp. 1−7.

Kronenfeld, B.J., Wong, D.W.S., 2017. Visualizing statistical significance of disease clusters using cartograms. Int. J. Health Geogr. 16, Article 19.

Krumbein, W.C., 1955. Statistical analysis of facies maps. J. Geol. 63 (5), 452−470.

Lavandier, C., Aumond, P., Gomez, S., Dominguès, C., 2016. Urban soundscape maps modelled with georeferenced data. Noise Mapping 3, 278−294.

Lin, J., Cromley, R., 2017. Inferring spatial scale change in an isopleth map. Cartogr. J. 54 (1), 48−60.

Little, E.L., 1971. Atlas of the United States Trees, Volume 1. Conifers and Important Hardwoods. U.S. Department of Agriculture, Forest Service, Washington, D.C. Miscellaneous Publication No. 1146. Map 26-W.

Loomis, F.B., 1961. Subsurface geology. In: Moody, G.B. (Ed.), Petroleum Exploration Handbook. McGraw-Hill Book Company, Inc, New York, 13−1 to 13−74.

Lukas, M., 2014. Cartographic reconstruction of historical environmental change. Cartogr. Perspect. 78, 5−24.

Mackay, J.R., 1955. An analysis of isopleth and choropleth class intervals. Econ. Geogr. 31 (1), 71−81.

Mane, K.K., Börner, K., 2004. Mapping topics and topic bursts in PNAS. Proc. Natl. Acad. Sci. 101 (Suppl. 1), 5287−5290.

Marks, P., 2013. Smartphone sensors offer crowdsourced weather map. New Sci. 218 (2917), 19.

Matti, J.C., Morton, D.M., Cox, B.F., Carson, S.E., Yetter, T.J., 2003. Geologic Map and Digital Database of the Yucaipa 7.5′ Quadrangle, San Bernardino and Riverside Counties, California. U.S. Department of the Interior, Geologic Survey, Tucson, AZ.

McKenzie, S.J.P., 2008. Disaggregating job seekers allowance statistics for Belfast using IKONOS satellite imagery. J. Maps 4 (1), 444−450.

Mendonça, A., Delazari, L., 2014. Testing subjective preference and map use performance: use of web maps for decision making in the public health sector. Cartographica 49 (2), 114−126.

Mendonça, A.L.A., Delazari, L.S., 2011. What do people prefer and what is more effective for maps: a decision making test. In: Ruas, E. (Ed.), Advances in Cartography and GIScience, Volume 1: Selection From ICC 2011, Paris. Lecture Notes in Geoinformation and Cartography. Springer, Berlin, pp. 163−181.

National Capital Planning Commission, 2001. Memorials and Museums Master Plan. National Capital Planning Commission, Washington, D.C.

Nordbeck, S., Rystedt, B., 1970. Isarithmic maps and the continuity of reference interval functions. Geografiska Annaler. Series B: Human Geography 52 (2), 92−123.

Norman, L., Wissler, C., Guertin, D.P., Gray, F., 2002. Digital soils survey map of the Patagonia Mountains, Arizona. U.S. Department of Agriculture, Soil Conservation Service, Washington, D.C.

Nuñez, J.J.R., 2013. Smartphone-based school atlases? Cartographica 48 (2), 126−133.

Oksanen, J., Halkosaari, H.-M., Sarjakoski, T., Sarjakoski, L.T., 2014. A user study of experimental maps for outdoor activities. Cartographica 49 (3), 188−201.

Olivero, J., Fa, J.E., Real, R., Farfán, M.A., Márquez, A.L., Vargas, J.M., et al., 2016. Mammalian biogeography and the Ebola virus in Africa. Mammal Review 47 (1), 24−37.

Opach, T., Rød, J.K., 2013. Cartographic visualization of vulnerability to natural hazards. Cartographica 48 (2), 113−125.

Palfrey, J., Gasser, U., 2008. Born Digital. Understanding the First Generation of Digital Natives. Basic Books, New York.

Passaro, S., de Alteriis, G., Sacchi, M., 2016. Bathymetry of Ischia Island and its offshore (Italy), scale 1:50.000. J. Maps 12 (1), 152−159.

Patterson, T., 2013. Mountains unseen: developing a relief map of the Hawaiian seafloor. Cartogr. Perspect. 76, 5−17.

Perdue, N.A., Lobben, A.K., 2016. Understanding spatial pattern cognition from tactile maps and graphics. Cartographica 51 (2), 103−110.

Pessel, G.H., Levorsen, J.A., Tailleur, I.L., 1978. Generalized isopach map of the Shublik Formation, Eastern North Slope Petroleum Province, Alaska. U.S. Department of the Interior, U.S. Geological Survey, Reston, VA.

Petrov, A., 2012. One hundred years of dasymetric mapping: back to the origin. Cartogr. J. 49 (3), 256−264.

Poulsen, E., Kennedy, L.W., 2004. Using dasymetric mapping for spatially aggregated crime data. J. Quantitative Criminol. 20 (3), 243−262.

Raisz, E., 1934. The rectangular statistical cartogram. Geogr. Rev. 24 (2), 292−296.

Raposo, P., Brewer, C.A., Sparks, K., 2016. An impressionistic cartographic solution for base map land cover with coarse pixel data. Cartogr. Perspect. 83, 5−21.

Saman, D.M., Cole, H.P., Odoi, A., Myers, M.L., Carey, D.I., Westneat, S.C., 2012. A spatial cluster analysis of tractor overturns in Kentucky from 1960 to 2002. PLoS ONE 7 (1), e30532.

Schobesberger, D., Patterson, T., 2008. Evaluating the effectiveness of 2D vs. 3D trailhead maps. A map user study conducted at Zion National Park, United States. In: Proceedings of the 6th ICA Mountain Cartography Workshop. Commission on Mountain Cartography, University of Ljubljana, Slovenia, pp. 201−205.

Schodde, R., Mason, I.J., 1999. The Directory of Australian Birds: Passerines. CSIRO Publishing, Melbourne.

Smith, D.B., Cannon, W.F., Woodruff, L.G., Solano, F., Ellefsen, K.J., 2014. Geochemical and Mineralogical Maps for Soils of the Conterminous United States. U.S. Department of the Interior, U.S. Geological Survey, Reston, VA. Open-File Report 2014-1082.

Soil Science Division Staff, 2017. Soil survey manual. In: Ditzler, C., Scheffe, K., Monger, H.C. (Eds.), Government Printing Office. Washington, D.C. U.S. Department of Agriculture Handbook, No. 18.

Speake, J., Axon, S., 2012. "I never use 'maps' anymore": engaging with Sat Nav technologies and the implications for cartographic literacy and spatial awareness. Cartogr. J. 49 (4), 326−336.

Stephens, S., 2017. Have you checked your location? U.S. Department of Defense, Air Force, Columbus Air Force Base, MS. <https://www.columbus.af.mil/News/Art/igphoto/2001778291/> (accessed 28.01.19).

Sternai, P., Herman, F., Fox, M.R., Castelltort, S., 2011. Hypsometric analysis to identify spatially variable glacial erosion. J. Geophys. Res. 116, F03001.

Stoorvogel, J.J., Bakkenes, M., Temme, A.J.A.M., Batjes, N.H., ten Brink, B.J.E., 2017. S-World: a global soils map for environmental modelling. Land Degradation Development 28 (1), 22–33.

Strickland, D.H., 2018. Edward I, Exodus, and England on the Hereford World Map. Speculum 93 (2), 420–469.

Tait, A., 2010. Mountain ski maps of North America: a preliminary survey and analysis of style. Cartogr. Perspect. 67, 5–18.

Theurer, C., 1959. Color and infrared experimental photography for coastal mapping. Photogram. Eng. 25 (4), 565–569.

Tobler, W., 2004. Thirty five years of computer cartograms. Ann. Assoc. Am. Geogr. 94 (1), 58–73.

Tobler, W.R., 1973. Choropleth maps without class intervals? Geogr. Anal. 5 (3), 262–265.

Ungar, S., Blades, M., Spencer, C., 1997. Teaching visually impaired children to make distance judgements from a tactile map. J. Visual Impairment Blindness 91 (2), 163–174.

U.S. Central Intelligence Agency, 2016. The World Factbook 2016-17. U.S. Central Intelligence Agency, Washington, D.C.

U.S. Department of Agriculture, Agricultural Research Service, 1990. USDA plant hardiness zone map. U.S. Department of Agriculture, Agricultural Research Service, Washington, D.C. <https://planthardiness.ars.usda.gov/PHZMWeb/Downloads.aspx> (accessed 08.04.19.).

U.S. Department of Agriculture, Forest Service, 2012. Chattahoochee National Forest Map. U.S. Department of Agriculture, Forest Service, Southern Region, Atlanta, GA.

U.S. Department of Commerce, National Oceanic and Atmospheric Administration, 2018. Houston Ship Channel, Alexander Island to Carpenters Bayou. U.S. Department of Commerce, National Oceanic and Atmospheric Administration, National Ocean Service, Coast Survey, Washington, D.C.

US Department of Commerce, National Oceanic and Atmospheric Administration, 2019. NWS Radar Mosaic. U.S. Department of Commerce, National Oceanic and Atmospheric Administration, National Weather Service, Silver Spring, MD. <https://radar.weather.gov/Conus/full.php> (accessed 08.04.19.).

US Department of Defense, Air Force Civil Engineer Center, 2014. Cleanup at Reese. US Department of Defense, Air Force Civil Engineer Center, San Antonio, TX. <https://www.afcec.af.mil/News/Photos/igphoto/2000830802/> (accessed 31.01.19.).

U.S. Department of Health and Human Services, Centers for Disease Control and Prevention, 2018. Ebola Virus Disease Distribution Map: Cases of Ebola Virus Disease in Africa Since 1976. U.S. Department of Health and Human Services, Centers for Disease Control and Prevention, Atlanta, GA. <https://www.cdc.gov/vhf/ebola/history/distribution-map.html> (accessed 26.01.19).

U.S. Department of Health and Human Services, National Library of Medicine, 2018. TOXMAP Environmental Health Maps. U.S. Department of Health and Human Services, National Library of Medicine, Bethesda, MD. <https://toxmap.nlm.nih.gov/toxmap/> (accessed 31.01.19.).

U.S. Department of the Interior, Fish and Wildlife Service, 2012. Skilak Lake. U.S. Department of the Interior, Fish and Wildlife Service, Kenai Fish & Wildlife Field Office, Soldotna, AK. <https://www.fws.gov/uploadedFiles/lake_Skilak.pdf> (accessed 22.02.19.).

U.S. Department of the Interior, Geological Survey, 2018. SCoRR: Sediment-Bound Contaminant Resiliency and Response Strategy. U.S. Department of the Interior, Geological Survey, Reston, VA.

U.S. Department of Transportation, Bureau of Transportation Statistics, 2014. Aviation and Highway Noise for the New York City Metropolitan Area: 2014. U.S. Department of Transportation, Bureau of Transportation Statistics, Washington, D.C. <https://www.bts.gov/content/aviation-and-highway-noise-new-york-city-metropolitan-area-2014> (accessed 29.05.19.).

U.S. Department of Transportation, Federal Aviation Administration, 2018. ATS Route. U.S. Department of Transportation, Federal Aviation Administration, Washington, D.C. <http://ais-faa.opendata.arcgis.com/datasets/acf64966af5f48a1a40fdbcb31238ba7_0?geometry=122.479%2C21.212%2C96.99%2C59.492> (accessed 28.01.19.).

U.S. Department of Transportation, Federal Highway Administration, 2017. Average Daily Long-Haul Truck Traffic on the National Highway System: 2015. U.S. Department of Transportation, Federal Highway Administration, Office of Freight Management and Operations, Washington, D.C.

U.S. Environmental Protection Agency, 2017. EnviroAtlas—Dasymetric Population for the Conterminous United States. U.S. Environmental Protection Agency, Washington, D.C. <https://edg.epa.gov/metadata/catalog/search/resource/details.page?uuid=%7BBBBDAEC9A-F9B6-490F-8CA0-40401C47DEBA%7D> (accessed 27.01.19.).

Uttal, D.H., 2000. Seeing the big picture: map use and the development of spatial cognition. Dev. Sci. 3 (3), 247–264.

van Breugel, P., Kindt, R., Lillesø, J.P.B., Bingham, M., Demissew, S., Dudley, C., et al., 2015. Potential natural vegetation map of Eastern Africa (Burundi, Ethiopia, Kenya, Malawi, Rwanda, Tanzania, Uganda and Zambia). Version 2.0. Forest and Landscape (Denmark) and World Agroforestry Centre (ICRAF). <http://vegetation-map4africa.org> (accessed 30.01.19.).

van Elzakker, C.P.J.M., 2001. Use of maps on the web. In: Kraak, M.-J., Brown, A. (Eds.), Web Cartography: Developments and Prospects, Taylor and Francis, London, pp. 21–37.

Verbeek, K., Buchin, K., Speckmann, B., 2011. Flow map layout via spiral trees. IEEE. Trans. Vis. Comput. Graph. 17 (12), 2536–2544.

Weir, M.J.C., 1997. A century of forest management mapping. Cartogr. J. 34 (1), 5–12.

Wood, W.R., 1993. Integrating ethnohistory and archaeology at Fort Clark State Historic Site, North Dakota. Am. Antiq. 58 (3), 544–559.

Yan, J., Guilbert, E., Saux, E., 2015. An ontology of the submarine relief for analysis and representation on nautical charts. Cartogr. J. 52 (1), 58–66.

Zeng, X., Mu, L., Liu, J., Yang, Y., 2015. Setting diverging colors for a large-scale hypsometric lunar map based on entropy. Entropy 17 (7), 5133–5144.

Map Components

A suitable, professional map should contain enough information (text and associated graphical content) to allow a map user to understand the desired message crafted by the map developer. Certainly, one can carelessly add graphical components to a media and call it a map. Yet in doing so, the map developer can unknowingly distract people's attention from the map's purpose. With cartographic principles in mind, and while interjecting a little creativity, a high-quality map product can be developed. Cartography, as one might recall from Chapter 1, Maps, is the art and science of developing maps. The term is derived from the words *carte* (map) and *graphia* (writing) and suggests that this art and science pertains to the practice of communication through mapping. In order to communicate a message effectively, a number of map components might be employed, and some thought should be expended on the design and placement of these (Fig. 3.1). Some map components are more technically defined as *map elements* or *data elements* (Yin and Huang, 2001); yet, we refer to all of these as map components in this chapter. While some maps

93

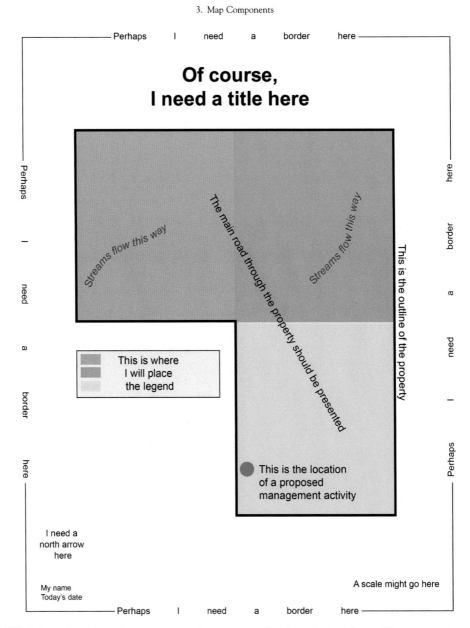

FIGURE 3.1 A hypothetical map, prior to the insertion of real symbols and text, illustrating where certain map components might be placed.

fail to include the necessary components suggested by cartographers (e.g., neat lines), we suggest that depending on the circumstance, some map components should be viewed as necessary, while others may be viewed as optional. And finally, we should note that the rules, the cartography, employed in the development of a map can vary across different societies (Harley, 1989); therefore, while we describe what we perceive to be essential map components, we do this from our perspective of North American natural resource management professionals.

The focus of this chapter is on the information that enhances the presentation of a landscape; those components necessary for assisting in the presentation of a clear message about the landscape. A map directs people's attention to a specific area, the land or water body around which a message is being conveyed. This area can be as small as a few hundred square meters or as large as the entire Planet Earth. Map components are placed on top and around the specific area portrayed, acting as *marginal information* that assists in the interpretation of both the landscape's location and the map's message (Akinyemi et al., 2013). Thus one way to evaluate a map is that it is composed of two sets of information, (1) the face of the map (the landscape) and (2) other nearby information or *marginalia*. The face or landscape is the main subject of the map, and the marginalia include the other map components, such as a legend, scale, north arrow, title, projection, and other map components. The efficiency of a person in reading and interpreting a map can be related to the amount of map components available, the complexity of information that map components provide, and how this information is allocated or spread across the printed map or digital screen (Edler et al., 2014). The spatial arrangement of map components can be described simply as the map's *layout* (Akinyemi et al., 2013). A good map has an organized and logical layout that supports the presentation of a complete and coherent story. In sum, the composition of the landscape and the marginalia are used to tell this story. The area focused upon by the map should be considered a map component, however, we do not discuss this point explicitly other than to suggest that some attention is necessary to framing the area of interest. Broader places beyond the area of interest need not be included in a map unless they contribute to an understanding of the context of space. For example, if one were to develop a map of the Francis Marion National Forest in South Carolina (Fig. 3.2), they may include the nearby city of Charleston since it can provide map users a place of reference of which they may be familiar.

Map Title

Reflective of the theme of the map, the title ideally provides the map user a concise and easily understood description of the map's purpose. Although not always provided, map titles are most often considered to be required map components. In many cases, map titles are used for various archiving purposes that involve storage (all types of media) and search-enabled processes. Therefore a title should contain the location of the place described graphically in the map along with other concise, pertinent information reflective of the general theme of the map. Furthermore, the title of a map should be apparent to the map user. In this regard, it also needs to be placed appropriately on the media (top, corner, bottom, etc.) so that it can be easily found, and it needs to be displayed in a manner that draws the attention of the map user. Various combinations of font, font size, and color (Fig. 3.3) can be employed in the design of the title. However, care should be taken to select the appropriate style or flair of the title so that the outcomes of these decisions complement the presentation of other features within the map. A flamboyant title paired with rather basic map features may result in an awkward map product.

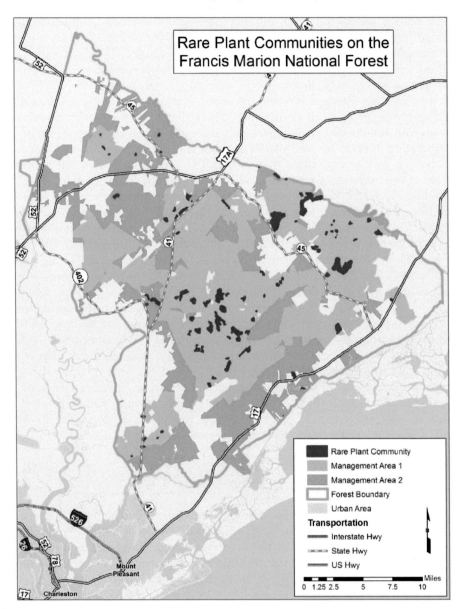

FIGURE 3.2   Map of the rare plant communities located on the Francis Marion National Forest, and the nearby city of Charleston. Source: *U.S. Department of Agriculture, Forest Service (2016a).*

Inspection 13

On this book's website (mapping-book.uga.edu), you can find a map developed by S.A. Mitchell in 1853 that illustrates the State of Louisiana, and the locations of cities, rivers, and streams that were understood at that point in time. Describe how the content,

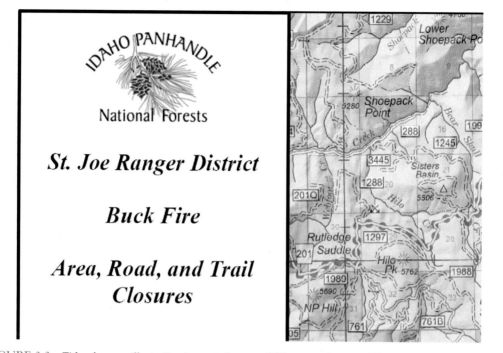

FIGURE 3.3   Title of a map illustrating impacts from a wildfire occurring on a US national forest. Source: *U.S. Department of Agriculture, Forest Service (2017a).*

placement, font, and artistry of the map title contribute to, or distract from, the message of the map.

A less formal map title may simply state the issue at hand, through a common font, using a standard color (e.g., black). For example, perhaps a thematic map was developed just to illustrate general ages of forests within a portion of a US national forest (Fig. 3.4). Here, a simple, general title style may be sufficient.

Orientation of the Landscape

Often, orientation of the landscape is indicated by a *north arrow* or a *compass rose*. A symbol representing landscape orientation should be viewed as a standard map component, as it provides map users a sense of direction. A north arrow (Fig. 3.5) may simply indicate the direction of north, without reference to other directions, yet it may also infer that the other directions are there; a southward direction is opposite of north on a map, an eastward direction is to the right of north, and a westward direction is to the left. A compass rose (Fig. 3.6) explicitly illustrates the *cardinal directions* (north, south, east, and west) and perhaps other finer divisions of these in a manner emblematic of the bloom of a flower (similar in pattern in all major and minor directions).

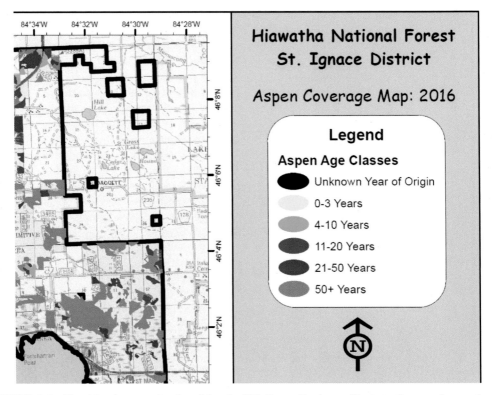

FIGURE 3.4   The title of a map developed by the U.S. Forest Service to illustrate the age classes of aspen (*Populus tremuloides*) on the St. Ignace District of the Hiawatha National Forest. Source: *U.S. Department of Agriculture, Forest Service (2016b).*

FIGURE 3.5   Four north arrows, some explicitly indicating the northward direction (N), and others implicitly suggesting the direction.

Reflection 11

Although we have noted that a north arrow should be viewed as a standard map component, sometimes it is lacking from a map. In what cases would you absolutely need to place a north arrow or compass rose on a map? In what cases might you not place a north arrow or compass rose on a map?

There are other types of "north" orientation that might also be displayed on maps (Fig. 3.7). *True north* represents a direction to the place on Earth where the lines of longitude all meet, and thus an arrow or compass rose illustrating true north is pointing to the North Pole. In this respect, true north is a geographic representation of orientation.

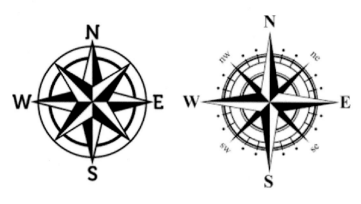

FIGURE 3.6 Two compass rose examples, with the cardinal directions (north, south, east, and west) only provided in one, and finer divisions provided in the other.

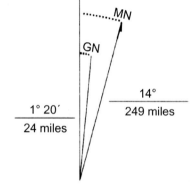

UTM grid and 2002 magnetic north declination at center of sheet

FIGURE 3.7 True north (★), grid north (GN), and magnetic north (MN) on the U.S. Geological Survey Pismo Beach, California Quadrangle map.

Magnetic north is a physical representation of orientation that is based on the Earth's magnetic field. The location of magnetic north on Earth is not exact, yet is close (about 11 degrees in latitude) to the North Pole. In addition, the magnetism of the Earth gradually shifts with time, and thus magnetic north will vary over time (Wei-Haas, 2019). For example, for the area around Athens, Georgia (United States), the direction to magnetic north has shifted about 1 degree in the last decade. The difference between true north and magnetic north (Fig. 3.8) can be described in terms such as *declination* or *variation*. For instance, imagine standing somewhere on the landscape being mapped, with a compass in hand. The compass needle will point to magnetic north. If magnetic north is 6 degrees to the west (left) of true north, the declination associated with this place is considered to be 6 degrees to the west. If, on the other hand, magnetic north is 10 degrees to the east (right) of true north, the declination of this place is considered to be 10 degrees to the east. *Grid north* ("GN" in Fig. 3.7) is related to a plane grid system, which itself is related to the coordinate and projection system used to represent a curved Earth on a flat surface (sheet of paper or digital device screen). Grid north is also a geographic representation of orientation, and at times may be very similar to true north.

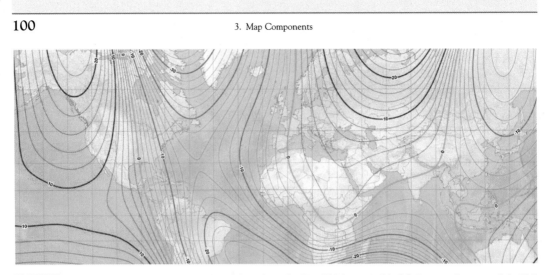

FIGURE 3.8   Declination around Earth (2019) based on the World Magnetic Model. Source: *Courtesy of the U.S. Department of Commerce, National Oceanic and Atmospheric Administration (2019).*

In the Northern Hemisphere, a northward direction, by default, is frequently assumed to be the top of a printed map and a southward direction is assumed to be the bottom of a printed map. But this is not always the case in printed maps, and so an explicit representation of orientation is necessary to avoid any potential confusion regarding this issue. In the Southern Hemisphere, printed maps also generally have a northward orientation at the top of a map and a southward orientation at the bottom. Similarities in the usage of orientation in both hemispheres of Earth may date back to the maps copied (or miscopied) from the book *Geographia*, written by Greco-Roman astronomer and Geographer Ptolemy, even though no original maps from this book exist today (Richardson, 2015). Small-scale maps of the world to which we have been accustomed (Fig. 3.9) may not necessarily need an indication of orientation if we understand the arrangement of the continents and the notion that lines of longitude all converge upon the poles. Even small-scale maps of countries, states, or provinces may not necessarily need this map component, but in these cases an in-depth knowledge of the political subdivisions will likely be required by map users to understand map orientation. Larger scale maps of small areas that are not commonly understood, such as a parcel of timberland in Escambia County, Florida (Fig. 3.10) will likely need an indication of orientation, simply because most people reading the map may be unfamiliar with the area.

A *graticule* is a spherical coordinate system, and uses the longitude and latitude lines of a map to provide a sense of orientation. A graticule includes evenly spaced meridians (longitudinal lines running north to south) and parallels (latitudinal lines running east to west) to provide map users spatial reference for features presented in a map (Edney, 1999). Technically, a graticule is represented by lines of latitude and longitude. For small-scale maps, the graticule can be described by whole numbers (e.g., they are placed 30 degrees apart); for large-scale maps, the graticule can be described by fine divisions of degrees, minutes, and seconds. When lines other than longitude or latitude are used to form a graticule, perhaps the system should simply be referred to as a *grid*. For example,

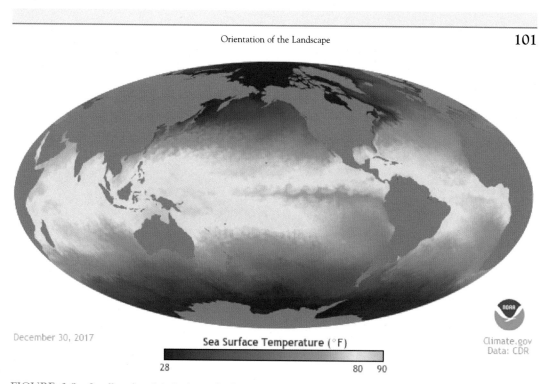

December 30, 2017

Sea Surface Temperature (°F)

Climate.gov
Data: CDR

28   80   90

FIGURE 3.9   Small-scale, global view of relative sea surface temperatures for various parts of the world. Source: *U.S. Department of Commerce, National Oceanic and Atmospheric Administration (2017).*

when the term *graticule* is used in other fields, it can represent a grid network of parallel and perpendicular lines that help one locate and establish reference of features within their respective landscape, which may be as large as outer space, or as small as a human polyp or tumor within the body of a human (Young et al., 2013).

Inspection 14

Access the map contained on the book's website (mapping-book.uga.edu) that represents a 7.5-minute topographic map of the Paisley, Oregon area. Describe both the graticule (its dimension and extent) and the other grids present on the map that might help people orient themselves. Then, access the Oregon Islands National Wildlife Refuge map, and compare the graticule present on this map to the graticule present on the topographic map.

Diversion 10

Open Google Earth, and navigate to your favorite place. Using an eye altitude of about 10,000 ft. (roughly 3000 m), enable the process to *view the grid*. What regular interval of latitude and longitude (in degrees) is presented as a graticule? The degrees of latitude and longitude should be the same.

With the pervasiveness of digital maps today, and the manner in which we are able to interact with them, an indication of landscape orientation can be quite important. As you

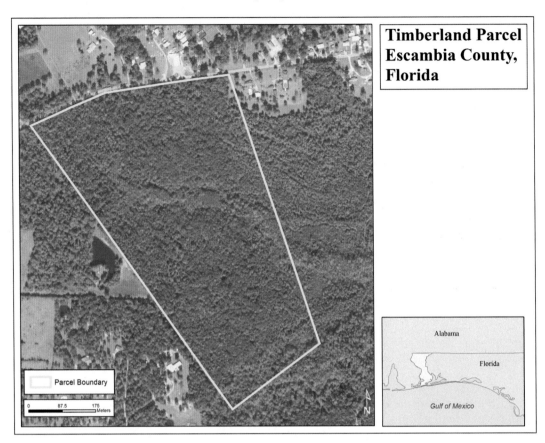

FIGURE 3.10   Large-scale map of a small timberland parcel in Escambia County, Florida, where a north arrow would be necessary.

may have observed, with handheld digital devices the orientation of a map may vary based on how the device is held. Hopefully, a north arrow or a graticule is present to lend assistance in this matter.

Scale

A scale is one of the most important components of a map because it allows a map user to judge absolute and relative sizes of mapped features, and approximate (or exact) distances between mapped locations. We discuss in more detail the uses of map scales to estimate areas and distances in Chapter 6, Map colors, and so we leave these endeavors for later. There are three general methods to express a scale. One is a *representative fraction* scale, which is a number without specific units (i.e., unitless) and indicates the relationship between distances on a map and actual distances on Earth (ground distances). For example, if a distance on a map is measured to be 2 in., and the actual ground distance is 2 miles, the ratio is 2 in.=2 miles. This is the model scale of the map, and once reduced it becomes

1 in. = 1 mile, or
1 in. = 5280 ft., or
1 in. = 63,360 in.

The representative fraction scale of the map would be presented simply as 1:63,360, a unitless relationship. This might be communicated orally as "1 to 63,360," which should imply that any one unit of measurement on the map (metric or English or otherwise) is equal to 63,360 of *those same units* on the ground, in real life. The second method to express a scale, the *equivalence scale*, presents words that represent units of measurement on both sides of the relationship, and often these units of measurement are different. For example, 1 in. = 1 mile and 1 in. = 5280 ft. are both considered equivalence scales. The third method for expressing scale on a map is to use a *graphical scale*. These employ bars or lines, accompanied by supplementary text, to help map users better understand scale through a firm example of horizontal distances. In addition, graphical scales are useful when maps are enlarged or reduced on photocopying machines. Graphical scales retain their value when stretched or compressed whereas numerical scales (representative fraction and equivalence) do not, since the graphical scale is enlarged or reduced proportionally along with the mapped features. Of course, there are many examples of the presence of all three types of scales on a map (Fig. 3.11).

Reflection 12
Think about the last physical, printed map that you viewed or used. What was the purpose of the map? What was the representative fraction or equivalence scale of the map?

When viewing or creating maps on digital devices (cell phones, tablets, and computers), the concept of scale is rather dynamic, and depends on the perspective of the map user at the time a map is viewed. The transfer of information across scales can be facilitated in a digital environment in a fraction of a second, permitting a *hyperlocal* manner of map interaction (Dalton, 2013). To illustrate map scale, some digital mapping applications use a graphical scale that changes size dynamically as a person zooms into (or out of) an area. In some digital applications, scale may be related to the height above the ground surface when viewing the landscape (as in Google Earth). Direct conversion to a representative fraction scale may be difficult due to the display properties of a digital device. One would need to measure a distance on the screen of the digital application and compare it to the actual ground distance.

Inspection 15
On a personal computer or other digital device, open Google Earth and navigate to the following set of latitude and longitude coordinates: 41.953246°N 72.510893°W. What message did the property owner convey to the world in October 2006?

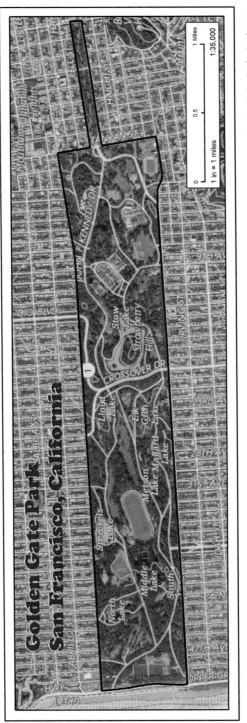

FIGURE 3.11 Representative fraction, equivalence, and graphical scales on a single map. Source: *Base map source is the U.S. Geologic Survey National Map.*

- At what eye altitude is the message no longer visible? Approximately what representative fraction scale is this?
- At what eye altitude can you understand the political unit (country, state, and county) in which the location resides? Approximately what representative fraction scale is this?

Symbols

Symbols on maps portray the existence and quality of many features associated with landscapes and water bodies. Symbols may be composed of lines, polygons, points, or other graphics, and presented using different colors, sizes, and enhancements that reflect the importance of the landscape or water body features they represent. The appearance of each symbol can be adjusted to increase or decrease the relative value of the associated feature. Symbols placed on maps should ideally describe the most important aspects of a landscape or water body as they relate to the overall message of the map. For example, a city bike map might portray streets as specifically having bike lanes, yet may also illustrate through the *symbology of the map* the different vehicular speed limits, number of lanes, and directionality of vehicular traffic of the roads (Wessel and Widener, 2015).

Some computer mapping programs allow extensive and intricate adjustments to the appearance of symbols, such as shadowing and transparency, to address the need to differentiate among features that may be similar (Huffman, 2018). Within some computer mapping systems, automated processes can also be employed to facilitate the development of *proportional symbol* maps, where the size of the symbol is a function of some quantitative aspect of a place, such as a city's population (Cano et al., 2013). Some symbols may portray to the map user that a location has been identified very precisely (Fig. 3.12), yet there may be cases where the places illustrated on a map have positional uncertainty, and therefore this imprecision might be better represented with other plain symbols or shapes (Piatti et al., 2009).

The use of common symbols on maps can be viewed, much like mathematics, as an international means of communication (Voženílek et al., 2014). Within specific fields of research or management, standard sets of symbols can facilitate effective cognitive interpretations of the ideas presented in maps, particularly when these are important enough to require extension across social, cultural, and geographical boundaries (Bianchetti et al., 2012). The standardization of symbols for various fields of study has been discussed for almost a century (Swartz, 1933), and has been attempted by various collective groups of people or organizations.

Diversion 11

Assume that you have been charged with the development of maps for a large area of land that has its primary purpose to serve as a refuge for wildlife. In association with this job, you will need to produce a series of maps that describe the land and water resources, and present these in paper and digital form to the public. Therefore, you have decided that a standard set of symbols would be of value to ensure consistency among the maps. Over the course of about 15 minutes, develop a set of guidelines for the presentation of the following five land and water body features: (1)

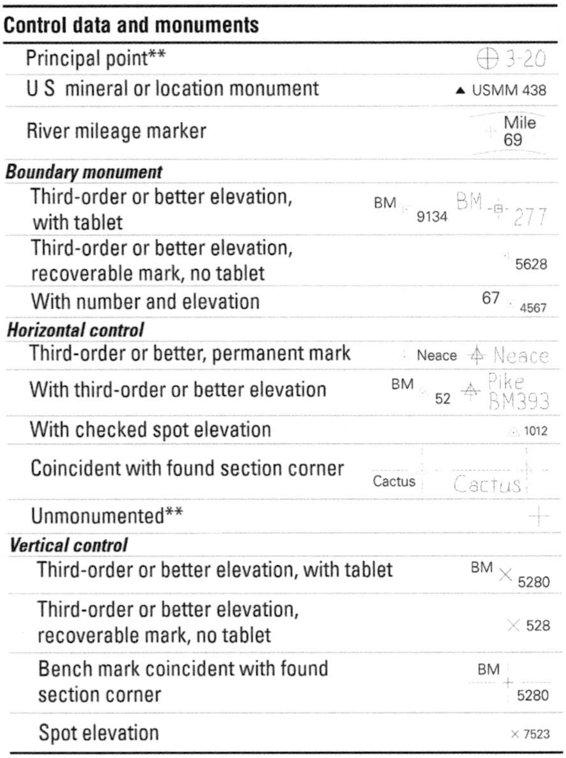

Control data and monuments	
Principal point**	⊕ 3-20
U S mineral or location monument	▲ USMM 438
River mileage marker	Mile 69
Boundary monument	
Third-order or better elevation, with tablet	BM 9134 BM -⊖- 277
Third-order or better elevation, recoverable mark, no tablet	5628
With number and elevation	67 4567
Horizontal control	
Third-order or better, permanent mark	Neace ⊕ Neace
With third-order or better elevation	BM 52 ⊕ Pike BM393
With checked spot elevation	⊕ 1012
Coincident with found section corner	Cactus Cactus
Unmonumented**	+
Vertical control	
Third-order or better elevation, with tablet	BM × 5280
Third-order or better elevation, recoverable mark, no tablet	× 528
Bench mark coincident with found section corner	BM + 5280
Spot elevation	× 7523

FIGURE 3.12 Symbols that might represent precise locations. *Source: U.S. Department of the Interior, Geological Survey (2013).*

permanent water bodies (lakes and ponds), (2) dry water bodies, (3) forests, (4) marshes, and (5) sand dunes. Assume that these features will be presented as polygons (closed areas) on maps. Include in these guidelines a consideration of these four characteristics: color of the boundary or outline of the polygons (where necessary), color of the inner area of the polygons (again where necessary), the style (solid lines, dashed lines, etc.) of the outlines of the polygons, and the texture of the inner areas of the polygons.

Sets of standard map symbols used by larger governmental and political organizations can often be located, but standards used by small land management organizations are much more difficult to locate, if these standards even exist. The U.S. Geological Survey, for example, published a standard set of map symbols for use in national topographic maps. An example of a small portion of this standard set is illustrated in Fig. 3.13. The U.S. National Park Service also utilizes a set of standard cartographic symbols to portray recreational information associated with national parks (Fig. 3.14). Given the large number of countries involved, symbols used to represent land-based systems on North Atlantic Treaty Organization (NATO) military maps have also been standardized (NATO Standardization Agency, 2017). However, in some fields such as emergency management, informal symbol conventions are used where standards and conventions have not been

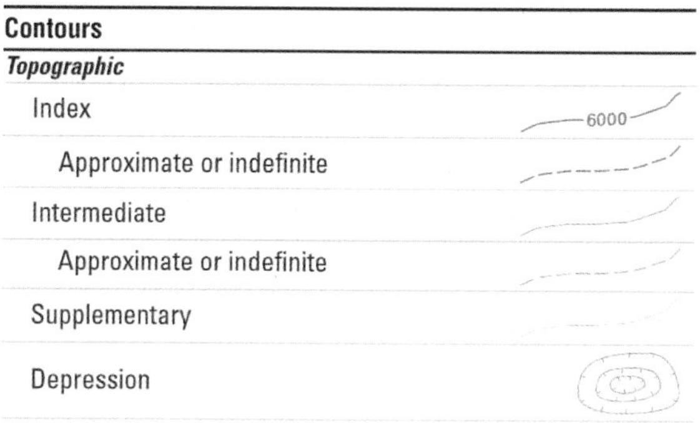

Contours	
Topographic	
Index	—6000—
Approximate or indefinite	
Intermediate	
Approximate or indefinite	
Supplementary	
Depression	

FIGURE 3.13 Standard symbols for representing contour lines (lines of consistent elevation) on topographic maps developed by the U.S. Department of the Interior, Geological Survey (2013).

Nature and wildlife

Bear viewing

Birding/wildlife viewing

Deer viewing

Fish hatchery

Fish ladder

Flower viewing

Sea lion viewing

Tidepooling

Viewing area

Waterfowl

Whale viewing

FIGURE 3.14 Standard nature and wildlife symbols used in maps developed by the U.S. Department of the Interior, National Park Service (2017).

widely adopted, perhaps because a wide variety of public and private organizations would need to collaborate in the development of the standards (Robinson et al., 2013). Sets of standard symbols (transportation, forestry, etc.) are also available within geographic information system (GIS) programs. For example, ArcGIS (ESRI, 2018) supports the use of sets of symbols through *style* files. As computing technology progresses, traditional cartographic symbols may need to be revised to allow effective visualization within digital devices (Lisitskii, 2016).

Inspection 16

Access the graphic standards document developed by the U.S. Department of the Interior, Fish and Wildlife Service (2001), which can be found through a search of the Internet or accessed through the website associated with this book (mapping-book.uga.edu). Navigate to page 2.18. Compare the graphic standards suggested for permanent water bodies (lakes and ponds), dry water bodies, marshes, and sand dunes to the symbology you suggested earlier for these same features. In what ways do the two models (yours and the U.S. Fish and Wildlife Service) differ?

Legend

As was suggested in the previous section, symbols can be associated with descriptions of landscape and water body features to form a visual language that may assist map users in interpreting a map (Rambaldi, 2005). A map *legend*, which defines the symbols employed, can further assist map users in making sense of the shapes and colors that are present on a map. For map users unfamiliar with symbols and features presented in maps, a legend (Fig. 3.15) acts as a key to interpreting them (Akinyemi et al., 2013). While symbols on a map often can be rather self-explanatory, the presentation of a formal list of the main symbols, through a legend, may be important.

Several important considerations can affect the quality and usefulness of a map legend. For example, the size of the legend is something that a map developer should carefully consider. The legend should not be so small as to be illegible and of no value to the map user, and should not be so large as to overwhelm other aspects of the map. The content of a map legend is another consideration that a map developer should carefully consider. Some maps contain features that should be obvious to map users; these may not need to be included in the map legend. Areas that are colored light blue and that represent water bodies, for example, might not need to be represented in a map legend, as the map developer may assume this form of representation of water would be obvious. The map legend, therefore, may not be all-inclusive of the symbols contained in the map, and may otherwise focus on a sparse subset of the more important symbols employed (Fig. 3.16). Further, since symbols representing common landscape features can occasionally be inconsistent from one map to the next, the legend acts to clarify these assignments. For the sake of consistency (and reflective of the professionalism of the map developer) the colors of the symbols contained in the map legend should be consistent with the colors employed upon the symbols in the map. The size of the symbols

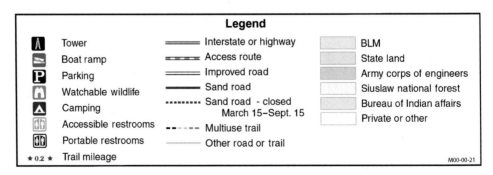

FIGURE 3.15   A legend associated the road systems, land uses, and beach access issues of the North Spit of Coos Bay, in Oregon. Source: *U.S. Department of the Interior, Bureau of Land Management (2017).*

Highways and roads

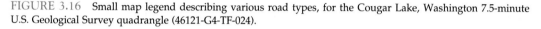

Interstate .... ⑤

U S .... ⑩⑪

State .... ⑦⑨

County .... ⑥

National Forest, suitable
for passenger cars 105   61

National Forest, suitable for
high clearance vehicles 1 0 5

National Forest trail 384

Primary highway ....

Secondary highway..................

Light-duty road

   Composition: Unspecified...

               Paved ....

               Gravel ....

               Dirt ....

Unimproved; 4 wheel drive 4WD

Trail ....

Gate; barrier............................

FIGURE 3.16   Small map legend describing various road types, for the Cougar Lake, Washington 7.5-minute U.S. Geological Survey quadrangle (46121-G4-TF-024).

contained in the map legend can at times be larger than the size of the symbols in the map, if it seems necessary to present them in a larger format to help prevent interpretation problems.

Inspection 17
Access the book's website (mapping-book.uga.edu) and view the *North Spit of Coos Bay map* that was developed by the U.S. Department of the Interior, Bureau of Land Management (2017). Although ideally a legend would be designed to illustrate all of the features presented in a map, sometimes this would require the development of a very large legend. Thus, some land and water body features presented within a map are not necessarily described in the map's legend. Which features on this map are not included in the map's legend?

Map legends are very useful for describing categorical data (e.g., land use classes) as well as for describing ranges of quantitative data (e.g., age of forests and annual amount of precipitation). With respect to quantitative data, a map legend can illustrate the discrete classes employed (Fig. 3.17) or the range of colors assigned to continuous data along a spectrum (Fig. 3.9). For example, the discrete classes of land ownership in the map mentioned in *Inspection 17* are obvious; state-owned lands are colored purple, privately owned lands are white, and national forest lands are green. With respect to a continuous spectrum of values, the *color ramp* employed can be an important aspect of an attempt to communicate the range of expected or observed values, from good to bad or from one end of a data range to another. For example, in the continuous range of values illustrated in Fig. 3.9, one can see that the map developer probably wanted cooler water surface temperatures to be illustrated through various shades of blue, a color people often associate with coldness, and warmer water surface temperatures to be illustrated through colors ranging from orange to yellow, which many people associate with warmth.

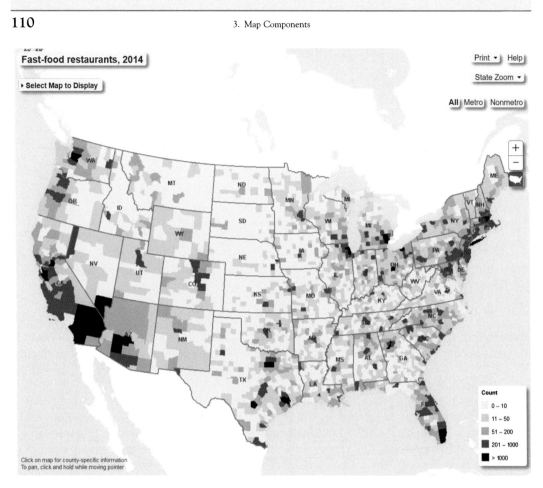

FIGURE 3.17 Digital map of fast-food restaurants per county in the United States, 2014. Source: *U.S. Department of Agriculture, Economic Research Service (2019)*.

Translation 11

Access from this book's website (mapping-book.uga.edu) the motor vehicle use maps of the Osceola National Forest in Florida (U.S. Department of Agriculture, Forest Service, 2017b). Without simply repeating the information provided in the legend of these maps, and from the perspective of someone who may want to drive through this national forest, what can the legend help the map user understand, in general, about where they may be going and what they may see along the way?

Unlike *map labels* and *annotation* (described next), which often need to be placed near their associated landscape features, map legends can be placed anywhere a map developer wishes. Therefore, in designing an aesthetically pleasing map, the placement of the legend should be carefully considered. Given a legend's potential size, the process of locating a suitable area for it on a printed or digital map can require significant thought, so that other important map features will not be obscured (covered by the legend).

A map legend can be composed of a fixed, standard set of items, as in the 15 Osceola National Forest motor vehicle use maps. Alternatively, a map legend can evolve dynamically over the course of developing a map, as final adjustments are made in the amount of information that is included (Rambaldi, 2005). Unless a map has been developed by a committee, through a participatory process, or developed under the guidelines of predefined protocols, the map developer has ultimate control over the design and presentation of the map legend. In addition, given advances in computing technology over the last decade or so, dynamically sized legends can be incorporated into digital maps through the use of programming logic (Donohue et al., 2013).

Labels and Text Annotation

The graphical marks made by adding text or label annotation to a map are often essential in communicating a message to the map user. Conceptually, *map labels* and *text annotation* are different from other symbols placed on a map; while they may refer to something of interest to the map user, the annotation is unlikely to be found when the landscape is viewed in real life from above (Skupin and Skupin, 2009). Labels and text annotation can consist of a variety of items:

- Political boundary names (country, state, province, city, county, etc.)
- Administrative boundary names (e.g., national forests, parks, or wildlife refuges)
- Common names of local communities
- Elevations and surveyed benchmarks
- Other place names or toponyms related to topography (e.g., names of mountains, hills, or mountain passes)
- Other place names created through human activity (e.g., churches, cemeteries, and gravel pits)
- Water bodies (e.g., oceans, bays, lakes, harbors, ponds, rivers, and streams)
- Utility corridors (e.g., pipeline, telephone, and electrical rights-of-way)
- Important recreation features (trails, parks, and scenic overlooks)
- Details of various land survey systems (e.g., the U.S. Public Land Survey System)
- Proposed management actions or activities (e.g., proposed forest harvests and proposed road development)

The artistry and associated techniques for placing text on maps are called *typography*. Typography is employed in order to enhance interpretation and readability of maps when printed or displayed on digital devices. Through typography, map developers can improve the utility of a map by maximizing the number of labels placed on a map without obscuring the labeled landscape features (or other nearby features), and therefore minimizing negative effects on a map's aesthetic quality (Rylov and Reimer, 2014).

Labels and text annotation should be clearly legible on a map through the use of strong contrasting styles. The importance of the relationships among various map features can be illustrated through typeforms (or typeface) which represent different font, font size, coloring, and style (e.g., italic and bold) combinations. Some common typeface conventions are used in label and text annotation efforts. Although not universally standard, names of water bodies are often presented on maps in blue text and italic font style. Names of higher level administrative or political units (countries, counties, etc.) are often displayed in capitalized letters, and names of places of perceptually lesser importance are presented using smaller case letters and smaller font sizes (Fig. 3.18). Labels and text annotation placed on maps provide both verbal and visual stimuli to the map user. When labels or text annotation are presented using a uniform typeface (font and size) regardless of each feature's importance, it may be difficult for map users to appropriately interpret the map. As Howarth (2015) suggests, a map developer should recognize the hierarchy of importance of places and features on a map, and make a list of the labels and text annotation that should be included on the map. Various text styles can then be employed to ensure that the labels and text annotation are (1) legible and (2) appropriately describe the hierarchy.

Diversion 12

A 2008 map of Georgia, the former Soviet Republic (United Nations, 2015), can be found on the book's website (mapping-book.uga.edu). From this map, develop a list of the various label and text annotation typeforms that are present. Once this has been completed, describe why you think the map developer may have assigned these typeforms to the text and annotation. Briefly describe the reasons why a hierarchy of relationships among the text and annotation might also exist.

Label placement efforts should also be carefully considered when developing a map. Generally speaking, one would want the name of a state or province to be placed inside

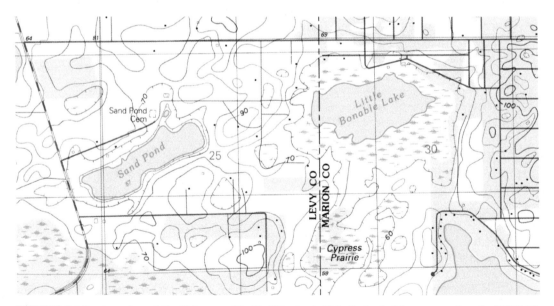

FIGURE 3.18   A portion of the Tidewater, Florida 7.5-minute U.S. Geological Survey quadrangle (29082-B5-TF-024) showing county names (Levy County and Marion County) in capitalized lettering along the county boundary, water bodies in italicized, blue lettering, and less important information (Sand Pond Cemetery, elevations, and Cypress Prairie) using smaller font sizes.

the boundary of the entity, one would want the name of a city to be placed near the point representing the center of the city, and one would want the name of a river to be placed along the line(s) representing the river's course. Fortunately, the placement of labels can be automated through a GIS or other online mapping software. Further, the presentation of labels and annotation digital maps is guided by rules for presenting text and annotation at various viewing scales. Regardless of how and when text and annotation are provided, there are a number of well-defined cartographic rules to consider for the placement of these on maps in order to achieve high levels of cartographic quality (Rylov and Reimer, 2014):

- The hierarchy (importance) of features should be reflected through the choice of font type.
- With respect to map orientation, the horizontal placement of labels is preferred.
- With respect to points, placement should be to the right, and slightly above the point.
- When describing land, labels should be placed completely on land.
- When describing water, labels should be placed completely on water.
- With respect to place names, these should be sufficiently spaced apart to avoid ambiguity.
- The overlap of two labels should be avoided.

These rules may be difficult to follow, depending on the scale of the map being developed and the heterogeneity of landscape or water body features that need to be placed on the map. Therefore, one should view these rules as professional advice for developing a map. Symbol and text placement, for example, are important in crowded maps as they may be impossible to avoid overlap, and may call for displacement, reduction, or generalization of the symbols and text (Korpi et al., 2013). Label placement for the rendering of maps on the screens of digital devices involves electronic processes that are adjusted as one increases or decreases the scale being viewed. Sometimes, the labels presented on maps within digital devices can be positioned or stretched automatically along the shape or area of the landscape features, scaled without incurring pixilation effects, and color-altered easily using computer processes (Skupin and Skupin, 2009).

Insets

Often within the borders of a map one will find both the landscape of interest, the central theme portrayed at a specific scale, and one or more smaller maps portrayed perhaps at smaller scales. These smaller maps are considered *insets*, and should be clearly distinguished from the main map, perhaps through the use of a border or frame (or the subject of "Neat Line" section). Insets can be locational in nature, illustrating a more general location of the information contained in the face of the map, and acting as an overview of the landscape. In this manner, an inset may allow the map user to orient themself by visualizing the mapped area in a broader geographic context. Called *locational insets* or *locator maps*, they might indicate where the land or water body resides with respect to the boundaries of other common entities such as a county, state, or country. In the "Map Title" section, we referenced a national forest map that can be found on the book's website: the aspen (*Populus tremuloides*) forest map, as of 2016, for the St. Ignace District of the Hiawatha National Forest. On this map is an inset (Fig. 3.19) which illustrates in green color the location of the St. Ignace District (the landscape portrayed in the main body of the map) with respect to other portions of the Hiawatha National Forest and some of the counties in the Upper Peninsula of Michigan. The inset, therefore, assists the map user in understanding the location of the St. Ignace District with respect to the broader landscape. Thus, using a smaller scale map, a locational inset may illustrate the broader area within which the main area of interest resides.

Some maps might also include other graphics, or *map insets*, that provide a broader reference to the larger geographic context within which the area of interest lies. For example, Fig. 3.20 provides two graphics that indicate where a specific U.S. Geological Survey 7.5-minute quadrangle map (Paisley, Oregon) (1) falls within the particular state (Oregon) in which it lies and (2) is situated with respect to the adjoining maps that cover adjacent 7.5-minute latitude and longitude areas. These graphics assist map users in understanding not only the names of nearby maps, but also where the area of interest is located with respect to a broader landscape.

An inset may also be used to provide greater detail about certain areas displayed in the main map. These types of *detail maps* illustrate a portion of the landscape at a larger scale than the main map that illustrates the area of interest, allowing a map user a more

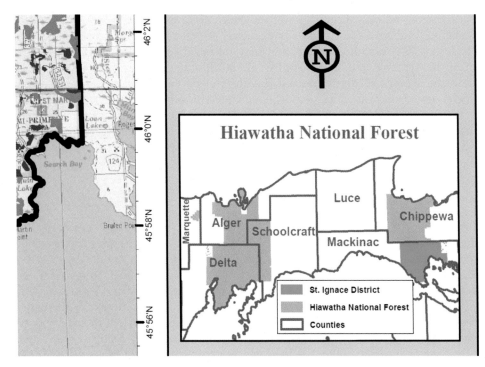

FIGURE 3.19 Locational inset contained within the aspen (*Populus tremuloides*) forest map, as of 2016, for the St. Ignace District of the Hiawatha National Forest. Source: *U.S. Department of Agriculture, Forest Service (2016b)*.

Oregon

■

Quadrangle location

FIGURE 3.20 General quadrangle location, and graphic describing adjoining quadrangles, for the Fremont Point, Oregon 7.5-minute U.S. Geological Survey quadrangle (42120-G7-TF-024).

1	2	3
4		5
6	7	8

1 Foster Butte
2 Summer Lake
3 Ana River
4 Pole Butte
5 South of Ana River
6 Shake Butte
7 Harvey Creek
8 Slide Mountain

Adjoining 7.5-minute quadrangles

informed view or perspective of an area of great interest. Fig. 3.21 is a map developed to illustrate the Harpers Ferry National Park (West Virginia), and in the lower right portion can be found a detail map that provides greater information on the area of the town at the junction of the Potomac and Shenandoah Rivers. The inset acts to convey at a scale larger than the main map area the location of the information center, the bookstore, and a trail in relation to other locations of interest (church and ruins) to recreationists or sightseers.

As you may have noticed, locational insets and detail maps can be presented at scales significantly different than the main map, to provide a reference or broader perspective or to focus attention on a specific place. The inclusion of a scale in both the main map and the inset(s) may therefore be of value to the map user. However, there are instances where an inset may be used to portray, at the same scale as the main map area, an area spatially separated from the landscape illustrated in the main map area. These types of *continuation insets* might be viewed as extensions of the main map that are meant to avoid the display of large expanses irrelevant to the message of the map. A good example is the various maps of the United States that show Hawaii within an inset, often situated very close (but incorrectly) to California. In these cases, the size of Hawaii may be presented at the same scale as the other states, but the vast Pacific Ocean that separates Hawaii from the continental United States is ignored (Fig. 3.22). An inset such as this allows Hawaii (or any other geographic location) to be placed anywhere on a map, and to be represented at the same scale as other areas portrayed on a map.

Translation 12

Examine the hypsometric map of Borneo (Fig. 3.23). Located in the upper-left corner of this map is a locational inset. How would you describe the purpose of the inset to another person who may be unfamiliar with map design? What land areas are included in the inset, and how does this coverage of the Earth differ from the main area of the map?

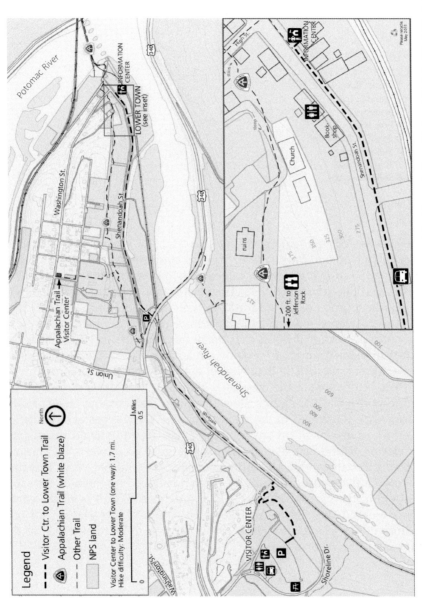

FIGURE 3.21 A larger scale inset, or detail map, located in the lower right corner of the Harper's Ferry National Park in West Virginia. *Source: Courtesy U.S. National Park Service through Wikimedia Commons.*

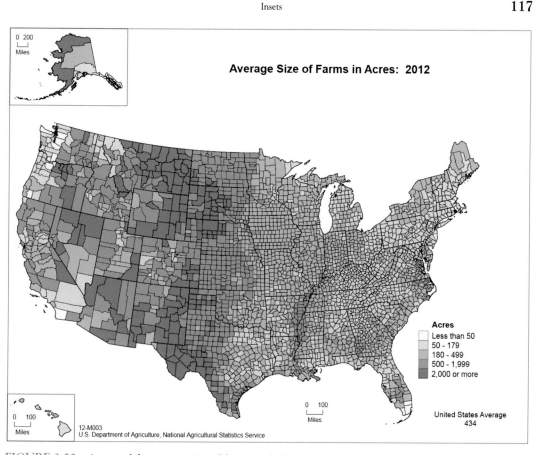

FIGURE 3.22   A map of the average size of farms in the United States in 2012, with two inset maps—one each for Hawaii and Alaska. Source: *U.S. Department of Agriculture, National Agricultural Statistics Service (2012).*

In some areas of the world, such as the United Kingdom, inset maps are separated from the main map to which they refer. Some of these might be referred to as *planning inset maps*, which are a subset of smaller scale policy or plan proposal maps that are developed to represent specific communities. These types of insets illustrate the locations where specific plan or policy proposals apply, and may provide more detail about individual communities situated within a broader city or district. Planning inset maps may technically be better described as *location diagrams* since they are usually provided in a manner separate from the smaller scale map to which they were associated.

When incorporating an inset into a map, in a manner similar to the incorporation of a legend into a map, a map developer should consider how the inset would best balance the overall cartographic quality of a map, given the other map components and marginalia that are present. As we suggested, insets typically are contained within specific borders or frames to separate them from the other parts of a map. This can allow a map user to understand that the information contained within the locational inset box

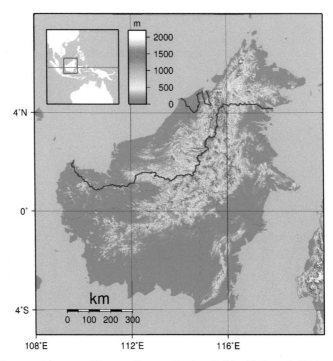

FIGURE 3.23   A hypsometric map of Borneo. Source: *Courtesy Sadalmelik through Wikimedia Commons.*

might be different (scale and theme) than the information contained within the main map area.

Neat Line

A *neat line* is a map component presented as a border, frame, or boundary between what represents the main area of the map itself, and the other parts of the page, screen, or media on which it resides, or commonly those areas that may not be printed (in other words, the margin of the media). A neat line may consist of one or more parallel lines of various widths, often connected in horizontal and vertical directions forming a decorative box or frame. The thickness of one or more lines used to create a neat line often reflects the importance of the information that they enclose (Peterson, 2015), but not necessarily, as in Fig. 3.24. There is some confusion as to whether a neat line, when used, represents only the outer boundary of the main map area, or whether it also represents the borders used to contain legends, insets, and other items that are contained inside a map, since the purpose of a neat line is to separate map components and facilitate an organized feel to the map (Peterson, 2015). Therefore, the terms *borders*, *frames*, and *neat lines* are used interchangeably in contemporary mapping processes.

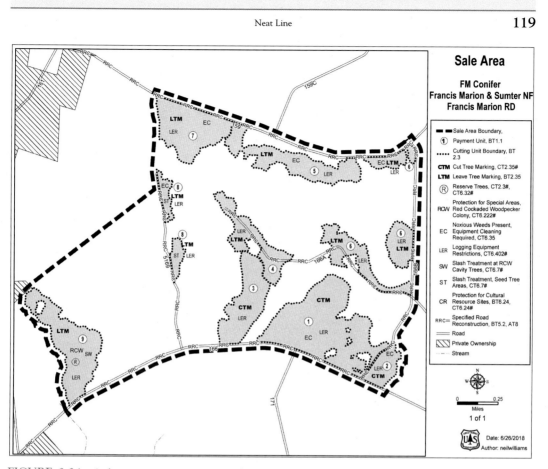

FIGURE 3.24   A forest management map indicating potential harvesting activities on the Francis Marion National Forest in South Carolina. Source: *Williams (2018).*

Reflection 13

Certainly, a neat line might be considered as the box drawn on a map to contain the entire contents of the map on printed or digital media. However, given the discussion earlier, what is your opinion of whether other boxes (borders and frames) drawn inside of maps, separating legends and insets, should be technically considered neat lines?

A neat line around an inset within a map may be visually useful, as it may improve the aesthetic quality of a map and may help map users focus on certain map details. However, a neat line is an optional map component, and not everyone is a proponent of incorporating these into a map. One argument against the use of a neat line is that avoidance of noncompulsory map components might liberate one's mind of rules and standard practices, and promote *outside the box* design thought (pun intended). A reduction of cartographic drudgery might then facilitate a work environment where cartographic artistry can flourish. Another argument against the use of a neat line is that it may leave an

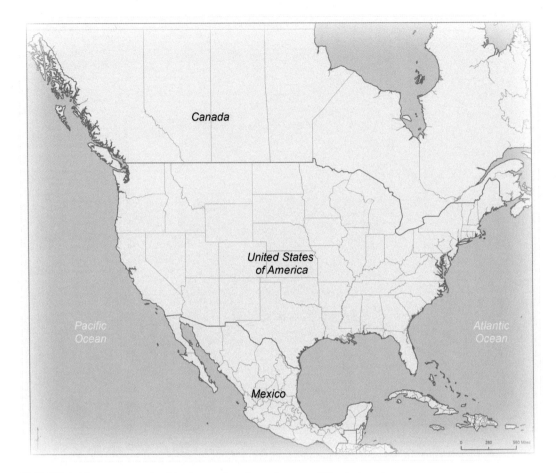

FIGURE 3.25   An example of feathering the edges of a map of North America.

impression in the map user's mind that the physical landscape ends at the edge of the neat line. To overcome this, a map could be designed to feather (gradually fade out) the landscape as it approaches the edge of the media on which it is presented (Fig. 3.25). Alternatively, a map could employ a *full bleed* approach, where the map continues to the edge of the media. Often this is the case when maps are presented on digital media.

Reference Information

In professional settings, for purposes of communicating information to decision-makers or the general public, a number of items of reference should be placed on maps. The date that a map was developed is one piece of reference information can significantly help people understand the temporal quality (or resolution) of the information presented. A map created in 1985 of New York City landmarks can be of significantly less value to tourists than a current map of city landmarks. In some circumstances, where incremental versions of specific maps need to be produced relatively frequently, the day and the time of day might also be provided to help map users understand which version is being viewed. Weather maps may represent a good example in this regard, especially during periods of severe weather events.

Inspection 18

Authors of maps, or the organizations within which they are employed, can be very cautious in communicating the caveats associated with map products. More information on this subject follows in "Warranties, Disclaimers, Copyrights, Caveats, and Discrimination Statements" section of this chapter. For this exercise, search the Internet for a map representing the *Building Benchmarking and Transparency Policies of the United States*. Hopefully, you will locate a map developed by the Institute for Market Transformation in Washington, D.C. When was the map last updated? To whom is this information of value? How is the date of the map of importance?

When the information presented in a map may be questioned, the *source* of the information might be noted. Adding sources to map annotation can support the legitimacy of the product, allowing a map user to conclude that the data used came from a reputable organization. A map developer may state that the data underlying or contributing to the mapped features was obtained from a state or federal agency, or some other source that suggests that the data may be current, accurate, or commonly acceptable. Some simple statements of map information sources include:

Source: U.S. Department of Agriculture, Forest Service
Source: U.S. Census Bureau

Other statements related to source material may direct map users to locations on the Internet where a substantial amount of metadata and background information might be accessible. Caution should be applied in these cases, as Internet addresses are not always stable, and the information you seek may, therefore, not be considered as permanent as information obtained through printed media. An example of a statement that would lead a map user to the source material for a map might be the following:

Full analysis information and accompanying computer code are available at: https://.....

The sources of data can also be important for understanding the relative quality of the map product (lineage and vintage of data), and for suggesting subtle caveats concerning map quality to the map user. A more general statement regarding the source of information provided on a map, and a deflection of sorts to the true sources of information, might include the following:

The information on this map was collected through a variety of secondary sources that included ____. Every effort was made to verify the quality of the information, however, concerned users of this map should contact the associated agencies before making decisions based on the information provided.

This map is based on the ____ base map. Other data presented in this map have been modified to be consistent with the base map.

Providing the specific name(s) of the map developer(s) on a map might represent a finer level of reference detail along these lines. A map of wetlands in Jackson County, Florida (for example) may appear to be more legitimate and official if it were produced by someone from the U.S. Geological Survey than if it were developed by Pete's Mapping Service, or worse, if no map developer or organization were listed. However, presenting the specific map developer's name may be more appropriate when a map is developed by one or two people, or a small team, rather than a larger collection of individuals. Forest management activity maps developed specifically for small landowners, government agencies (Fig. 3.24), or companies by consultants or analysts, may likely identify the map developer.

Translation 13
Access the map contained on this book's website (mapping-book.uga.edu) that describes Cameroon's Forest Estate. Based on the information provided in the *data and map production* section of the map, how would you describe the stature and perceived legitimacy of the map? Based on this information, try to develop a statement to support or refute the notion that the contents of the map represent the highest authority in identifying the locations of hunting areas and protected zones.

Other reference information that may be of value to a map include the projection and coordinate systems employed. Projection and coordinate systems are described in more detail in the next chapter. They assist in the transformation of a spherical surface (the Earth) to a flat representation (the map), and provide one the ability to reference places on the surface of land (or water) through positional data (coordinates). The physical location of the digital files associated with the map might also be provided as reference information. For instance, a map may indicate somewhere in the marginal area a piece of text such as

c:/MAPS/January_2018/Sophies_Forest_Reserve.mxd

Of course, unless otherwise accessible to a larger audience, this type of reference information would only be of value to the map developer and close associates, pointing them to the physical computer (or network) location of the file used to generate the map. Placement and aesthetic quality of these reference items on a map are a reflection of the cartographic skill of the map developer; they should be located in indiscrete parts of the map, out of the way, but still accessible. Unfortunately, when maps are rendered on digital devices, many of these pieces of reference information can be lacking.

Warranties, Disclaimers, Copyrights, Caveats, and Discrimination Statements

Sometimes a map will contain information meant to protect the map developer and the publisher of the map from action (legal and otherwise) taken by users of the map. In these rare but serious cases, some map users may feel that they have been led astray by the inherent inaccuracies in a map, and may argue that they incurred damages as a result of relying upon the accuracy of the map. For example, a map may have indicated the location of a natural gas pipe, and a land developer may have used the map to avoid the pipe, but unfortunately struck the pipe during construction activities, causing an explosion. Naturally, the map (and map developer) may be blamed for the unfortunate event, whether true or not. Often a *warranty* will state that the map developer or publisher is offering no guarantee, expressed or implied, of integrity on the validity or accuracy of the information presented in the map. The length and detail of a map warranty can vary, as one can see in the following examples:

> *While every effort has been made to ensure the accuracy of the information in this map, ____ accepts no responsibility for errors or omissions or map features or annotation. ____ also accepts no responsibility for damages alleged to have been suffered as a result of the use or misuse of the information presented in the map.*

> *No warranties, expressed or implied, are provided by ____ concerning the validity, accuracy, or completeness of the information presented in the map.*

> *This map was compiled by ____ using data known to be reasonably accurate. However, some level of error is inherent in all maps, therefore this map is distributed "as is", without warranties of any kind, expressed or implied.*

> *This information on this map is made available as a public service, and is to be used only for reference purposes. The ___ provides this map AS IS, without warranty of any kind regarding accuracy, validity, completeness or fitness of use.*

> *The ____ does not warrant the accuracy of this map, therefore no decision involving a risk of economic loss or physical injury should be made in reliance of the information it contains.*

Statements related to data limitations and inconsistencies within a map can serve to warn people of limitations associated with uses of the map. A *disclaimer*, therefore, represents a denial of responsibility of the map developer with respect to actions taken by the map user based on the quality or content of the map. Disclaimers can also relate to the vintage of the information used, implying that more current data may one day become available. They may also refer to potential problems in the interpretation of map products caused by spatial data that is not seamlessly (spatially) integrated. The exclusion of information, or the recognition that certain data was not used in the development of a map, may also be communicated through a disclaimer.

> *Some data records could not be georeferenced.*

> *Some information was deliberately excluded for reasons of safety and privacy.*

> *The ____ was unable to verify the accessibility of all recreational sites noted on this map.*

> *The ___ is not responsible for any inaccuracies in the data presented in this map, and does not endorse any further interpretations or derived products that used this information.*

> *This map uses data from multiple public sources, ____ staff cannot verify the accuracy of all features.*

Similarly, *warnings* associated with data uncertainty or feature omission may be communicated through a statement on a map.

This map is not suitable for uses involving land titles or land claims.

This is a thematic map to be used for ____ purposes only.
This map contains no authoritative positional information.

Maps developed by ____ are currently being validated.

Warnings, disclaimers, and warranties are all very similar, and are designed to limit the legal liability of the map developer and their employer. Often, maps may also contain statements of caution, in the form of *caveats*, to the users of maps. These statements are also meant to warn map users of potential map interpretation pitfalls. For example, on a small-scale hydrologic map, a caveat might state the following:

... the data provided does not illustrate streams that are less than one mile in length.

In thematic maps that illustrate current environmental conditions across a landscape, perhaps based on a published model or a set of rules, a caveat to the users of the map might suggest that

An increase in positional precision is required to more accurately examine the influence of ____ on ____.

Translation 14

Access the drought map of the United States, the January 20, 2019 version of which is located on this book's website (mapping-book.uga.edu). Alternatively, the current version of the United States drought map may be located through a search of the Internet, using *United States Drought Monitor* keywords. The official website for this work is managed by the National Drought Mitigation Center at the University of Nebraska-Lincoln. Once the map has been located, describe in general terms, what the disclaimer on this map suggests to people who may use the map.

Often a map developer will allude to the fact that they were aware that the information contained therein may not be totally accurate, and that the data may have needed to be improved or updated. In these cases, the map developer may provide statements that resemble a disclaimer, such as the following:

Feedback and corrections on the information contained in this map are welcomed.

Please send any corrections to this map to ___.

This map is subject to, and undergoes, quarterly revisions.

Inspection 19

Access the Apalachicola National Forest FY 2019 prescribed burn planning map from this book's website (mapping-book.uga.edu). In general terms, without repeating what the map has to offer, what are the caveats and warnings noted on this map?

With respect to maps, a *copyright* is the exclusive legal right of an entity, the originator of a work or one to which the rights have been assigned, to print and publish the work. Along with this intellectual property protection, these entities also have the right to authorize other entities (people and organizations) to do the same. Unless works are explicitly stated to be in the public domain (where there is no copyright protection, or where the copyright has expired), permission to use the work is usually required. Our tactile map in Chapter 2, Map Types, is an example where we received permission to reuse another's work. Most of the maps in this book are either considered to be in the public domain, or they have been developed by the book's authors. Many very useful and aesthetically pleasing maps were not included in this book due to copyright restrictions. Copyright laws vary by country, which may affect international uses of maps. The Creative Commons system allows authors to assign copyright licenses to their works, which may allow varied uses of these published works when the original author is given credit in the manner that the license requires. When provided on maps, a copyright statement might look like the following three examples:

Copyright © 2018 University of _____, All Rights Reserved

© 2013–2018 Center for _____, University of _____

Copyrights reserved

Occasionally a discrimination policy is provided on a map that serves two purposes: it provides a defense for the map developer's organization against claims related to certain types of intolerable behavior, and it enables the map developer's organization to communicate to its employees and the public that it will not tolerate these types of behavior.

It is the policy of _____ to maintain a work environment free of discrimination and harassment. The _____ prohibits discrimination and harassment against any person because of age, sex, race, religion, veteran status, disability, handicap, sexual orientation, or gender identity. Discrimination or harassment against _____ will not be tolerated.

Reflection 14
Imagine that you are the owner of a small natural resource management consulting firm. Your work often entails developing maps that accompany resource assessments your firm conducted for towns, cities, and counties in the area where you reside. On the maps you develop and distribute to the public, what types of warranties, disclaimers, copyrights, sources, caveats, or discrimination statements would you provide? For what reasons would you place these on your maps?

As with reference information, the placement and aesthetic quality of these items on a map are a reflection of the cartographic skill of the map developer; they should be located in indiscrete parts of the map, out of the way, but still accessible.

Backdrop

Various geographic databases can be used as the *backdrop*, or base information for the map. For example, in addition to thematic information concerning the types of forest management activities an organization might propose (e.g., the location of seed tree harvests or proposed thinnings of trees), forest management maps may provide an aerial image of the landscape in the background. The addition of an aerial image can allow a map user to clearly see the houses, businesses, and roads near a property, and thus judge their significance in relation to the proposed activities. While other cartographic symbols can be used to represent the location of houses, businesses, or roads on a map, an aerial image can more firmly ground the map to mental images or memories people may have of the area. Therefore, this type of backdrop allows map users to further place in perspective the landscape resources found in the area of interest. Other types of base maps may be used as backdrops on maps to provide similar landscape perspective for the user of the map (Fig. 3.26). For example, it is not uncommon for digital road or weather maps to contain light gray or green images (*canvases*) of county, state (provincial), or national boundaries. The use of backdrops in this manner may also include a subtle recognition of the major roads within the area of interest of a map.

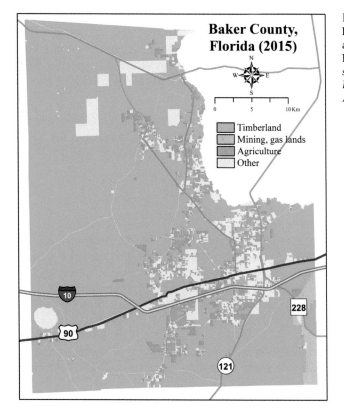

FIGURE 3.26   General thematic map of land uses in Baker County, Florida, 2015, accompanied by a base map of national highways. Source: *Background base map source: National highway network from the U.S. Department of Transportation, Federal Highway Administration (2013).*

Reflection 15

Consider the most recent digital map you viewed, either on your cellular phone, tablet, or personal computer. Aside from the general theme of the digital map, what was presented as a backdrop to the thematic data to provide context for the map? Do you think that the backdrop was of value to the map?

When incorporating a backdrop into a map, in a manner similar to the incorporation of a legend or an inset into a map, a map developer should consider how the inset would best balance the overall cartographic quality, given the other map components and marginalia that are present.

Concluding Remarks

Each map, depending on its audience and purpose, will require a certain set of map components. An object (north arrow) to define map orientation is often a standard item, unless a map developer assumes that map users will understand landscape orientation through convention (north is always at the top of the map) or through knowledge of well-known, accompanying geographic features. An indication of the scale of the map is often necessary to allow a map user to understand the relative sizes of landscape features, and also to judge distances between places on the map. A scale reflects the dimension of the model, where *large scale* suggests that the map is closer to the real-life dimension or size of the landscape or water body than *small scale*, as if viewing the landscape from further away. Legends and map annotation may also be necessary to add insight or to alleviate confusion among the features presented. Similarly, with warranties, disclaimers, and other reference information, map developers may need to adhere to organizational standards while also informing the map user of any issues that may affect important potential outcomes, should they use the map as a basis for making decisions. The amount of detail one can include on a map can be considerable, therefore map developers may be faced with design considerations that limit the amount of information presented on the map.

Diversion 13

Using the Talladega National Forest data made available for this exercise through the book's website (mapping-book.uga.edu), along with your preferred computer mapping software, develop a thematic map that includes each of the map components noted in this chapter. Consider an informal hierarchy as you make this map; some information you may wish to include in the map may be more important than other information. As you develop the map, pay attention to the emphasis (color, font, placement, etc.) you are placing on each map component, given your opinion of the hierarchy and importance of each.

Hopefully, an appreciation of the need for various map components has been gained through the discussions contained in this chapter. A balance between too little and too much information on a map must be struck, as a paucity of information may add little value to the user and be of low intellectual complexity, while an excessive amount of information displayed on a map may result in too much visual noise and may affect its legibility (Touya et al., 2016). An excess of information on a map may also affect the cognitive ability of a person to process and remember information (Bestgen et al., 2017). From different perspectives, map developers and users each need to make sense of, and be able to reason with, the components found on a map (Gregg, 2001) as most people who engage with maps think critically about the symbols and colors that are displayed (Barbour, 2001). An awareness for the good design of maps is therefore important (Lee, 1995).

References

Akinyemi, F., Kibora, P.M., Aborishade, P., 2013. Designing effective legends and layouts with a focus on Nigerian topographic maps. Cartographica 48 (1), 1–12.

Barbour, D., 2001. Maps for fun and profit: one small company's experience. Cartogr. J. 38 (1), 81–85.

Bestgen, A.-K., Edler, D., Miller, C., Schulze, P., Dickman, F., Kuchinke, L., 2017. Where is it (in the map)?: Recall and recognition of spatial information. Cartographica 52 (1), 80–97.

Bianchetti, R., Wallgrün, J.O., Yang, J., Blanford, J., Robinson, A.C., Klippel, A., 2012. Free classification of Canadian and American emergency management map symbol standards. Cartogr. J. 49 (4), 350–360.

Cano, R.G., Kunigami, G., de Souza, C.C., de Rezende, P.J., 2013. A hybrid GRASP heuristic to construct effective drawings of proportional symbol maps. Comput. Oper. Res. 40 (5), 1435–1447.

Dalton, C.M., 2013. Sovereigns, spooks, and hackers: an early history of Google geo services and map mashups. Cartographica 48 (4), 261–274.

Donohue, R.G., Sack, C.M., Roth, R.E., 2013. Time series proportional symbol maps with Leaflet and jQuery. Cartogr. Perspect. 76, 43–66.

Edler, D., Bestgen, A.-K., Kuchinke, L., Dickmann, F., 2014. The effects of grid line separation in topographic maps for object location memory. Cartographica 49 (4), 207–217.

Edney, M.H., 1999. Reconsidering enlightenment geography and map making: reconnaissance, mapping, archive. In: Livingstone, D.N., Withers, C.W.J. (Eds.), Geography and Enlightenment. University of Chicago Press, Chicago, IL, pp. 165–198.

ESRI, 2018. About ArcGIS, the Mapping and Analytics Platform. ESRI, Redlands, CA. <https://www.esri.com/en-us/arcgis/about-arcgis/overview> (accessed 02.02.19.).

Gregg, M., 2001. Making sense of maps: what part of the map are we talking about? Pa. Geogr. 39 (2), 30–37.

Harley, J.B., 1989. Deconstructing the map. Cartographica 26 (2), 1–20.

Howarth, J.T., 2015. A framework for teaching the timeless way of mapmaking. Cartogr. Geogr. Inf. Sci. 42 (Suppl. 1), 6–10.

Huffman, D.P., 2018. The power of appearances. Cartogr. Perspect. 90, 71–81.

Korpi, J., Haybatollahi, M., Ahonen-Rainio, P., 2013. Identification of partially occluded map symbols. Cartogr. Perspect. 76, 19–31.

Lee, J., 1995. Map design and GIS – a survey of map usage amongst GIS users. Cartogr. J. 32 (1), 33–39.

Lisitskii, D.V., 2016. Cartography in the era of informatization: new problems and possibilities. Geogr. Nat. Resour. 37 (4), 296–301.

NATO Standardization Agency, 2017. NATO Joint Military Symbology. NATO Standardization Agency, Brussels. APP-6(D).

Peterson, G.N., 2015. GIS Cartography: A Guide to Effective Map Design, second ed. CRC Press, Boca Raton, FL.

Piatti, B., Bär, H.R., Reuschel, A.-K., Hurni, L., Cartwright, W., 2009. Mapping literature: towards a geography of fiction. In: Cartwright, W., Gartner, G., Lehn, A. (Eds.), Cartography and Art. Lecture Notes in Geoinformation and Geography. Springer-Verlag, Berlin, pp. 177–192.

Rambaldi, G., 2005. Who owns the map legend? URISA J. 17 (1), 5–13.

Richardson, W.A.R., 2015. Asian geographical features misplaced south of the Equator on sixteenth-century maps. Terr. Incogn. 47 (1), 33–65.

Robinson, A.C., Pezanowski, S., Troedson, S., Bianchetti, R., Blanford, J., Stevens, J., et al., 2013. Symbol Store: sharing map symbols for emergency management. Cartogr. Geogr. Inf. Sci. 40 (5), 415–426.

Rylov, M.A., Reimer, A.W., 2014. A comprehensive multi-criteria model for high cartographic quality point-feature label placement. Cartographica 49 (1), 52–68.

Skupin, A., Skupin, M., 2009. On written language in works of art and cartography. In: Cartwright, W., Gartner, G., Lehn, A. (Eds.), Cartography and Art. Lecture Notes in Geoinformation and Geography. Springer-Verlag, Berlin, pp. 207–222.

Swartz, D.J., 1933. A study for variation in map symbols. Elem. School J. 33 (9), 678–689.

Touya, G., Hoarau, C., Christophe, S., 2016. Clutter and map legibility in automated cartography: a research agenda. Cartographica 51 (4), 198–207.

U.S. Department of Agriculture, Economic Research Service, 2019. Food Environment Atlas, Fast-Food Restaurants, 2014. U.S. Department of Agriculture, Economic Research Service, Washington, D.C., <https://www.ers.usda.gov/data-products/food-environment-atlas/go-to-the-atlas/> (accessed 15.04.19.).

U.S. Department of Agriculture, Forest Service, 2016a. Revised Land Management Plan, Francis Marion National Forest. U.S. Department of Agriculture, Forest Service, Region 8, Atlanta, GA, R8-MB 149 A.

U.S. Department of Agriculture, Forest Service, 2016b. Hiawatha National Forest, St. Ignace District, Aspen Coverage Map: 2016. U.S. Department of Agriculture, Forest Service, Hiawatha National Forest, Gladstone, MI.

U.S. Department of Agriculture, Forest Service, 2017a. St. Joe Ranger District, Buck Fire, Area, Road, and Trail Closures. U.S. Department of Agriculture, Forest Service, Idaho Panhandle National Forests, St. Joe Ranger District, St. Maries, ID.

U.S. Department of Agriculture, Forest Service, 2017b. Motor Vehicle Use Map, Osceola National Forest. U.S. Department of Agriculture, Forest Service, Southern Region, Atlanta, GA.

U.S. Department of Agriculture, National Agricultural Statistics Service, 2012. Census of Agriculture, 2012 Census Ag Atlas Maps - Farms. U.S. Department of Agriculture, National Agricultural Statistics Service, Washington, D.C., <https://www.nass.usda.gov/Publications/AgCensus/2012/Online_Resources/Ag_Atlas_Maps/Farms/> (accessed 15.04.19.).

U.S. Department of Commerce, National Oceanic and Atmospheric Administration, 2017. Data Snapshot Details: Sea Surface Temperature (SST). U.S. Department of Commerce, National Oceanic and Atmospheric Administration, Silver Spring, MD, <https://www.climate.gov/maps-data/data-snapshots/data-source-sea-surface-temperature> (accessed 21.12.18.).

U.S. Department of Commerce, National Oceanic and Atmospheric Administration, 2019. World Magnetic Model Out-of-Cycle Release. U.S. Department of Commerce, National Oceanic and Atmospheric Administration, National Centers for Environmental Information, Asheville, NC. <https://www.ncei.noaa.gov/news/world-magnetic-model-out-cycle-release> (accessed 24.05.19.).

U.S. Department of the Interior, Bureau of Land Management, 2017. North Spit Coos Bay Map. U.S. Department of the Interior, Bureau of Land Management, Washington, D.C.

U.S. Department of the Interior, Fish and Wildlife Service, 2001. Graphics standards. U.S. Department of the Interior, Fish and Wildlife Service, Washington, D.C. <https://www.fws.gov/external-affairs/marketing-communications/printing-and-publishing/pdf/graphicstandardsjuly01.pdf> (accessed 22.05.19).

U.S. Department of the Interior, Geological Survey, 2013. Topographic Map Symbols. U.S. Department of the Interior, Geological Survey, Reston, VA.

U.S. Department of the Interior, National Park Service, 2017. NPS Map Elements: Updated April 18, 2017. U.S. Department of the Interior, National Park Service, Washington, D.C.

U.S. Department of Transportation, Federal Highway Administration, 2013. National Highway Network. U.S. Department of Transportation, Federal Highway Administration, Washington, D.C.

United Nations, 2015. Georgia. United Nations, Department of Field Support, Geographic Information Section, Rome. <http://www.un.org/Depts/Cartographic/map/profile/georgia.pdf> (accessed 21.12.18.).

Voženílek, V., Morkesová, P., Vondráková, A., 2014. Cognitive aspects of map symbology in the world class atlases. Procedia - Soc. Behav. Sci. 112, 1121–1136.

Wei-Haas, M., 2019. Magnetic north just changed. Here's what that means. Natl. Geogr. February 4, 2019.

Wessel, N., Widener, M., 2015. Rethinking the urban bike map for the 21st century. Cartogr. Perspect. 81, 6–22.

Williams, N., 2018. Sale Area, FM Conifer. U.S. Department of Agriculture, Forest Service, Francis Marion and Sumter National Forests, Columbia, SC.

Yin, P.-Y., Huang, Y.-B., 2001. Automating data extraction and identification on Chinese road maps. Opt. Eng. 40 (5), 663–673.

Young, H.T.M., Carr, N.J., Green, B., Tilley, C., Bhargava, V., Pearce, N., 2013. Accuracy of visual assessments of proliferation indices in gastroenteropancreatic neuroendocrine tumours. J. Clin. Pathol. 66 (8), 700–704.

Map Reference Systems

Map reference systems help us understand where we are with respect to other places on, under, or above the surface of a planet such as Earth. In understanding where a place is located, we require reference points, often known places on the surface of the Earth. Then, we need a process for describing the relative position of other places, with respect to the reference points, when mapping landscape features to a flat surface, paper or digital device (Greenwood, 1964). The collection of tools for transforming a nonflat surface to a flat surface, and the means by which we decide to understand where we are on this surface, compose the reference system. A reference system for describing places on Earth may thus be composed of a map projection, a coordinate system, and a datum. No map reference system is perfect, and each have their advantages and limitations. Cartographers and engineers have developed many map reference systems over the last several centuries to explain how one might best represent on a flat surface the features of an approximate sphere (Arnaud, 2013).

Complicating our need to develop and use a map reference system to describe places on Earth is the fact that Earth is an oblate spheroid or oblate ellipsoid (hereafter, an *ellipsoid*). Our planet is slightly flattened at the poles, resulting in an equatorial bulge. But of course Earth is more complicated than this, as on its surface there are numerous mountains, depressions, and flat plains. When describing the shape of Earth, scientists and geographers often refer to the *geoid*, the shape of the planet with respect to a theoretical sea level surface, or mean sea level surface. This is the shape of the Earth at sea level where oceans and seas exist, and the shape of the Earth under land (but at sea level elevation) where land now exists. Rather than a perfectly round ball, our geoid is a complicated shape (King-Hele, 1967), and this is important because the geoid is used as a basis for measuring surface elevations. We often utilize a static version of a geoid to represent the shape of Earth; however, the surface of our planet continuously changes, and these need to be continually monitored (Sun et al., 2019). A number of reference ellipsoids have been developed to approximate Earth's geoid. We discuss these very briefly when we provide a brief description of the reference points, or *datums*, on which coordinate systems are based.

Coordinate Systems

A coordinate system uses a pair (X,Y) or a triplet (X,Y,Z) of numbers to represent a point on the ground, a point under or above the ground, or similarly a point within or at the surface of a water body. X usually represents a horizontal east-west direction, Y usually represents a horizontal north-south direction, and Z usually represents a vertical direction. The units employed can be degrees, meters, feet, or any other quantitative value to reflect the difference in landscape position from some known location (e.g., the equator, a meridian, a base line, and a geoid). A coordinate system allows for distances, directions, and areas to be computed using this information. As we noted earlier, the type of map projection system employed can influence the types and amounts of errors that then exist in a map. Therefore, the choice of coordinate and map projection systems are both important in developing a common language for describing the position of features on maps. Examples of coordinate systems include latitude and longitude, Universal Transverse Mercator (UTM), and the state plane systems of the United States. After our discussion below of common coordinate systems, we provide a section in this chapter that describes several map projection systems.

Latitude and Longitude

One of the most common global coordinate systems involves the use of latitude and longitude measurements. Latitude (a north-south reference) and longitude (an east-west reference) are components of a coordinate system that describes each place on Earth as it relates to the equator and a prime meridian. The equator divides the Earth into north and south halves, and *latitude* measurements (degrees) begin there and proceed toward the appropriate pole. In the northern hemisphere, higher latitudes represent areas further north. In the southern hemisphere, higher latitudes represent areas further south. A

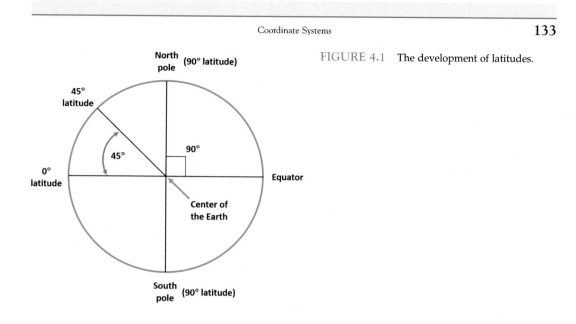

FIGURE 4.1   The development of latitudes.

latitude is developed by projecting an angle from the center of the Earth to the surface of the Earth, using the equator as the base edge of the angle (Fig. 4.1). The poles are located at 90° angles from the center of the Earth thus, their latitude is 90° north or south of the equator. Latitudes of the same magnitude represent the same distance north or south of the equator. The lines they form can be visualized as rings around the Earth that get smaller as one moves closer to a pole. These rings or lines drawn around the Earth at a constant latitude are called *parallels*. Parallels are therefore parallel to the equator. While any constant degree of latitude around the Earth may be considered a parallel, the Tropic of Cancer and the Tropic of Capricorn are two of the more commonly known parallels, at 23.5° north (Cancer) and south (Capricorn) of the equator. The longest of the parallels around the Earth is located at 0° latitude; with increases in latitude, the size of the parallels (circles around the Earth) becomes smaller (Fig. 4.2).

North-south meridian lines can also be drawn on a model of the Earth, running from the north pole to the south pole, representing essentially one-half of an ellipsoid (Fig. 4.3). When extended around the Earth, each complete loop (consisting of two lines of latitude) divides the planet in half, as if cutting a melon through its center. The *longitude* is the angle formed by moving east or west of a reference meridian, a line formed by drawing a straight north-south line from pole to pole through an arbitrary point such as the Royal Observatory at Greenwich, England. This meridian is also commonly called the *prime meridian*, or the *zero meridian*. The prime meridian is the one to which other meridians are referenced (they lie east or west of it), up to 180°, or half-way around the world. For example, New Orleans, Louisiana, is located approximately 90° west of the prime meridian, or one-quarter of the way around the Earth. Istanbul, Turkey, is located approximately 29° east of the prime meridian (Fig. 4.4). Lines of longitude are not parallels, as they begin and end at a pole. The continuation of a longitude line around the other side of the Earth results in a different longitude value with respect to the origin. As we suggested, these north-south reference lines on maps can simply be called *meridians*, and interestingly every

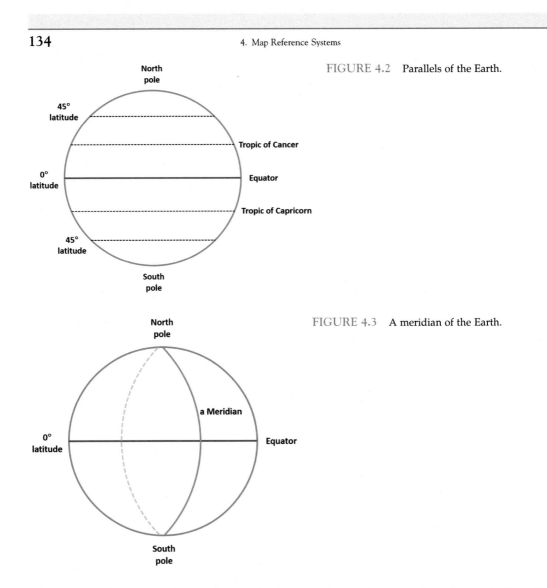

FIGURE 4.2   Parallels of the Earth.

FIGURE 4.3   A meridian of the Earth.

place along these lines, from pole to pole, has midday at the same time (Frye, 1895). Other local coordinate systems, such as the state plane or the Public Land Survey System systems used in the United States, employ shorter meridians that are perhaps as long (in a north-south direction) as one or two states or provinces (or perhaps less). Here though, within the latitude and longitude coordinate system, a meridian represents a north-south reference line that extends from pole to pole.

Reflection 16

Imagine that you needed to develop a broad range of maps, perhaps ranging in scale from very local representations of land parcels to worldwide representations of forest cover. When would it matter that the geographic coordinates of an area are (or are not) displayed on these maps?

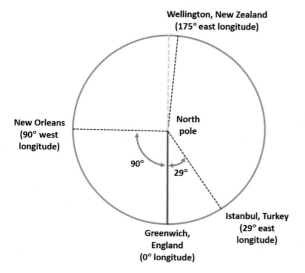

FIGURE 4.4 An azimuthal view of longitudes with respect to the prime meridian in Greenwich, England.

Great circles are the largest circles that can be drawn on spheres. Great circles of Earth are formed by an imaginary plane that intersects the center of the Earth. The resulting intersection of the plane with the surface of the Earth defines the great circle. Technically, there are an infinite number of these (Williams and Ridd, 1960) even though only a few located at specific longitudes and latitudes are discussed. For example, some great circles are composed of lines of longitude that connect the two poles and divide the Earth equally into two halves. One of the great circles divides Earth equally into northern and southern hemispheres, where the latitude is 0°.

Translation 15

Imagine you find yourself one day engrossed in a game of trivia, perhaps a televised game show. A question arises: *What is the common name for the great circle of Earth found at 0° latitude?* Of course you know this answer, which is …

A coordinate system that employs latitudes and longitudes uses *geographic coordinates*. The grid on a map that is formed with lines of latitude and longitude that are spaced apart at an assumed, specific degree interval is considered a *geographic grid* or *graticule*. Latitude and longitude lines, along with great circles, are characteristic elements of a graticule (Kessler, 2018). A graticule can of course be developed for many other uses, such as medical assessments of human arteries (Morgan and Adams, 1974), but in mapping the graticule represents the meridians and parallels that are used to represent the locations of landscapes and water bodies. On a map, particularly a small-scale map, one should be able to recognize the meridians (north-south reference lines) and parallels (east-west reference lines). If one were to think about coordinates in a two-dimensional graph, positions along horizontal lines (east-west) would be represented by movement along the *X*-axis, and positions along vertical lines (north-south) would be represented by movement along the *Y*-axis. *X,Y* coordinate pairs are, therefore, used to reference north-south (with respect to the equator) and east-west (with respect to the prime meridian) positions on a map. For example, the *X,Y* coordinates representing the location of the *Grizzly Statue* at the University of Montana are:

46.860076° north and 113.986805° west

Diversion 14

Using your preferred mapping program, what set of longitude (*X*) and latitude (*Y*) coordinates might you use to describe the position of the following features?

a. The Eiffel Tower in Paris
b. The Empire State Building in New York
c. The Four Corners Monument associated with Utah, Colorado, New Mexico, and Arizona

As you can see in the example above, degrees of latitude and longitude may be divided very finely since a single degree on the surface of the Earth can represent a very long distance, depending on how far or how close a place is to the equator. At the equator, for example, one degree of longitude is approximately 111 km (69 miles) wide. At 45° latitude (approximately where Minneapolis, Minnesota, is located) one degree of longitude is approximately 79 km (49 miles) wide. In communicating general locations of places, rounded latitude and longitude coordinate pairs may suffice, yet for exact locations, finely divided latitude and longitude coordinate pairs may be necessary.

Diversion 15

Using your preferred mapping program again, how far (meters or feet, kilometers or miles) is the exact location of the *Grizzly Statue* at the University of Montana (noted above) from 47° north, 114° west?

Latitude and longitude can be represented in terms of *decimal degrees*, as we shown above in the *Grizzly Statue* example. Latitude and longitude can also be represented in terms of *degrees*, *minutes*, and *seconds*. In this sense,

- One degree can be divided into 60 minutes.
- One minute can be divided into 60 seconds.

While degrees are denoted with the degree symbol (°), minutes are denoted with one apostrophe symbol ('), and seconds are denoted with two apostrophes ("). For example, the latitude 43°52′45″ is 43 degrees, 52 minutes, and 45 seconds. This form of a coordinate can be converted to decimal form as a fraction of a whole degree. Obviously, this latitude is greater than 43° and less than 44°. Therefore, the range of the decimal degrees is somewhere between these two. Since there are 60 seconds in each minute, there are an additional 0.75 minutes (45 seconds, or 45/60) to add to 52′. The latitude is, therefore, 43°52.75′. Further, since there are 60 minutes in one degree, there are an additional 0.8792 degrees (52.75 minutes, or 52.75/60) to add to 43°. Thus, the decimal degree equivalent of 43°52′45″ latitude is 43.8792°. To summarize,

43°52′45″ = 43°52.75′ = 43.8792°

Diversion 16

The location of *Harkness Tower* on the campus of Yale University can approximately be described by the following latitude and longitude coordinates: 41°18′34″ north, 72°55′46″ west. Without using a mapping program, what are the decimal degree equivalents of these coordinates?

Plane Coordinates

Even though the Earth is an ellipsoid, it is so large that for smaller geographical areas we can pretend that it is a flat plane, and when collecting measurements to develop maps based upon this assumption, we can employ *plane surveying* techniques (Sincovec, 2008). Plane coordinate systems used in maps employ perpendicular north-south and east-west lines as a grid. In these systems, the planar surface, or flat grid, is often referred to as a *rectangular coordinate system*. Since the curvature of the Earth can propagate distortion in maps using plane coordinate systems, the distortions may be minimized if the systems employed are limited in geographic scope. As opposed to plane surveying, *geodetic surveying* may be necessary for larger areas and distances to describe features on the surface of the Earth, where corrections are made to take into account the curvature of land or water.

The *state plane system* of the United States is a grid-based plane coordinate system that acts as a compromise between the plane and geodetic surveying. Systems like these are adaptations of the practice of using latitudes and departures, concepts that have been employed by surveyors for over 5000 years (Dracup, 1977). In these systems, a latitude is simply a measurement in a north-south direction (not necessarily a degree), and a departure is a measurement in an east-west direction. The US state plane system began in the 1930s, and 125 separate plane coordinate systems have since been developed (Fig. 4.5). The state plane system began with a basic ellipsoid representation of Earth projected onto a surface using a conformal projection (Lambert Conformal or Transverse Mercator) where shapes and angles are preserved. Small zones were then created where the distortion from the act of projection was small enough to be considered within the level of surveying accuracy assumed in the 1930s (Sincovec, 2008). Within the US state plane system, states with boundaries that are generally longer in the north-south direction use the Lambert Conformal Conic map projection, whereas states with boundaries that are generally longer in the east-west direction use the Transverse Mercator map projection (these are described later in this chapter). The exceptions include California and other large states (e.g., Alaska) that were subdivided into pieces that do not represent the general shape of the entire state boundary. Modifications to the US state plane system are being considered, and perhaps by the year 2022 further subdivisions of some states might occur (Baumann, 2019).

Plane coordinate systems can be applied to small areas using *Cartesian coordinates* which are signed, numerical coordinates that are referenced to two fixed, perpendicular reference lines (*Cartesian axes*). A Cartesian coordinate system can also be used to determine the location of points drawn on a plane or flat surface (Wyoming Department of Transportation, 2015). In the basic representation of these systems, the coordinates are considered signed (+, −), as they may represent positive or negative *northings* (latitude measurements) and *eastings* (departure measurements) around an origin, depending on the position of a point with respect to the origin. The origin of the axes is some assumed place

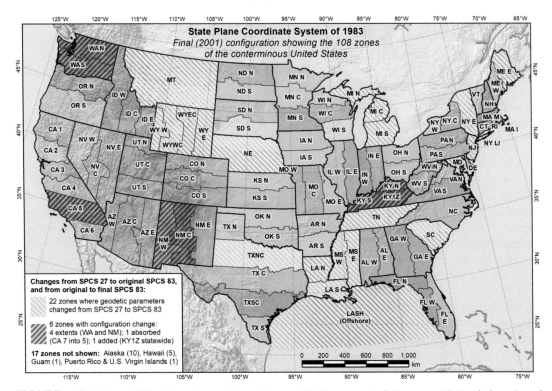

FIGURE 4.5 Map of US state plane systems. Source: *US Department of Commerce, National Oceanic and Atmospheric Administration (2018).*

where the two perpendicular reference lines cross (coordinate pair 0, 0). Therefore, eastings are the *X*-coordinates, and northings are the *Y*-coordinates in the *X,Y pairs* of coordinate values. As mentioned earlier, a three-dimensional Cartesian coordinate system may include a third axis (*Z*), forming a triplet of data that represent the horizontal position and the associated elevation. The concepts of northings and eastings are not specific to small area Cartesian coordinate systems. As you will learn, the worldwide UTM system also uses these concepts to define places on Earth.

In attempts to avoid the use of negative *X,Y* coordinates in some coordinate systems, a *false origin* may be employed. In using a false origin, the values of the *X,Y* position that describe the original origin position of the coordinate system are shifted to a different location, perhaps outside of the area covered by the coordinate system. The area covered by each state plane coordinate system in the United States, for example, has a false origin that is located either outside or on an edge of the area covered. In moving the origin, the *Y*-axis of the coordinate system is shifted far to the left and beyond the edge of the geographic zone covered, so that all eastings (*X*-coordinates) are positive in nature (no longer signed). In addition, the *X*-axis of the coordinate system is shifted far to the south, either directly to the bottom of the system (if the lower boundary is horizontal) or below the bottom of the system. Associated with this adjustment to the origin of a coordinate system are a *false easting* and a *false northing*. The false easting and false northing are the difference in *X*- and *Y*-coordinate values between the original origin of the coordinate system and the shifted origin of the system. A false easting is a value added to all *X*-coordinates to eliminate negative coordinates, while a false northing may be added to all *Y*-coordinates for the same purpose (Snyder, 1987).

Universal Transverse Mercator System

Most industrialized countries use the UTM coordinate system for their authoritative maps (Buchroithner and Pfahlbusch, 2017), and many organizations (public and private) use the UTM system as the common geodetic coordinate system for databases that they develop and manage. Although often suggested as having been developed by the United States as a worldwide plane coordinate system (Dracup, 1995), the origin of the UTM system can be attributed to parallel or perhaps earlier developments by geographers in Germany (Buchroithner and Pfahlbusch, 2017).

In the UTM system, the Earth is divided into 60 zones, each six degrees wide at the equator. Each zone is widest at the equator, and each zone tapers to a point at the poles, such that all zones come together at the poles. The UTM coordinate system utilizes the metric system for its measurements. Within the UTM coordinate system, the *northing* coordinate (*Y*) represents the number of meters north of the equator in the northern hemisphere, and the number of meters north of the south pole in the southern hemisphere. The *easting* coordinate (*X*) is more difficult from some to comprehend. One first needs to imagine a line that passes directly through the middle of each zone, from north to south. Coordinates along this line are given a *false easting*, or value such as 500,000 m. Then, areas to the east and west of this imaginary line are measured or located relative to the false easting, or middle of each zone. When land is mapped it is recorded with reference to a specific UTM zone. As an example, a set of UTM coordinates representing the *Grizzly Statue* at the University of Montana are:

5,193,946.9 m north and 272,340.9 m east (UTM Zone 12)

UTM zones are not as wide as the false easting employed. Therefore, if necessary when large area databases are developed, some land areas in an adjacent UTM zone (to the left) can be included without the easting values becoming negative (falling below 0 m).

Diversion 17
Open Google Earth or some other Internet-based mapping program and navigate to the *Space Needle* in Seattle. What is the latitude and longitude of this structure in degrees, minutes, and seconds? What is the latitude and longitude of the structure in decimal degrees? What are the UTM coordinates (zone, northing, and easting)?

Inspection 20
View the portion of the U.S. Geological Survey Tidewater Quadrangle map (Florida) shown in Fig. 4.6. What are the approximate ranges of UTM coordinates (minimum and maximum northings and eastings) for the area represented in the map?

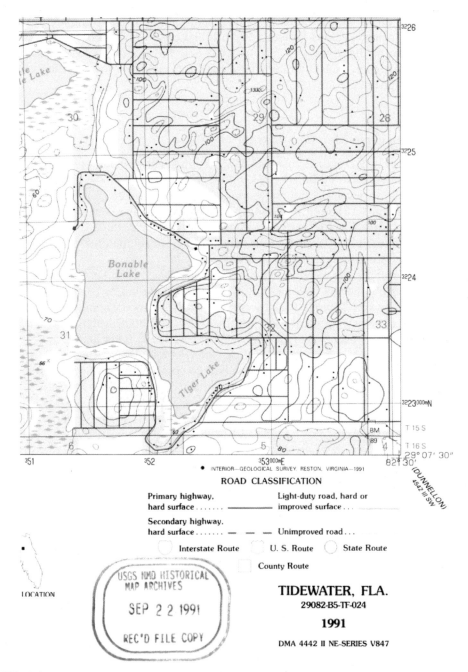

ROAD CLASSIFICATION

Primary highway,
hard surface . . . .   Light-duty road, hard or
                                  improved surface . . .

Secondary highway,
hard surface . . . . — — —   Unimproved road . . .

◯ Interstate Route   ◯ U. S. Route   ◯ State Route

◻ County Route

TIDEWATER, FLA.
29082-B5-TF-024

1991

DMA 4442 II NE-SERIES V847

LOCATION

FIGURE 4.6   A portion of the Tidewater, Florida, 7.5-minute U.S. Geological Survey quadrangle (29082-B5-TF-024).

When using the UTM system in the southern hemisphere a false northing of 10,000,000 m is applied. Since the equator represents the baseline of this coordinate system, all positions south of the equator would naturally be assigned negative coordinates. However, if the baseline is shifted 10,000,000 m south (the false northing), northing values near the south pole would be close to 0 m, while northing values near the equator would be close to 10,000,000 m.

Inspection 21

Located on the book's website (mapping-book.uga.edu) is a publication *Plano distrital de Saneamento Báscio e de Gestão integrada de resíduos sólidos* ("District sanitation plan and solid waste integrated management") for the city of Brasília that was developed by Serenco Serviços de Engenharia Consultiva Ltda (2017). View the map located on Page 32. From this map, describe both the false easting and false northing employed, and describe the extent of the landscape illustrated through the range in UTM values that are given.

Inspection 22

Using Google Earth or another Internet-based mapping program, what is the UTM northing value for Mount Kenya? What is the UTM northing value for nearby Nanyuki Airport? These places are only about 38 km apart. Why are the northing values so different?

Metes and Bounds Surveys

In general, a metes and bounds survey is a description of a property that uses carefully measured directions, distances, and angles. This information can be used to draw a map, can be presented as a narrative of a map, and may become the legal description of a piece of land. *Metes* refers to the measures of distance and direction, while *bounds* refers to local boundary descriptions. Metes and bounds systems have been used on every continent to describe both urban and rural lands (Libecap and Lueck, 2011). In places where land ownership is not well described by other means, or in areas where land ownership was already established due to previous ownership systems (e.g., the Spanish land system found in parts of California, or the French arpents in Louisiana) a metes and bounds survey may still be employed. In the eastern United States, various state-based metes and bounds systems were created during the period of European colonization. The State of Georgia, for example, employed two metes and bounds survey systems that are inherent in the land ownership system today in the eastern part of the state: the Royal headright system and the state headright system. The State of Louisiana employed two metes and bounds systems prior to implementing the Public Land Survey System (Taylor, 1951).

While stipulations and guidance provided by a broader organizational program (e.g., Georgia's state headright program) may have guided the progress of metes and bound surveys, compared to other types of systems (UTM and those noted below), the implementation of metes and bounds surveys is considered to be decentralized. In some cases, individuals located and marked their property, only later were these carefully surveyed. Property boundaries may be very irregular in certain metes and bounds surveys, and the descriptions of their boundaries were often problematic and disputed, since they may have referenced natural landmarks, trees, mounds of rocks, or other structures that may have changed, or may not exist today (Libecap and Lueck, 2011). Property boundaries described with metes and bounds surveys may have further been influenced in size and shape by rules governing how much water access each individual could claim. For example, certain *long lots* along waterways in Michigan and other places in the United States have narrow access to a shoreline, yet extend a considerable distance back into the upland areas.

Diversion 18
Access the book's website (mapping-book.uga.edu) and print the map associated with this diversion exercise. Imagine that you are one of the first inhabitants of the place denoted in the map, and that you have collectively decided to develop a system for the allocation of land to everyone who is there. Assume that you have no technology to assist in this effort. What rules would your metes and bounds system employ? Draw an example on the map.

Other Coordinate Systems

In the United States, the Public Land Survey System (PLSS) is used in some states as a coordinate system for subdividing and describing lands. The Land Ordinance of 1785, championed by Thomas Jefferson, described the system to be employed with further instructions for implementing the system developed over the next several decades by the federal government (White, 1983). States that include Alabama and Ohio, and most conterminous states west of these two, along with Alaska and Florida, employed the PLSS on lands that were considered to be in the public domain (i.e., lands generally not previously surveyed). Texas uses a system very similar to the PLSS, and parts of northern Maine also use a similar system. The PLSS was said to have been modeled around the early efforts in the United States to create towns in the wilderness. In the late 1700s and early 1800s, Dutch investors in New York State divided land into 6 to 10 mile square tracts within which parcels and village centers would have been placed, enacting a survey process that integrated a practical system of land subdivision with proactive town planning efforts (Wycoff, 1986). These early efforts were appealing and effective due to their simplistic and regular characteristics. Technically, one could consider the PLSS a metes and bounds system, since it is indeed used to describe properties with carefully measured directions, distances, and angles. However, given its stature (more than half of the United States) regularity of shape, and the use of base lines and meridians, it is often considered different than the metes and bound surveys it supplanted.

The PLSS system has 37 Principal Meridians and associated base lines. Conceptually, *tracts* 24 miles square were first surveyed, then within the tracts *townships* 6 miles square were surveyed. Within the townships, 36 *sections*, each one-mile square were then surveyed (Fig. 4.7). Given the curvature of the Earth, challenges related to the terrain, and surveying mistakes, these guidelines may not have been executed perfectly (Fig. 4.8). Townships are denoted by their distance (north or south) from their associated *Base Line* and by their distance (east-west) from their *Principal Meridian*. For example, a township

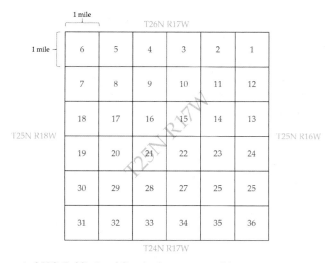

FIGURE 4.7   A theoretical U.S. Public Land Survey System township.

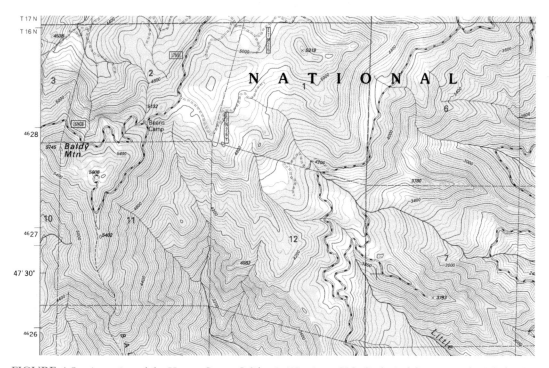

FIGURE 4.8   A portion of the Happy Camp, California 7.5-minute U.S. Geological Survey quadrangle (41123-G4-TF-024), showing the upper-right portion of T16N R6E (Sections 1, 2, 11, and 12) and the upper-left portion (Sections 6 and 7) of T16N R7E.

described as T23N R12E would be located 23 townships north of the base line and 12 townships east of the meridian. In other words, it is located in the 23rd row of townships above the base line, and in the 12th column to the east of the meridian. In this model, *T* references a *township*, *N* references *north*, *R* represents *range* (distance east or west of the meridian), and *E* represents *east*. Within each of the 37 systems, no two townships have the same township and range designation.

Translation 16

Imagine that you are about to conduct a survey northern spotted owls in Trinity County, California. You are working with some summer interns who have little experience as of yet in following and using maps. On the map that you are using on this particular day is the notation T34N R9W. Explain to your coworkers what this notation means.

The 36 sections within a township are all numbered in the same manner, except in some areas of Ohio where the PLSS system was first tested. Sections are numbered in a serpentine manner, where Section 1 is located in the upper-right corner of a township, Section 6 in the upper-left corner, Section 31 in the lower-left corner, and Section 36 in the lower-right corner (Fig. 4.7). Under normal procedures, Section 36 was generally the first section surveyed within a township, and those sections above it (north to Section 1) followed next. Section 35 was then surveyed along with those above it, and so on. In some cases, due to problems related to the curvature of the Earth, due to the terrain, or due to surveying error, the sections along the western or northern edges of a township may be too small or too large than the system prescribed. Sections are further subdivided into quarter sections, half sections, quarter—quarter sections, and other parts and pieces that form a legal description of the area. For example, the following notation might be used to describe the parts of a section:

SE 1/4 = southeast quarter of a section
NE 1/4 SW 1/4 = northeast quarter of the southeast quarter of a section
N 1/2 SW 1/4 NW 1/4 = north half of the southwest quarter of the northwest quarter of a section

The smallest portion of a subdivided section is located on the left side of these land descriptions, and the largest portion is on the right side. In the final example above, one would first look to the northwest quarter of the section, and within this area, look to the southwest quarter, then to the northern half ("C" in Fig. 4.9) to locate the area described. If theoretically a section is one square mile, it is 640 acres, given that there are 43,560 ft^2 in one acre. However, the size of a section may differ due to the curvature of the Earth and errors in the original survey.

1 mile $\times$ 1 mile = 5280 ft $\times$ 5280 ft
1 mile $\times$ 1 mile = 27,878,100 ft^2
1 mile $\times$ 1 mile = 640 acres

Therefore, theoretically a quarter of a section of land contains 160 acres (one-quarter of 640 acres), and a quarter of a quarter of a section contains 40 acres (one-quarter of

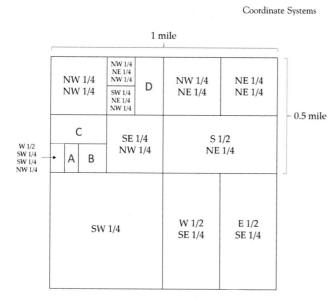

FIGURE 4.9 A theoretical U.S. Public Land Survey System section and several potential subdivisions. E 1/2 SW 1/4 SW 1/4 NW 1/4 = A, SE 1/4 SW 1/4 NW 1/4 = B, N 1/2 SW 1/4 NW 1/4 = C, E 1/2 NE 1/4 NW 1/4 = D.

160 acres). A complete *legal description of land* surveyed under the PLSS system might involve the following language:

NW 1/4 SE 1/4 Section 23 T34N R5W

Here, the land described is the northwest quarter *of the* southeast quarter *of* Section 23 *of* Township 34 north, Range 5 west. As you may see, the smallest subdivision of land is on the left-hand side of the legal description.

Diversion 19
From the following legal descriptions of land that use the US PLSS, determine how much land is theoretically described by each.

1. Section 23 T34N R5W
2. SE 1/4 Section 23 T34N R5W
3. NW 1/4 SE 1/4 Section 23 T34N R5W
4. SE 1/4 NW 1/4 SE 1/4 Section 23 T34N R5W
5. W 1/2 SE 1/4 NW 1/4 SE 1/4 Section 23 T34N R5W

In eastern Canada, land parcels were delineated using metes and bounds systems such as the French seigneurial system (blocks of land granted to seigneurs and later subdivided into strip farms) and the British rectangular township system (Parson, 1977). In many areas of eastern Canada, as in the eastern United States, the resulting irregular pattern of land parcels is evident today. In much of western Canada, the Dominion Land Survey (DLS) was employed as a land subdivision and survey system, primarily to promote agricultural development of lands that were not reserved prior to the establishment of the system. Enacted by the Dominion Lands Act of 1872, the DLS uses seven meridians and is applied primarily to Manitoba, Saskatchewan, Alberta, and parts of British Columbia.

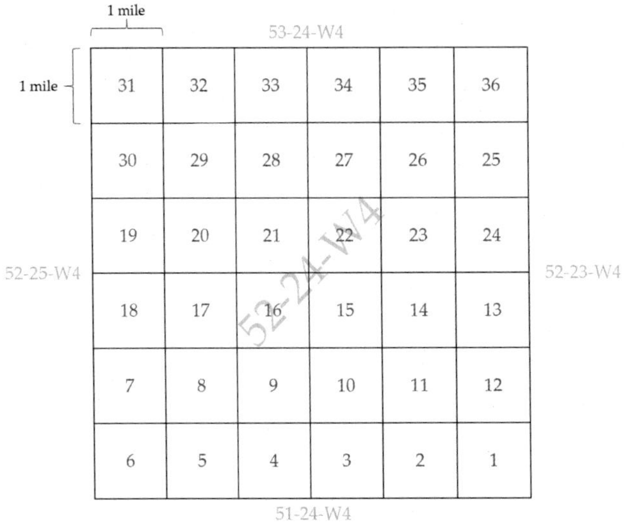

FIGURE 4.10 A theoretical Dominion Land Survey township.

While the dimensions of the townships and sections are similar to the PLSS, this system differs from the PLSS in that townships are numbered only in a northward direction from each baseline, and the sections are numbered in a serpentine manner with Section 1 being located in the southeast corner, and Section 36 being located in the northeast corner (Fig. 4.10). Legal designation of sections involves placing the section number before the township and range number, and adding the meridian information to the end of the label. For example, a section of land in Manitoba might be described as *17-6-7-W1*, which suggests that it is Section 17 of Township 6, Range 7 in lands west of the Prime Meridian used in Manitoba (W1). The northeast quarter of this section might be described as *NE17-6-7-W1*, where simply the *NE* designation was attached to the beginning of the legal description. Sections can be divided into 16 legal subdivisions of theoretically 40 acres each, and each of these can be divided into quarters of land that are 10 acres each. The legal subdivisions are also numbered in a serpentine manner from the southeast corner of a section to the northeast corner (Fig. 4.11). A legal subdivision might be described as *LSD7−17-6−7-W1* (legal subdivision 7 within Section 17, etc.), and a quarter of a legal subdivision might be described as *SELSD7−17-6−7-W1* (southeast quarter of legal subdivision 7 within Section 17, etc.).

Datum

In representing land or sea resources on a map through a coordinate system, a reference model of the Earth's surface is necessary. For mapping purposes, an *ellipsoid* is a model that is used to describe the form of the Earth (Arnaud, 2013), and a *datum* fixes the location of an ellipsoid to the Earth. Therefore, a datum is a fundamental basis for mapping, navigation, and surveying work (Baumann, 2019). In general, while the term *datum* suggests

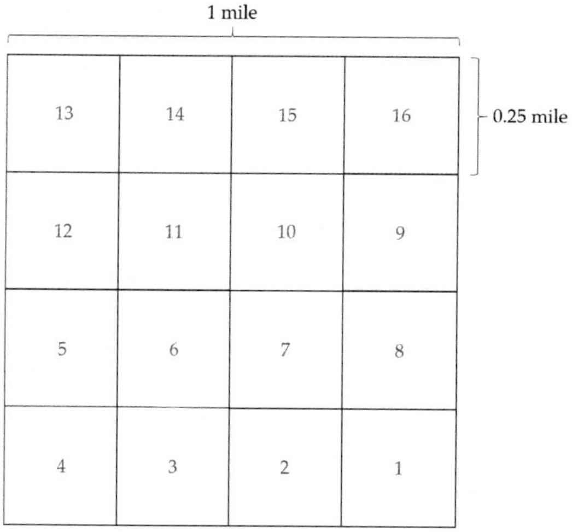

1 mile

13	14	15	16
12	11	10	9
5	6	7	8
4	3	2	1

0.25 mile

FIGURE 4.11 A theoretical Dominion Land Survey section and its 16 legal subdivisions of 40 acres each.

the singular form of perhaps a wider and broader set of *data*, in geographic terms, it is used to reference the starting point or position from which all other measurements emanate. Therefore, to become a datum, a mathematical model must both describe an ellipsoid and have a base station, or starting point, that provides location and orientation (Sincovec, 2008). Hundreds of local datums have been developed by various countries, utilizing their national systems of reference points (Baumann, 2019); several global datums have also been developed. Further, there are horizontal and vertical datums, therefore when elevations are of interest, a vertical datum would provide the base upon which all elevations on the map are also referred (Loomis, 1961).

A datum is based on a description of the shape of the Earth, and therefore the land or water body position for a specific set of geographic coordinates may differ based on the datum employed. One common global datum is the World Geodetic System of 1984 (WGS 84). WGS 84 acts as both an ellipsoid model and a datum, and is used today by the U.S. NAVSTAR global positioning system (GPS). For broad areas, WGS 84 has been suggested as one of the best geodetic models, or datums, to represent places on Earth (National Geospatial-Intelligence Agency, 2014). The North American Datum of 1983 (NAD 83), the North American Datum of 1927 (NAD 27), the South American Datum of 1969 (SAD 69), the European Datum of 1950, the Japanese Geodetic Datum of 2011 (JGD2011), and the European Terrestrial Reference System 1989 (ETRS 89) are all considered local datums. Each of these uses Greenwich, England, as its prime meridian. However, they may reference different ellipsoids: ETRS 89, NAD 83, and JGD2011 reference the GRS 1980 ellipsoid, NAD 27 references the Clarke 1866 ellipsoid, SAD 69 references the GRS 67 modified ellipsoid, and the European Datum of 1950 references the International 1924 ellipsoid. The North American Vertical Datum of 1988 (NAVD 88) is a datum providing vertical control for surveying efforts. A new global datum, the North American-Pacific Geopotential Datum of 2022 (NAPGD 2022), is also being developed. This datum is based on global navigational satellite system (GNSS, or GPS) measurements and an estimate of the Earth's center of mass, and may better represent Earth's gravity field (Baumann, 2019).

Map Projection Systems

A *map projection* is a systematic transformation of coordinates from a model of Earth to a map (Ghaderpour, 2016). In making a map, particularly of a large area, one challenge is to transform the curved surface represented by a sphere's graticule (lines of meridians and parallels) to a flat surface represented by a piece of paper or the screen of a digital device (Ipbuker and Bildirici, 2005). This is achieved through a projection of the sphere onto the flat surface (Robinson, 1974). Map projections are thus mathematical systems for describing how to conduct this effort and transform three-dimensional coordinates on Earth to two-dimensional coordinates on a flat surface (paper or digital device). In doing so, a map projection may attempt to maintain certain properties of the curved surface of the Earth so that distortions and errors are limited (National Geospatial-Intelligence Agency, 2014). But, one of the great challenges of mapping land and water bodies is to do this without significantly distorting other aspects of the place being mapped (Deetz, 1918). What a map projection does to areas, shapes, distances, and directions are the main concerns (Kuniansky, 2017). Geographers have been studying the advantages and disadvantages of map projections for quite some time. Through these comparisons, map projections have been classified based both on their manner of representing the Earth (on a cone, cylinder, etc.) and their ability to preserve geometric properties.

Much of our early mapping history comes from European endeavors that illustrated trade routes and political power; however, Asian cartographers were simultaneously active (or perhaps even on the forefront) in the development of maps using similar map projections (Gang, 2007). As will be seen in the discussion that follows, a variety of systems and map projections have been devised in an attempt to solve this perplexing problem. The popularity of certain map projections with cartographers has been shown to intensify or wane as time progresses. For example, an earth-globe azimuthal map projection, displaying the continents within two circles that represent each hemisphere (Fig. 4.12), was used extensively in the 17th century (Livieratos, 2008), yet it is not a popular map projection choice today.

Small-scale maps of the world have the most possible projections from which to choose (Robinson, 1974). For small-scale maps, the selection of a projection is often based on the intended use of the map and the distortions that will result. These maps may have straight or curved parallels or meridians, depending on what the map developer has decided. However, for small-scale maps, map users may prefer projections that utilize straight parallels as opposed to curved parallels, and meridians that are elliptical as opposed to sinusoidal in shape (Šavrič et al., 2015). For large-scale maps, where the representation of area is important, projections that minimize distortions in land area would seem more appropriate. Where the direction of water or airflow is important, a projection that minimizes the distortion in angles would seem appropriate (Snyder and Voxland, 1989). Thus, both the scale of the map and perceptions concerning the magnitude of geographic feature distortion should be considered when selecting a projection system. To address these concerns, country-specific map projection systems, such as the Romanian national stereographic projection system (Stereo70), have been developed for larger scale maps (as used in Buterez et al., 2016).

A *conformal projection* is one where small features on Earth essentially retain their original form or shape after the projection has been applied. A conformal map projection

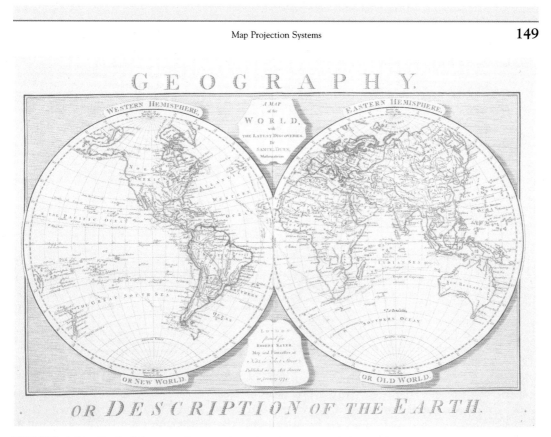

FIGURE 4.12 Earth-globe map. Source: *Dunn (1774), courtesy of the Lionel Pincus and Princess Firyal Map Division of The New York Public Library.*

preserves the relative angles between the model of Earth and the mapped features (Ghaderpour, 2016). However, areas further away from the center of a map may be distorted, and shapes of landscape features may be oddly represented as the region represented increases in size (Snyder and Voxland, 1989). These problems may be insignificant when developing or using large-scale maps of small land areas. So, while conformal maps may distort the sizes of landscape features, they represent true shapes of landscape features on maps of small areas; therefore, they are advantageous if one is concerned about the utility of using rectangular coordinate systems (Robinson, 1974). Conformal systems include the Mercator, Lambert's conformal conic, and the stereographic map projection systems, among others (Table 4.1).

When an *equal area projection* is used, the sizes of landscape features on a map are represented relatively well in relation to their true areas on Earth. In other words, equal area map projections attempt to preserve the relative sizes of areas of mapped features (Ghaderpour, 2016; Snyder and Voxland, 1989). This type of map projection does not completely remove distortions in areas when the Earth is projected onto a plane (Yildirim and Kaya, 2008). However, for small areas of 1000 ha (about 2500 acres) or less, the use of an equal area map projection (e.g., Albers, Behrmann, Bonne, Goode, and sinusoidal and

TABLE 4.1 Characteristics of a small set of map projection systems.

Conformal	Equal area	Equidistant
Bipolar Oblique Conic Conformal	Albers Equal Area Conic	Azimuthal Equidistant
Lambert Conformal Conic	Behrmann	Equidistant Conic
Mercator	Bonne	Plat Carrée
Oblique Mercator	Eckert IV	
Stereographic	Eckert VI	
Transverse Mercator	Equal Earth	
	Goode Homolosine	
	Lambert Azimuthal Equal Area	
	Lambert Cylindrical Equal Area	
	Lambert Equal Area Conic	
	Mollweide	
	Sinusoidal	

others) can present land areas with a precision of 1/10,000,000 or better, which suggests a 1 m^2 difference from true area for every 1000 ha of land (Yildirim and Kaya, 2008). Interestingly, when developing small-scale maps, the further one moves away from the projection's origin, angles and shapes of landscape features can become distorted using this type of map projections (Snyder and Voxland, 1989).

When an *equidistant projection* is used, an attempt is made to preserve relative distances along major lines of longitude and latitude. It has been suggested that this type of projection can only be used successfully on maps of limited areas (Snyder and Voxland, 1989). The azimuthal equidistant and equirectangular map projections, among others, are representative of this class.

Reflection 17
Think about and select an important type of map that you often use. Given that there is the problem of mapping the curved surface of the Earth onto a flat surface of paper or digital device, which one of the properties (angles, distances, shapes, areas) would you least like to see distorted? Why?

In constructing a map projection, the features of a sphere are mathematically and graphically altered as they are placed onto the surface of a plane, cylinder, cone, or other solid. The characteristics (shape, extent, spacing) of meridians and parallels illustrated on a map are directly related to the method of map projection. Whether the meridians and parallels are represented as straight or curved lines reflects how the sphere (Earth) was transposed onto a flat surface. The choice of map projection can, therefore, be important, particularly when a map needs to express shapes, areas, distances, or directions correctly. At least 200 map projections have been developed (Schmunk, 2019; Jung, 2019). In the following sections of this chapter, we describe a small set of these map projections. When creating a small-scale map of large areas, perhaps the decision of which map projection to employ should be made first (Snyder and Voxland, 1989).

Cylindrical Projections

Cylindrical projections represent Earth as a rectangle. With this type of map projection, the sphere (the Earth) is projected onto a cylinder. Imagine wrapping a piece of paper around a tennis ball, without collapsing or crumpling the paper. Only one line of tangency typically occurs, and this is the place where the paper touches the ball. Along this line of tangency, characteristics of the surface of the ball may be easily transferred to the paper (Fig. 4.13). However, all other characteristics of places on the surface of the ball must be projected onto the paper. When applied to the Earth, the result of these map projections illustrates meridians as straight lines running north and south, and parallels as straight lines extending east and west. The meridians and the parallels are also positioned at right angles to each other. A secant projection intersects the sphere along two standard parallels, as conceptually the cylinder may then rest below the surface of the sphere and be present above the surface of the sphere.

The *Mercator map projection* has existed for over 400 years and is considered a conformal map projection system (National Geospatial-Intelligence Agency, 2014). When using this map projection, the meridians (C) and the parallels (B) are displayed as being perpendicular and straight (Fig. 4.14). The meridians are equally spaced, while the parallels are unevenly spaced. On maps that use this projection system, the poles are implied to be located at the ends of the meridians; therefore they are not represented as single points on the maps. The cylindrical model touches the Earth along the equator (E). This is the central line of tangent. Other areas of the Earth are then projected onto the cylinder. With the

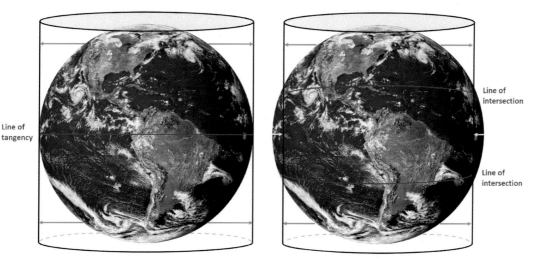

FIGURE 4.13 Conceptual model of tangent (left) and a secant (right) cylindrical map projections. Source: *Image of Earth courtesy National Aeronautics and Space Administration (2000).*

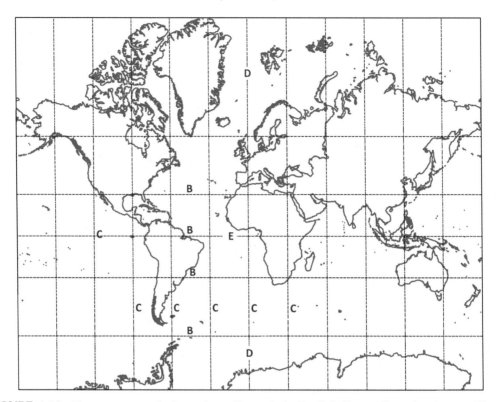

FIGURE 4.14   Mercator map projection, using a 30° graticule. Parallels=B, meridians=C, prime meridian=D, and equator=E.

Mercator projection, the sizes of land areas are distorted (made larger) as one moves toward the poles, due to east-west stretching of the Earth as it is converted from a sphere to a cylinder, and a corresponding north-south stretching to preserve the angles between areas. Thus, places such as Antarctica and Greenland are larger than normal on maps using this projection system, and the sizes of places near the equator are fairly accurately represented. Some have cautioned on the use of the Mercator map projection, as this variation in scale can be excessive (Goode, 1925; Robinson, 1974).

The Transverse Mercator map projection is a conformal system that uses a (any) meridian as the central line of tangency in the cylindrical projection, and therefore it is an adaptation of the Mercator map projection. The Transverse Mercator map projection is used in association with the UTM coordinate system, which describes the positions of land areas on Earth within 6° longitude (north-south) strips, or UTM zones. An *Oblique Mercator map projection* is also an adaptation of the Mercator map projection. An oblique Mercator map projection is considered conformal, and the central line of tangency can be any great circle that has an oblique angle with respect to the equator (Snyder and Voxland, 1989). The *Space Oblique Mercator map projection* is a dynamic, conformal map projection that facilitates continuous mapping of the Earth's surface. This map projection was inspired by the

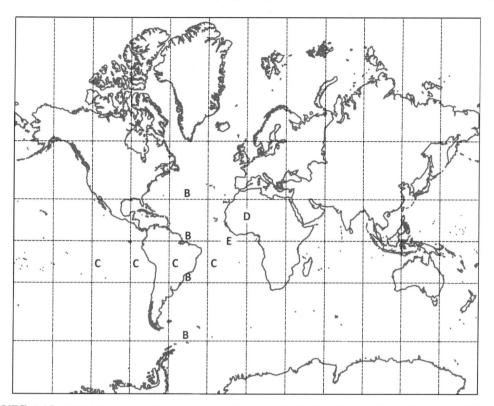

FIGURE 4.15   Web Mercator projection, using a 30° graticule. Parallels=B, meridians=C, prime meridian=D, and equator=E.

development of the Landsat program in the early 1970s, and was designed to help map the landscape conformably within each narrow swath of the Earth covered by the oblique angle of orbit of Landsat satellites. Given the ground track of the Landsat satellites, the central line of tangency is actually a circular arc around the Earth (Snyder, 1981).

Internet-based mapping applications use a popular mapping and visualization system, the *Web Mercator* map projection (Fig. 4.15). Here, the meridians (C) and the parallels (B) are represented by straight lines. The meridians are spaced equally apart on these maps, while the parallels are spaced at unequal lengths, and the meridians and parallels intersect at right angles. On maps that use this projection system, the poles are implied to be located at the ends of the meridians; therefore, they are not represented as points on the maps. The Web Mercator system is technically similar to the Mercator system, yet is a simplified spherical variant, and therefore is not truly conformal and not recognized as a geodetic system (National Geospatial-Intelligence Agency, 2014; Battersby et al., 2014). The map projection system does exemplify characteristics of the standard Mercator system, which allows simpler and faster computations and facilitates continuous panning and zooming on digital devices. The Web Mercator system is considered a *Spherical Mercator* map projection method because rather than being based on a model of the shape of the

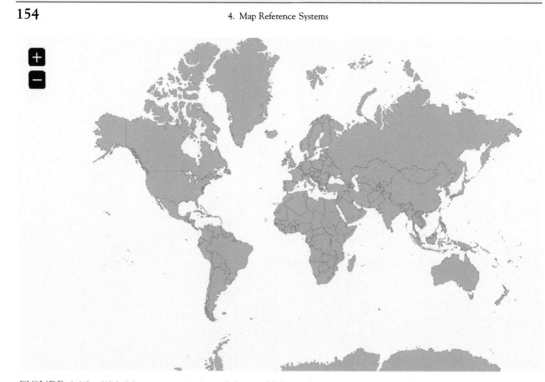

FIGURE 4.16   Web Mercator projection of the world from the interactive *Learn About Your Destination* Internet site. Source: *U.S. Department of State (2017)*.

Earth (such as the WGS 84 ellipsoid model), it is based simply on a sphere. While the easting coordinates may be similar between Web Mercator (spherical) and Mercator (ellipsoidal) systems, the northing coordinates can differ by up to 40 km in higher latitudes (National Geospatial-Intelligence Agency, 2014). Thus, the Web Mercator projection is nonconformal, and error increases with increasing distance from the equator (Battersby et al., 2014). Although information on Internet-based map projections is often lacking (Favretto, 2014), the Web Mercator system should be recognizable (Fig. 4.16), and one might remember that when using this map projection, there will be a cartographic versus geodetic trade-off in locational accuracy.

Similar to the Mercator map projection, the *Miller Cylindrical map projection* (Fig. 4.17) presents the meridians (C) as straight lines flowing north and south, and the parallels (B) as straight lines flowing east and west. The meridians are equally spaced, while the parallels are unevenly spaced. Further, as in other cylindrical projections, the poles are implied to be located at the ends of the meridians; therefore they are not represented as points on the maps. In contrast to the Mercator map projection, the Miller cylindrical map projection is not considered conformal, but a compromise system. While there may be no distortion along the equator, places located at higher latitudes (toward the north or south poles) are distorted; however, distortion in area and scale near the poles is less than what might be observed when using the Mercator map projection (Snyder and Voxland, 1989).

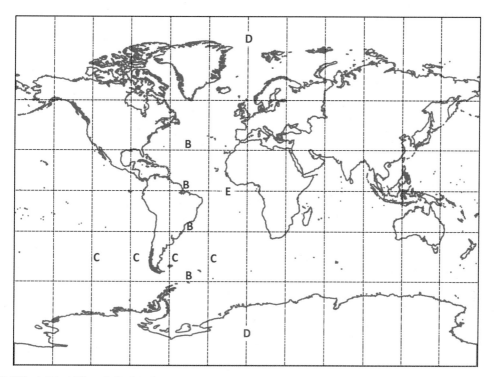

FIGURE 4.17 Miller Cylindrical map projection, using a 30° graticule. Parallels=B, meridians=C, prime meridian=D, and equator=E.

In an *Equidistant Cylindrical map projection*, the meridians and parallels intersect at right angles, and the meridians and parallels are equally spaced. Again, in contrast to the Mercator map projection, the equidistant cylindrical map projection is not considered conformal, yet as the name implies, it is equidistant in nature. When the equator is selected as the standard parallel, the map projection becomes the *Plat Carrée map projection* (Snyder, 1982). A Plat Carrée map projection is a simple cylindrical projection that is both azimuthal and equidistant about a given point (Botley, 1951). With this map projection, the tangent line of the Earth to the cylinder is along the equator, and therefore there is only one standard parallel. Both meridians (C) and parallels (B) are straight lines in the map projection (Fig. 4.18) and spaced apart at equal lengths. The Plat Carrée map projection is composed of the most simply constructed graticule of any type of map projection, and is the simplest and most limiting form of the Equidistant Cylindrical map projection (Snyder, 1982). With this map projection, distortion of both shape and area increases with distance from the equator. However, scale is true along the meridians (i.e., it is equidistant) and along the equator, and scale is constant along the parallels. Yet as with other cylindrical projections, distortions in world maps can be severe at higher latitudes (Botley, 1951). The Plate Carrée projection has been employed, among many other uses, in the illustration of global land masses (Bauer-Marschallinger et al., 2014), ground tracks of the orbits of GPS satellites (Oloffson et al., 2009), and a global vegetation index (Goward et al., 1994).

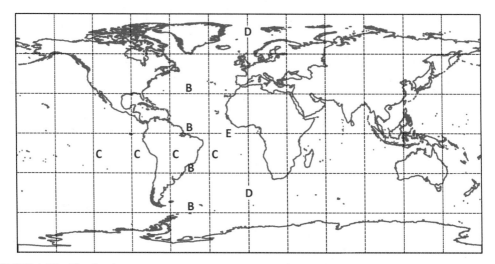

FIGURE 4.18   Plat Carrée map projection, using a 30° graticule. Parallels=B, meridians=C, prime meridian=D, and equator=E.

Further, the map projection has been of value in illustrating more focused areas of Earth, such as in the development of the medium-scale bathymetric map *Seafloor Map of Hawai'i* (Patterson, 2013).

Translation 17
Imagine you are developing a map of Puerto Rico to illustrate the forest damage incurred by Hurricane Maria in September 2017, and you have decided to use the Plat Carrée map projection for this effort. You are having dinner one night with some friends and conversation evolves to your current job and the tasks you need to complete. Of course you are excited to tell your friends about the map of Puerto Rico that you are making. You casually mention the map projection and this prompts some interest. In very simple terms (two sentences), what is a Plat Carrée map projection?

As a cylindrical projection, the meridians (C) of the *Lambert Cylindrical Equal Area* map projection (Fig. 4.19) are represented as straight lines flowing north and south, and the parallels (B) are represented as straight lines flowing east and west. The meridians are equally spaced, while the parallels are unequally spaced. As with other cylindrical map projections, the central line of tangency is the equator (E), and there is significant shape distortion near the poles but no area distortion. Again, in contrast to the Mercator map projection, the Lambert Cylindrical Equal Area map projection is not considered conformal, yet as the name implies, it is equal area in nature. The standard parallel, where there is no distortion, is also the equator.

Diversion 20
On the Internet, locate the *Compare Map Projections* website graciously hosted by Jung (2019). As of 2019, the website address is https://map-projections.net/index.php.

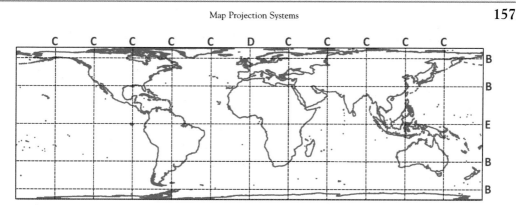

FIGURE 4.19 Lambert Cylindrical map projection. Parallels=B, meridians=C, prime meridian=D, and equator=E.

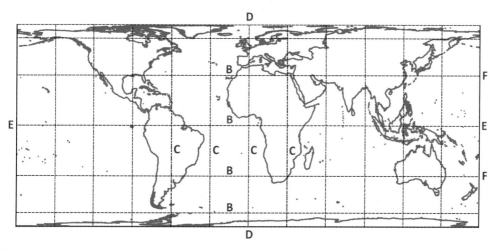

FIGURE 4.20 Behrmann Cylindrical Equal Area map projection. Parallels=B, meridians=C, prime meridian=D, equator=E, and standard parallels=F.

Compare, via a list of projections, the Mercator map projection and the Lambert Cylindrical map projection. Using the silhouette map comparison, in terms of the location of continents, how do these two map projections differ and how might they be similar?

The *Behrmann map projection* was introduced by Walter Behrmann in 1910 and is similar to the Lambert Cylindrical Equal Area map projection. The Behrmann map projection is not considered conformal, but equal area in nature. The central line of tangency is the equator. For maps that use the Behrmann map projection (Fig. 4.20), the meridians (C) and the parallels (B) are drawn as straight lines that intersect at 90° angles. The standard parallels (F), where there is no distortion, are located at 30° north and 30° south of the equator (E). The meridians are equally spaced across the map, while the parallels are unequally spaced. On a positive note, the sizes of areas are not distorted when using this model (i.e., it is an equal area projection), yet distances are distorted almost everywhere except along the equator. The distortion of shapes is at its minimum in the areas near the parallels. As you may find, the Behrmann map projection excessively reduces the size of the polar regions (Robinson, 1974). Although this system may be of value in mapping the entire world, it has often been employed as a map projection system for smaller areas where the distortions may be minimal. For example, it has been employed in the development of maps of snake distributions in Central and South America (Guedes et al., 2018) and floristic conditions in Africa (Droissart et al., 2018).

Pseudocylindrical Projections

Pseudocylindrical projections represent the world in a manner slightly different than cylindrical projections; the parallels are straight lines that still flow east to west, yet the meridians are elliptical and only somewhat parallel, curved to indicate possible convergence on the north and south poles. For example, Max Eckert (1868–1938) presented in 1906 the *Eckert IV map projection* and the *Eckert VI map projection*, which are considered both pseudocylindrical and equal area; they have often been used to illustrate climate maps and other themes of interest to society. The National Geographic Society once used the Eckert IV map projection for their wall maps of the world, until they transitioned in 1988 to the Robinson map projection. In the Eckert IV map projection system (Fig. 4.21), the central meridian (D) is half as long on the map as the equator (E) (Snyder, 1987; Snyder and Voxland, 1989). The lines representing the poles are equal in length to the central meridian. The parallels (B) are perpendicular to the central meridian; where they intersect, the

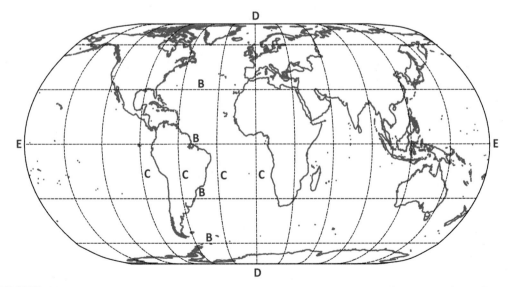

FIGURE 4.21   The Eckert IV map projection system, using a 30° graticule. Parallels=B, meridians=C, prime meridian=D, and equator=E.

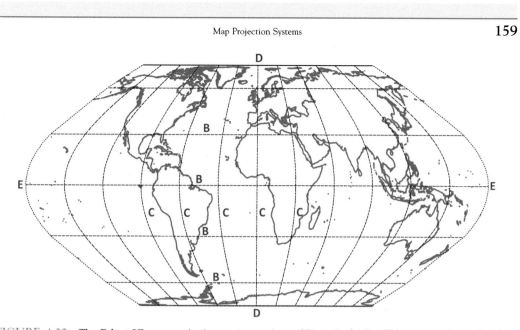

FIGURE 4.22   The Eckert VI map projection system, using a 30° graticule. Parallels=B, meridians=C, prime meridian=D, and equator=E.

meridians (C) are equally spaced along these lines. The parallels themselves are unequally spaced in order to represent areas correctly and are separated more widely at the equator than in other latitudes. The meridians are represented as elliptical arcs, and the latitudes of true scale are located at 40°30′. Distortion is free along the central meridian, and land areas are essentially stretched at the equator and flattened above 40°30′ (Snyder, 1987). Thus, north-south distances are elongated among the lower latitudes along the equator (Robinson, 1974). The Eckert VI map projection (Fig. 4.22) is considered both pseudocylindrical and equal area, and is similar to Eckert IV except that the meridians are represented as sinusoidal curves, and the latitudes of true scale are located at 49°16′ distortion is free here along the central meridian (Snyder and Voxland, 1989).

The *Goode Homolosine map projection* was developed by J. Paul Goode (1862–1932) as a pseudocylindrical, equal area system for the display of worldwide maps (Goode, 1925). With this map projection (Fig. 4.23) the meridians (C) are curved lines while the parallels (B) are straight lines. The poles (A) are represented as several distinct points due to the polycylindrical nature of the projection. There are several central meridians (D), and the longitudinal location of these may be different for each continent in order to minimize distortions (Snyder, 1982). Tobler (1962) characterized this map projection as a composite of the Mollweide map projection and sinusoidal equal area map projections (both described below). Goode's Homolosine map projection was designed to be applicable to areas of limited expanse in east-west directions, and was once noted as an improvement over the Mercator map projection system for representing areas of the Earth's surface (Ekblaw, 1928). As an example of the interrupted nature of this map projection system, Fig. 4.24 illustrates a visible Earth map developed by the National Aeronautics and Space Administration (NASA) program of the US government.

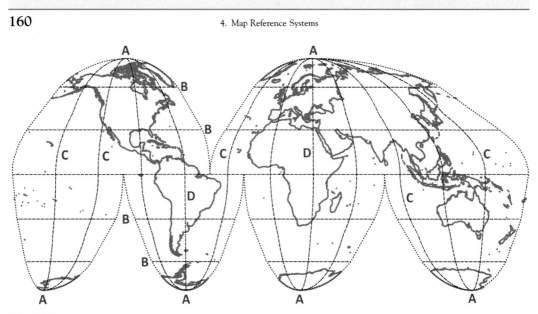

FIGURE 4.23   Goode's Homolosine map projection, using a 30° graticule. Poles=A, parallels=B, meridians=C, prime meridian=D, and equator=E.

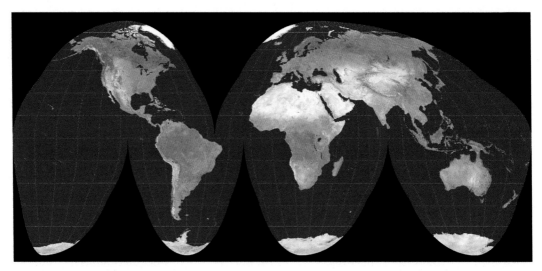

FIGURE 4.24   The visible Earth, 2005, using the Goode's Homolosine map projection, where the reticule is 15° in latitude and longitude. Source: *Courtesy NASA through Wikimedia Commons.*

Reflection 18
Can you think of four specific purposes or themes that might require you to develop a map that uses Goode's homolosine map projection? What are they?

Carl Mollweide (1774–1825) presented a pseudocylindrical, equal area projection in 1805 that has been used as an inspiration for other map projections (Snyder, 1987). The

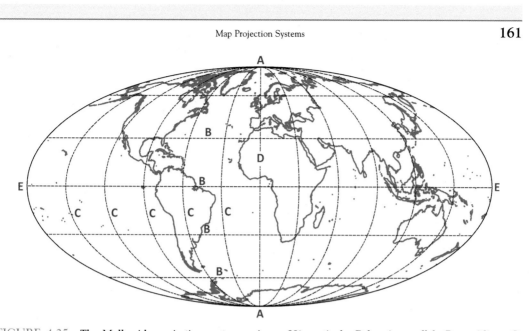

FIGURE 4.25   The Mollweide projection system, using a 30° graticule. Poles=A, parallels=B, meridians=C, prime meridian=D, and equator=E.

Mollweide map projection (Fig. 4.25) is bounded by an ellipse, and the equator (E) is twice as long as the central meridian (D). The central meridian (D) is represented as a straight, vertical line, and the meridians (C) on either side are represented as circular arcs, equally spaced along the equator. The parallels north and south of the equator (B) are represented by elliptical arcs in this system, and are unevenly spaced apart as one deviates northward or southward from the equator (Snyder, 1987). The poles (A) are presented as points in this projection system. Scale is true at latitudes of 40°44′ north and south of the equator and distortion is free only where these latitudes intersect the central meridian. Maps frequently use the Mollweide map projection to illustrate worldwide phenomena or themes, such as National Geographic Society maps of oceans, or other specific worldwide themes such as variations in magnetic fields observed in the upper layer of the lithosphere (Fig. 4.26), the outer, rigid shell of the Earth (the *crust* and part of the *upper mantle*). However, the map projection has been suggested to contain too much distortion in the higher latitudes (Robinson, 1974), even though areas represented with this projection are less than 1% different than those portrayed by the Albers Equal Area Conic map projection described below (Kuniansky, 2017).

The *Sinusoidal map projection* compresses the polar and high latitudes (Robinson, 1974) (Fig. 4.27). This map projection is considered to be an equal area projection. The sinusoidal map projection, an old map projection system, can be used with one or more central meridians. The use of multiple central meridians would suggest that the *interrupted form* of the map projection system is being used (Snyder, 1982). Where they are present, the central meridians are represented as a straight vertical lines, while other meridians are represented as equally spaced sinusoidal curves. The meridians converge at points representing the poles. When a single central meridian is used, it is half as long in length as the equator

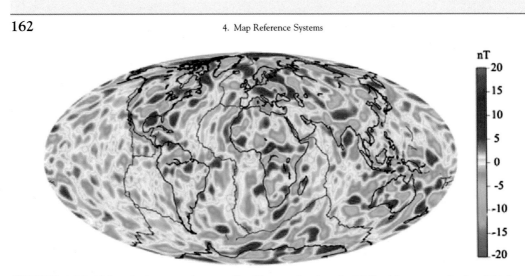

FIGURE 4.26 Lithospheric magnetic anomalies illustrated using the Mollweide map projection. The legend indicates areas with positive and negative magnetic fields. Source: *Courtesy Terrance Sabaka et al./NASA GSFC through Wikimedia Commons.*

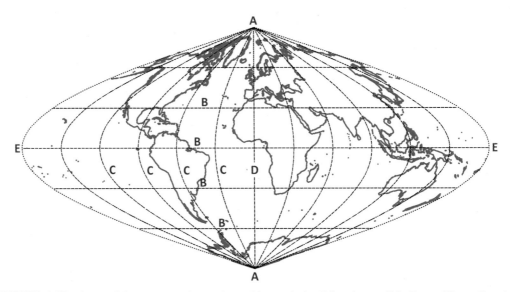

FIGURE 4.27 Sinusoidal map projection, using a 30° graticule. Poles=A, parallels=B, meridians=C, prime meridian=D, and equator=E.

(Snyder and Voxland, 1989). With this map projection system, distortion can be great at higher latitudes near the poles; however, the use of the interrupted form can help reduce these effects.

In 1963 Arthur Robinson presented a projection system that was intended to be visually appealing for worldwide maps, even though as a global system it makes compromises in several areas to limit distortions. His system was used by *National Geographic* and others, and is still used today, as a standard reference for maps of global scope. However, the *Robinson map projection* system is not considered equivalent, conformal, equal area, equidistant, nor azimuthal. The system attempts to minimize errors related to distance, direction, shape, and area that occur when the Earth is represented on a flat surface. The projection system is based on a set of transformation coordinates (Robinson, 1974) rather than mathematical formulas. With the Robinson projection (Fig. 4.28), lines of latitude and longitude are relatively evenly spaced across a map. The meridians (C) are generally concave, although the central meridian (D) is a straight vertical line. The equator and the parallels (B) are straight horizontal lines in this projection system. The poles are not represented as points, but as straight lines, 0.5322 times as long as the equator. Directions are true along the central meridian and along the east-west parallels. Scale of land areas is somewhat correct around 38° north and south of the equator (E), where the lines of tangency meet when projecting the Earth (an ellipsoid) onto the grid. Some countries that lie further from the equator, such as Russia, Canada, and Greenland, are represented larger than their actual size, but Greenland is generally compressed in size in this projection system. Distortion of land areas thus becomes worse beyond 45° latitude. Due to the manner in which the Earth is represented on maps, with distortions minimized throughout most land areas, the Robinson projection system continues to be a popular choice among map developers as it might be perceived as providing an acceptable balance between shape and area distortion (Robinson, 1974). As an example, a measure of social well-being (happiness) was assessed for each country (Helliwell et al., 2017) and outcomes were displayed as a thematic map indicating the relative standing of each country (Fig. 4.29).

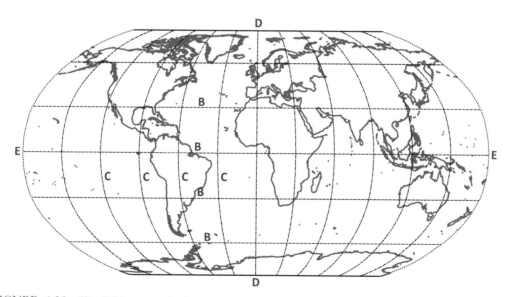

FIGURE 4.28 The Robinson projection system, using a 30° graticule. Parallels=B, meridians=C, prime meridian=D, and equator=E.

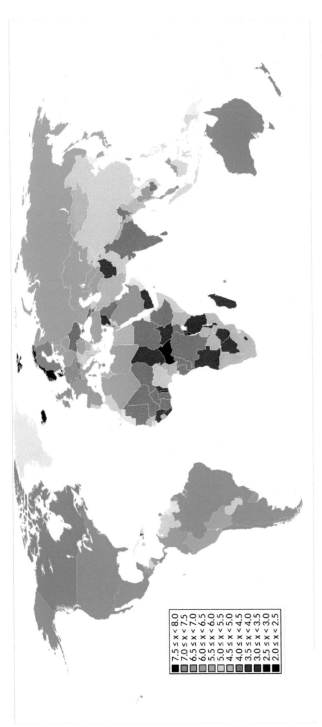

FIGURE 4.29 World Happiness Report scores (2017) depicted for the world using the Robinson map projection. Source: *Courtesy Jolly Janner through Wikimedia Commons.*

Diversion 21

On the Internet, locate the *Compare Map Projections* website graciously hosted by Jung (2019). As of 2019, the website address is https://map-projections.net/index.php. Compare, via a list of projections, the Behrmann map projection and the Robinson map projection. Using the silhouette map comparison, in terms of the location of continents, how do these two map projections differ and how might they be similar?

A new pseudocylindrical projection has also been recently developed as an alternative to the Robinson map projection (and others) to provide aesthetic appeal via an equal area map projection system (Šavrič et al., 2019). The *Equal Earth map projection* maintains the true relative sizes of features, using unequally spaced, yet straight parallels (east-west lines) and equally spaced, yet curved meridians (north-south lines). In contrast to the Robinson map projection, global features far from the equator are not exaggerated, and in contrast to the Mollweide and Sinusoidal map projections, the poles are not represented as points (Šavrič et al., 2019).

Conic Projections

With *conic projections* (Fig. 4.30), the Earth is projected onto a cone. The parallels are represented as curved arcs, and the meridians as straight lines that converge upon the appropriate pole (north or south). Conceptually, a conic projection rests a cone over a sphere (the Earth), and features within some limit of the surface of the sphere (not the entire sphere) are projected conformally onto the surface of the cone. After projecting features onto the cone, the cone is unrolled, revealing the mapped surface. A tangent projection is where the cone simply touches the surface of the sphere once, resulting in a single *standard parallel*, or line of intersection with the sphere. A secant projection intersects the sphere along two standard parallels, as conceptually the cone may then rest below the

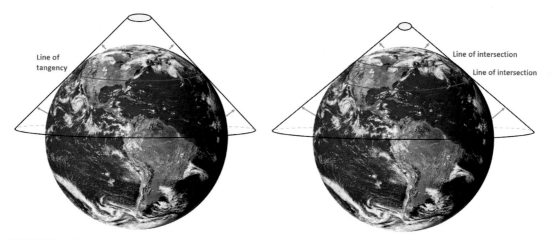

FIGURE 4.30   Conceptual model of tangent (left) and a secant (right) conic map projections. Source: *Image of Earth courtesy National Aeronautics and Space Administration (2000).*

surface of the sphere and be present above the surface of the sphere. These types of map projections are important for maritime and aerial navigation, military, and cadastral (surveying) mapping needs (Tutić, 2010). The line of least distortion in the tangent case of a conic projection is the circle formed where the cone just touches the edge of the sphere (Earth). The lines of least distortion in the secant case of a conic projection are the two circles formed where the cone intersects or cuts through the sphere (Earth). A considerable amount of official world mapping efforts use either conic or cylindrical projections (Borisov et al., 2015); however, the conic projections are perhaps the most easily constructed for temperate landscapes on Earth (Snyder, 1978).

The *Albers Equal Area Conic map projection* was presented by Heinrich Christian Albers (1773–1833) in 1805. Given that the Earth is superimposed on a cone, not by a single tangent line but through the intersection of two secants, the projection contains two standard parallels. Along these standard parallels, however defined, there is no distortion. The Albers Equal Area Conic map projection represents the poles ("A" in Fig. 4.31) as arcs or circles, depending on how much of the world is contained in the map. The parallels (B) are concentric arcs that are unevenly spaced as one moves north or south. The meridians in this model (C) are linear. Scale is constant along each parallel, a characteristic of all normal conic projections; there is also no angular distortion along the parallels; variations in scale along the meridians do cause angular distortion. The spacing of the parallels is such

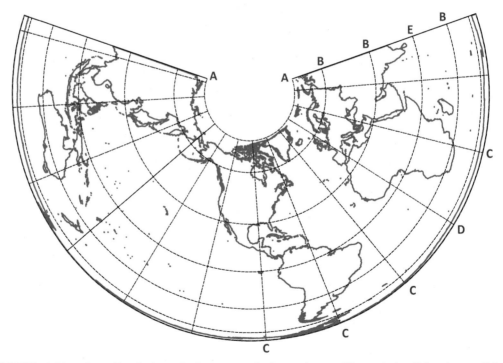

FIGURE 4.31 Albers Equal Area Conic map projection, using a 30° graticule. Poles=A, parallels=B, meridians=C, prime meridian=D, and equator=E.

that areas representation is correct along the meridians, making this an equal area map projection (Snyder, 1978), yet shapes may be distorted somewhat. The meridians are conceptually the radii of circles and are equally spaced, meeting the parallels at right angles in the model (Snyder, 1987). There is only one central meridian, chosen by the map developer. The two standard parallels should be selected so that scale error in the middle of the map (between the two parallels) is minimized. The two standard parallels for U.S. Geologic Survey maps of the conterminous United States are 29.5° and 45.5° latitude north of the equator, but a map developer can chose others. For Canada, these will likely be higher in latitude, perhaps 50° and 70° north. The Albers Equal Area Conic projection has been said to preserve both areas and distances of continental areas located in midlatitudes of Earth (25° to 45° north or south) that are also oriented in an east to west manner (Kuniansky, 2017). The map projection has been used for maps that illustrate landscapes with predominant east-west expanses, and for maps where equal area is important. An example of the use of the Albers Equal Area Conic map projection is NASA's map of nighttime illumination levels (Fig. 4.32).

Diversion 22

On the Internet, locate the *Compare Map Projections* website graciously hosted by Jung (2019). As of 2019, the website address is https://map-projections.net/index.php. Compare, via a list of projections, the Robinson map projection and the Albers Equal

FIGURE 4.32   United States at night, 2012, developed by the National Aeronautics and Space Administration (NASA) using the Albers Equal Area Conic map projection, WGS 84 datum, standard parallels of 29.5°N and 44.5°N, and a central meridian of 96°W. Source: *Courtesy NASA Earth Observatory through Wikimedia Commons.*

Area Conic map projection. Using the silhouette map comparison, in terms of the location of continents, describe how these two map projections differ and how might they be similar.

The conformal conic map projection described by Johann Heinrich Lambert in 1772 is one of the most popular map projections employed today (Grafarend and Okeke, 2007). The meridians of the *Lambert Conformal Conic map projection* system (Fig. 4.33) are represented by straight lines (C) that on one end of the map (here, the outer edge) are tacitly suggested to intersect at a common point located beyond the limits of the map developed (Deetz, 1918). On this map, the north pole (A) is represented where the meridians would all meet. This is a secant projection, and two standard parallels (east-west lines) are selected where conceptually, and both these intersect the surface of the Earth. For example, for some of the official cartography of Serbia, the two standard parallels selected are 38°30′ and 49°00′ north (Borisov et al., 2015). The meridians (C) and the parallels (B) on maps using this projection intersect at nearly right angles and can be considered linear graticules. Land or water bodies mapped between the parallels are the artifact of the case where the cone employed is conceptually below the surface of the sphere, and the resulting scale is slightly too small. Land or water bodies mapped outside the parallels are the artifact of the case where the cone employed is conceptually above the surface of the sphere, and the resulting scale is slightly too large. However, scale error when using this system is suggested to be lower than when using a polyconic map projection system (Deetz, 1918). Often, the two standard parallels are both either above or below the equator.

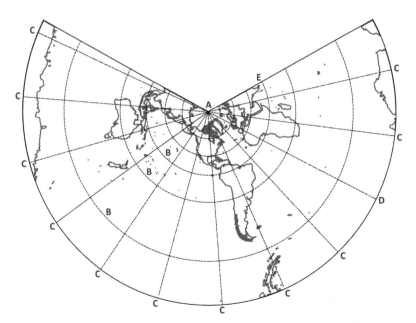

FIGURE 4.33 Lambert Conformal Conic map projection, using a 30° graticule. Poles=A, parallels=B, meridians=C, prime meridian=D, and equator=E.

However, if they happen to be chosen to represent areas near the equator and are symmetrical about the equator, this map projection essentially becomes the Mercator map projection (Snyder, 1982). Even though distortions in areas and distances may be sensitive to the two standard parallels that are chosen (Kuniansky, 2017), scale distortion is minimal for areas near the two standard parallels. And, as its name implies, the map projection is considered conformal, and angles are preserved at each point (Snyder and Voxland, 1989). The map projection has been used in practice for some time; for example, during World War I the French used this map projection as a basis for their battle maps (Deetz, 1918). The Lambert Conformal Conic map projection is used today for official state cartographic products of countries such as Croatia (Tutić, 2010) and others. Areas with wide longitude (east-west direction) and narrow latitude (north-south direction), such as the United States (Fig. 4.34), may be best suited for a conformal conic map projection such as this. If the two standard parallels are shifted further toward a pole, where the higher parallel is near the pole, this projection system may be useful in the development of polar charts (Skopeliti

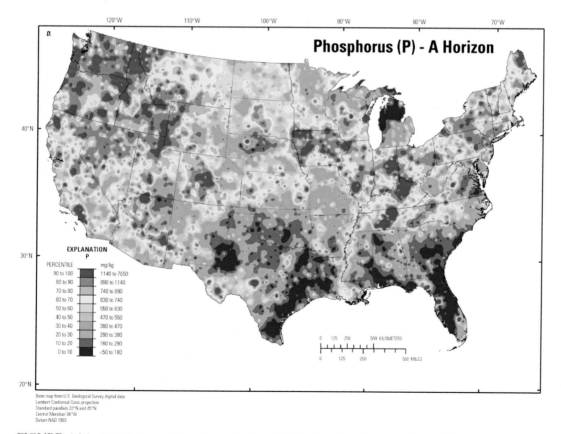

FIGURE 4.34 Distribution of phosphorus in the soil A horizon for the conterminous United States using a Lambert Conformal Conic projection. Source: *Smith et al. (2014).*

and Tsoulos, 2013). The map projection system has also been used in the development of maps of Mars, Mercury, and Earth's Moon (Snyder, 1982).

The *Equidistant Conic map projection* can be traced back to Ptolemy (c.CE 150), later enhanced beginning in about the 16th century, and refined to its present form in 1745 by French map maker De l'Isle (Snyder, 1978, 1987). The Equidistant Conic map projection (Fig. 4.35) is based on one or two standard parallels, and consists of straight meridians (C) and curved parallels (B). As in other conic projection models, when one standard parallel is used, this map projection is a *tangential model*. When two standard parallels are used, this map projection becomes a *secant model* (similar to the Albers Equal Area Conic map projection). The map projection is considered neither conformal nor equal area, but equidistant. The equally spaced meridians and the parallels meet at right angles on maps that employ this projection system (Snyder, 1987). The spacing of the arcs of latitude (parallels) is such that the result is a correct scale along the meridians (Snyder, 1978). Scale is also constant along each parallel, but it changes as one moves to other parallels. Along the one or two standard parallel lines, shapes are truly depicted, yet shapes become distorted the further one moves away from the standard parallels. Directions are also true along the standard parallels, and distances should be true along the meridians and along the standard parallels. This map projection system is most useful for small countries and landscapes with a predominant east-west expanse (Snyder, 1987).

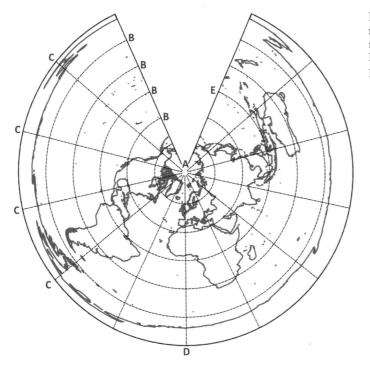

FIGURE 4.35 Equidistant Conic map projection, centered on the north pole, using a 30° graticule. Poles=A, parallels=B, meridians=C, prime meridian=D, and equator=E.

Translation 18

During the same dinner engagement described in Translation 17, one of your friends, who completed a geography course in college, asks why you did not use an equidistant conic map projection for the map of hurricane damage. Other people at the dinner table are enthralled with the conversation, but unaware of the difference. In very simple terms (two sentences), how does an equidistant cylindrical map projection differ from an equidistant conic map projection.

Polyconic projections represent parallels as nonconcentric circular arcs, centered on a straight line that represents a central meridian (Snyder and Voxland, 1989). The *polyconic map projection* (Fig. 4.36) is considered neither conformal nor equal area (Snyder, 1982). It has a central meridian that is represented as a straight line, but other meridians are represented as complex curves (Snyder and Voxland, 1989). Distortion is minimized along the central meridian, where scale is also true (Snyder, 1982). The standard parallel is the equator and is also represented as a straight line. All other parallels are represented as arcs that can be described where cones of different sizes would meet to form a tangent at that latitude. The fact that many cones are theoretically involved in

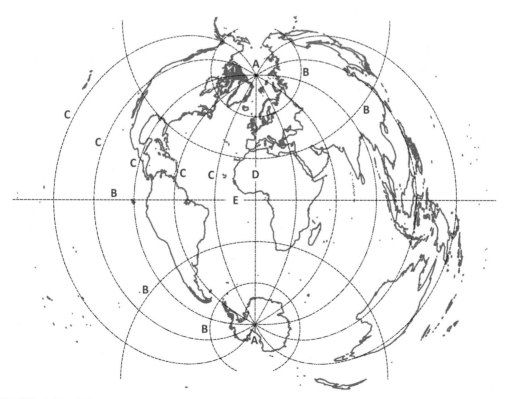

FIGURE 4.36   Polyconic map projection, using a 30° graticule. Poles=A, parallels=B, meridians=C, prime meridian=D, and equator=E.

this map projection process gave rise to the *poly* part of the name. The US government use the polyconic map projection for nautical and topographic maps through about the 1950s, afterward favoring instead the Transverse Mercator and Lambert Conformal Conic map projections due to their association with the state plane coordinate system (Snyder, 1982).

The *Lambert Equal Area Conic map projection* is similar to the Albers Equal Area Conic map projection, except that one of the poles is represented as a point rather than an arc. The map projection system preserves the relative sizes of areas; thus it is considered an equal area map projection. However, in contrast to the Lambert Conformal map projection, the Lambert equal area map projection is not conformal, and thus angular distortion exists everywhere except along a standard parallel (Snyder and Voxland, 1989).

Pseudoconic Projections

Pseudoconic projections are similar to conic projections, except now the meridians are represented as curved lines that converge upon the poles. Here, one might consider these to be oval projections, where both the meridians and the parallels are curved (Tobler, 1962). For example, the *Bonne map projection* (Fig. 4.37) has its central meridian (D) represented as a straight, vertical line, yet other meridians (C) are curved. The parallels (B) are represented by concentric circular arcs and the poles (A) are represented by points. The Bonne map projection is often attributed to Rigobert Bonne (1727–95), but was actually developed in the 16th century by others, and the map projection is perhaps more frequently used for atlas maps of continents rather than the entire Earth, but this does not necessarily exclude its application to maps illustrating the entire world (Snyder, 1987). In this system, the center of the projection is often represented by the center of the region being mapped. Scale is true along the central meridian and along parallels in this pseudoconic system (Snyder, 1987). The Bonne map projection is considered an equal area map projection and was once preferred because of the equal area property (Robinson, 1974). The map projection system is related to the sinusoidal map projection, where the standard parallel is 0°, or the equator (Snyder, 1987). As an example of its use, many 19th century topographic maps of central Europe were developed using this map projection system (Timár and Mugnier, 2010). The shape and display of a map using this projection, particularly for the entire globe (Fig. 4.38), may require some adjustment on behalf of the map user, depending on the chosen center of the projection.

> **Diversion 23**
> On the Internet, locate the *Compare Map Projections* website graciously hosted by Jung (2019). As of 2019, the website address is https://map-projections.net/index.php. Compare, via a list of projections, the Bonne map projection and the Albers Equal Area Conic map projection. Using the silhouette map comparison, in terms of the location of continents, how do these two map projections differ and how might they be similar?

The *Bipolar Oblique Conic Conformal map projection*, an adaptation of the Lambert Conformal Conic map projection, was designed for a 1:5,000,000 scale map of the

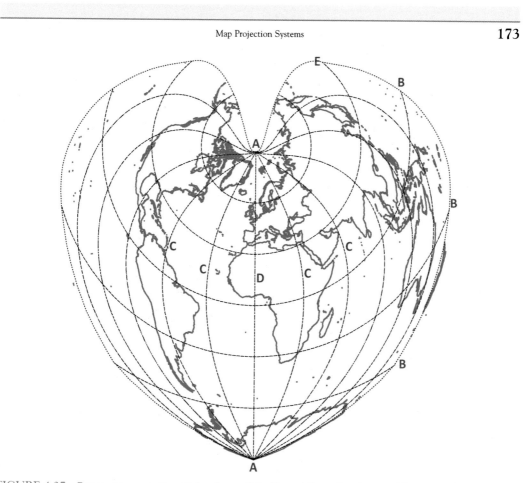

FIGURE 4.37   Bonne map projection. Poles=A, parallels=B, meridians=C, prime meridian=D, and equator=E.

Americas by the American Geographical Society in 1941 (Snyder, 1987). The map projection system involves the use of two side-by-side conic projections, spaced 104° apart, one to address each continent. This projection was once used for maps made by the U.S. Geological Survey until they decided to switch to the Transverse Mercator map projection. Although developed for the Americas, the Bipolar Oblique Conic Conformal projection has been used to develop maps in other areas of the world, such as maps of vegetation conditions in southeast Asia and India (Blasco et al., 1996).

Azimuthal Projections

When an *azimuthal projection* is used, an attempt is made to preserve relative directions and angles from a central point on a map. This type of map projection might be used when the direction from the central point on a map needs to be accurately described. However, azimuthal projections may only be of value for maps that focus on a central point; when using an azimuthal projection, the direction from a given central point to any

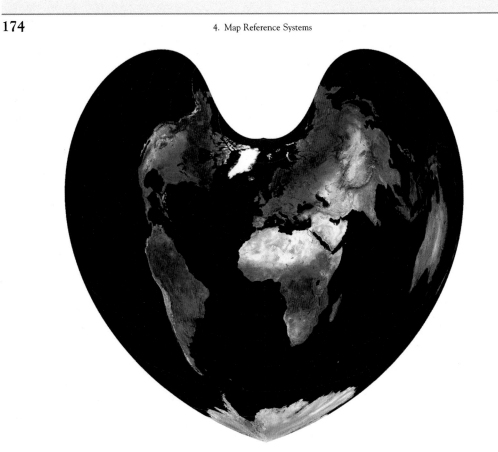

FIGURE 4.38   A Bonne map projection of the Earth. Source: *Courtesy Mdf through Wikimedia Commons.*

other place on the map is correct (Snyder and Voxland, 1989). In these map projections, the meridians are represented as straight lines, inclined differently to reflect their true position at each latitude, and the parallels are represented as a set of concentric circles (Snyder and Voxland, 1989).

The *Gnomonic map projection*, perhaps the oldest map projection system, is an azimuthal projection that represents the surface conditions of sphere (Fig. 4.39) from the point of view of the center of the sphere (Williams, 1997). The point at which the sphere touches the plane of the map (the point of tangency) can be anywhere on Earth. When using the Gnomonic map projection (Fig. 4.40), many of the great circle arcs are presented as straight lines (Williams and Ridd, 1960). A Gnomonic map projection is a panoramic image of sorts that is based on a central perspective (Barazzetti et al., 2014). Shapes, distances, and areas are increasingly distorted in a radial fashion from the center point of the map. This map projection is considered neither conformal nor equal area (Snyder and Voxland, 1989). Directions are accurate from the perspective of the center of the map. This map projection can illustrate areas about 60° from the center point; therefore maps that have one of the poles as the center tangent point are limited in describing conditions around the equator. Maps of Antarctica often employ a Gnomonic map projection. These types of map

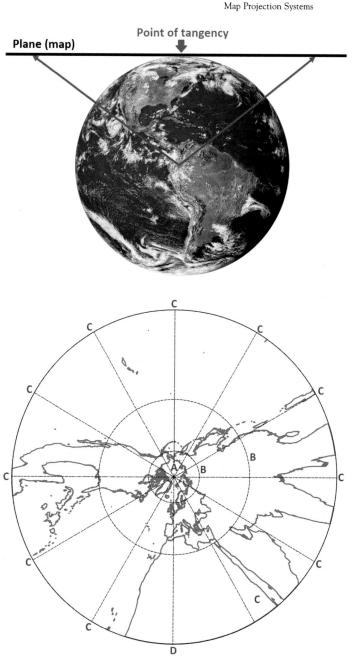

Plane (map)

Point of tangency

FIGURE 4.39 Conceptual model of the Gnomonic map projection using Earth as viewed from the Apollo 17 mission. Source: *Image of Earth courtesy National Aeronautics and Space Administration (2000).*

FIGURE 4.40 A Gnomonic map projection focusing on the north pole, using a 30° graticule. Pole=A, parallels=B, meridians=C, and prime meridian=D.

projections are also of use in the study of bird migration routes (Gudmundsson and Alerstam, 1998).

The *Azimuthal Equidistant map projection* is also quite old; it may have been used by ancient Egyptians in the development of star charts, and it was used by Gerardus

Mercator in his world map of 1569 (Snyder, 1987). As its name implies, this map projection is considered to be equidistant. The aspect of the graticule can center on a pole, the equator, or some other central point. Depending on the setting for the graticule, the meridians may all be straight lines (using a polar aspect), or only the central meridian may be (using the other aspects). Only with the equatorial aspect will a parallel (the equator) be represented as a straight line (Snyder and Voxland, 1989). If one were to view the Earth from afar, with a certain center point (or aspect), the Azimuthal Equidistant map projection would contain all you see, but also include the other side of the Earth that you cannot normally see due to the curvature of the sphere (Fig. 4.41). Thus, in contrast to the Gnomonic map projection where less than half of the Earth is displayed, the entire Earth can be displayed using this map projection system. For example, in Fig. 4.41, the poles (A) are both apparent, the parallels (B) are represented by curved arcs as they circle the Earth, and the meridians (C) are also represented as curved arcs. The parallels and meridians in this model are spaced apart by 30°. The only linear graticules are the Prime Meridian (D) and the equator (E). When using the Azimuthal Equidistant map projection, shapes are all distorted, except those that might be located in the center of a map. The directions implied by the map are true from the center point outward, thus from the center point of a map, all mapped locations can be found at the correct azimuth. However, when using this projection system, areas are increasingly distorted as one moves outward from the center of

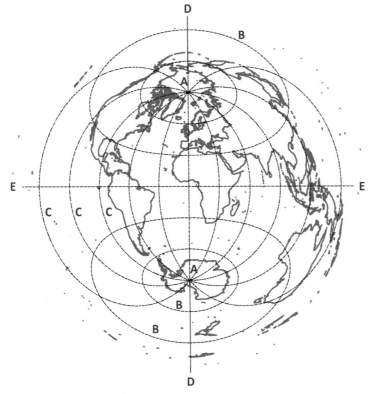

FIGURE   4.41   Azimuthal Equidistant   map   projection, using a 30° graticule. Poles=A, parallels=B, meridians=C, prime meridian=D, and equator=E.

the map. For polar-centric maps, where the center point is one of the two poles, distances are maintained from the center point outward on a map that uses this projection, while distances along the parallels (the concentric circles) contain some distortion. Polar applications are therefore useful applications of this projection system. At a very broad scale, and when centered on the north pole, the emblem of the United Nations utilizes the azimuthal equidistant map projection. The map projection has also been used to illustrate places at a more local scale, such as for depicting major land management areas within the state the Colorado (Fig. 4.42). Maps indicating human disease issues have also employed this projection (Jarcho, 1945). However, maps of the Earth are not the only application of this map projection method. For example, an Azimuthal Equidistant map projection has been applied to landscapes as small as the human heart (Edelman et al., 2014) and places as far away as asteroids in outer space (Nyrtsov et al., 2015).

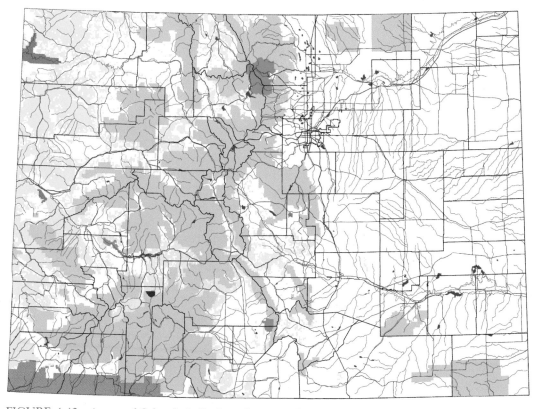

FIGURE 4.42 A map of Colorado indicating select major land management organizations and management areas: U.S. Forest Service (light green), U.S. Bureau of Land Management (yellow), U.S. National Park Service (dark green), Powderhorn wilderness (red), American Indian reservations (pink), national monuments or historic sites (brown), and major water bodies (blue). Also indicated are county boundaries (black lines) and major roads (reddish lines). The Azimuthal Equidistant map projection is centered on −105.7167° longitude and 39.1333° latitude. Source: *Courtesy David Benbennick through Wikimedia Commons.*

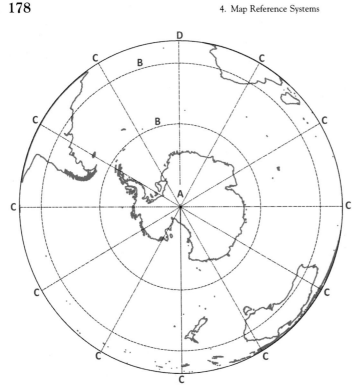

FIGURE 4.43 An Orthographic map projection focusing on the south pole. Pole=A, parallels=B, meridians=C, and prime meridian=D.

Similar to the azimuthal equidistant map projection, an *Orthographic map projection* (Fig. 4.43) can employ a pole, the equator, or some other place as the center point of the projection. An Orthographic map projection is neither conformal nor equal area, but the projection has the appearance of a globe (Snyder and Voxland, 1989). Similar to the Gnomonic map projection, about half of Earth can effectively be displayed when using this map projection system. Distortions in shapes and areas can be severe near the edges of the map, and therefore while perhaps aesthetically pleasing, measurements from these maps are not very useful (Snyder, 1982).

A *stereographic map projection* (Fig. 4.44), in contrast to other azimuthal projections mentioned here, is considered conformal in nature. Similar to other azimuthal projections, about one-half of Earth can effectively be represented on these maps. Depending on the aspect (pole, equator, other place), one or more meridians may be represented as straight lines, and perhaps one parallel (the equator when using an equatorial aspect) may be represented as a straight line. Some of the earliest known world maps and star maps used a stereographic map projection. A map developed by Jacques de Vaulx in 1583 illustrated one hemisphere of the world centered on Paris and the other hemisphere centered on the opposite point (Snyder, 1982), somewhere in the south Pacific Ocean, east of New Zealand.

With the *Lambert Azimuthal Equal Area map projection*, the entire Earth can also be projected from the perspective of a single point (e.g., point A, the north pole, in Fig. 4.45). The meridians (C) emanate as radii from the central point and are drawn with straight lines.

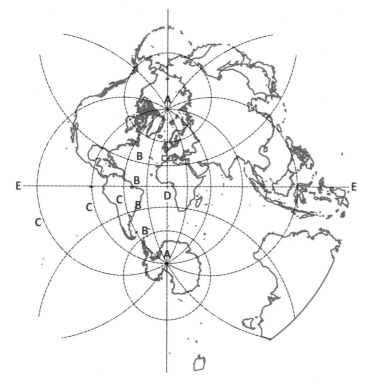

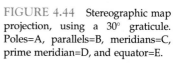

FIGURE 4.44 Stereographic map projection, using a 30° graticule. Poles=A, parallels=B, meridians=C, prime meridian=D, and equator=E.

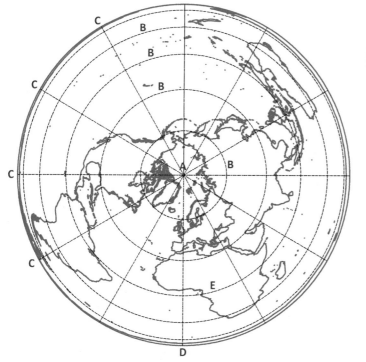

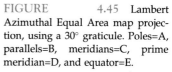

FIGURE 4.45 Lambert Azimuthal Equal Area map projection, using a 30° graticule. Poles=A, parallels=B, meridians=C, prime meridian=D, and equator=E.

The parallels (B) are concentric circles spaced apart unevenly, along the locations of the graticule (in this case, every 30°). This system is considered to be an equal area map projection, as its name implies. As with the others mentioned in this section, about one-half of Earth can effectively be represented on these maps, and depending on the aspect (pole, equator, other place), one or more meridians may be represented as straight lines, and perhaps one parallel (the equator when using an equatorial aspect) may be represented as a straight line. The equatorial aspect rendition of this map projection system is often used for Eastern and Western Hemisphere atlas maps (Snyder and Voxland, 1989).

Other Map Projections

Alphons Van der Grinten (1852−1921) received a US patent in 1904 (Patent no. US 751226 A), for a model (*Van der Grinten map projection I*) that displays the entire Earth on a plane surface using a circular outline. In this projection (Fig. 4.46), both the central meridian (D) and the equator (E) are straight lines that cross each other in a perpendicular manner in the center of the map. The other meridians (C) and parallels (B) are represented as various arcs of circles, and as a result, there is a large amount of distortion as one moves north or south to the polar areas. While the meridians are equally spaced along the equator, the parallels are unequally spaced along the central meridian. The poles are represented as points (A) rather than as lines in other systems such as the Robinson map projection. Due to the

FIGURE 4.46 The Van der Grinten projection system. Poles=A, parallels=B, meridians=C, prime meridian=D, and equator=E.

extreme exaggeration in the polar regions, Van der Grinten suggested that these could be left off of maps to reduce confusion that may arise within the mind of the map user (Reeves, 1904). Therefore, maps using the Van der Grinten projection system can be truncated and limit the display of these areas, unfortunately for those eager to see Antarctica and areas north of Greenland on maps. In the Van der Grinten map projection system, areas located near the equator are true to scale, but otherwise, the map projection system is neither considered conformal nor equal area (Snyder, 1982). The Van der Grinten projection is said to be a blend of the Mercator map projection, to reduce distortions of areas and distances, and the Mollweide map projection, to reduce angular distortion (Goode, 1905). This map projection was used by the National Geographic Society for some of their world maps for around 60 years, until 1988 when it was replaced by the Robinson map projection.

The *Winkel Tripel map projection* was developed by Oswald Winkel in 1921 as a compromise that acts to minimize area, distance, and angular distortions produced in the map projection process (Ipbuker and Bildirici, 2005). Therefore it contains a moderate level of distortion of each. The word "tripel" means triple in German, thus the name of the map projection reflects the fact that three sources of distortion are involved in the compromise. Technically, the Winkel Tripel map projection (Fig. 4.47) is a modified azimuthal map projection and considered neither equal area nor conformal (Ipbuker and Bildirici, 2005). Parallels (B) and meridians (C) are slightly curved, except the equator (E), the Prime Meridian (D), and the poles. The Winkel Tripel has been used by the National Geographic Society for wall maps of the world, replacing the Robinson map projection in 1998. The

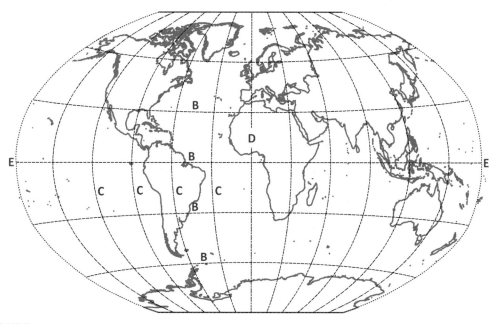

FIGURE 4.47 Winkel Tripel map projection, using a 30° graticule. Parallels=B, meridians=C, prime meridian=D, and equator=E.

map projection can thus be found in world maps such within the National Geographic Atlas of the World (Bush, 2015), and in other texts such as the Oxford Atlas of the World (Buckingham, 2018). General map readers in one research study ranked the Winkel Tripel map projection third in preference, after the Robinson and *Plate Carrée* map projections (Šavrič et al., 2015).

Map Grids or Indexes

As a map reference system, *grids* are often used to assist a map user in understanding and finding coordinates and locating landscape features; they also represent a systematic structure that divides a map into smaller geographic areas (Edler et al., 2014). On a map, a grid (Fig. 4.48) can be developed through the placement of evenly spaced north-south and east-west lines that divide the landscape into pieces of equal size. A grid need not be specifically associated with a projection or coordinate system, it simply can be the creation of the cartographer with the sole purpose of assisting the map user in locating features. Using this system, one division of the axes along the frame of a map may be represented by letters (A–Z), while the other may be represented by numbers (1–n). In conjunction with this

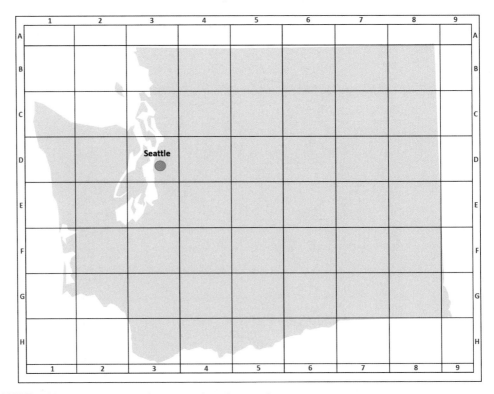

FIGURE 4.48   A grid reference for a map of Washington State.

system, an index could refer people to combinations of the lettering and numbering grid where certain landscape features might be located. For example, in Fig. 4.48 the index might indicate the city of Seattle is located within grid cell D-3. An index such as this would be of additional benefit to the map user for efficiently locating places even while one or more other geographic reference systems might be employed (Greenwood, 1964). Unfortunately, this type of grid and indexing system would only be applicable for the map on which it is presented, since it generally has no relation to a broader geographic system.

Inspection 23

Using the Internet, search for the *Washington State Highway Map*, developed by the Washington State Department of Transportation (2015). Once this website has been located, find a link that allows you to view the entire map. Within grid cell C-5, what cities and highways do you find?

Map grids can also be formed through divisions of latitude and longitude, or from distances implied by the UTM system. In these cases, the grid may be formally presented by the extension of parallels or meridians drawn across a map, or may only be implied by notation in the marginalia (outer edge) of a map. The scale of the map plays an important role in the amount of space between grid marks, or in *grid separation*. Often, on 1:24000 scale topographic maps developed by the U.S. Geological Survey, a 1000 m grid line separation is used, and these are drawn consistent with 1000 m northing and easting intervals of the UTM coordinate system.

Depending on the map, one may find 10 or more grid lines oriented north-south, and a similar number oriented east-west. A more dense set of grid lines may add unnecessary detail to the map, whereas a less dense set of grid lines may not provide enough reference for a map user to appropriately locate a landscape feature. For example, in urban areas a smaller grid spacing may, interestingly, not be as useful as a larger grid spacing due to the complexity of landscape features present within a map (Edler et al., 2014).

Concluding Remarks

Many of the coordinate and projection systems noted here have been used in practice. Some are antiquated, some are relatively new (e.g., Web Mercator, Equal Earth map projections). All map projections provoke deformations in mapped distances, angles, shapes, and areas (Borisov et al., 2015). The criteria for selecting a map projection may involve an assessment of area and angular distortions, the portrayal of shapes, and the portrayal of the meridians and parallels (Skopeliti and Tsoulos, 2013). Choosing a map's coordinate and projection system requires an understanding of the intended purpose of a map. For example, a map developer might consider whether the purpose is to represent an area correctly at a large or small scale, whether the purpose is to use the map for navigation at large or small scales, or whether the purpose is to describe precisely land ownership of land. A thoughtful consideration of these concerns can inform the selection of coordinate and projection systems.

Reflection 19

Imagine you have a job whereby you provide mapping support for a large agency that manages land across a broad area. At some point in time, you are asked to develop a map that illustrates the tree species composition of forests through a thematic map of tree species categories. Of course, it is suggested that this map will represent the entire country in which you live. Of the coordinate and map projection systems described in this chapter, which (1) coordinate system and (2) the map projection system would you choose? Why did you choose these?

Knowledge of coordinate and map projection systems was once deemed important for all map developers and map users. Yet unfortunately in our digital age, with widespread delivery of maps through electronic means, the level of knowledge of these systems seems to have diminished, and today, speed and usability of an interactive digital mapping program may be of more interest to a large segment of society than the geodetic accuracy that appropriate coordinate and map projection systems may provide (Favretto, 2014).

References

Arnaud, J.-L., 2013. Production of georeferenced data—use, cost and accuracy. e-Perimetron 8 (2), 101−105.

Barazzetti, L., Previtali, M., Scaioni, M., 2014. Simultaneous registration of gnomonic projections and central perspectives. Photogrammetric Record 29 (147), 278−296.

Battersby, S.E., Finn, M.P., Usery, E.L., Yamamoto, K.H., 2014. Implications of Web Mercator and its use in online mapping. Cartographica 49 (2), 85−101.

Bauer-Marschallinger, B., Sabel, D., Wagner, W., 2014. Optimisation of global grids for high-resolution remote sensing data. Comput. Geosci. 72, 84−63.

Baumann, J., 2019. Moving from static spatial references systems in 2022. ArcUser 22 (1), 34−37.

Blasco, F., Bellan, M.F., Aizpuru, M., 1996. A vegetation map of tropical continental Asia at scale 1:5 million. J. Vegetation Sci. 7 (5), 623−634.

Borisov, M., Petrović, V.M., Vulić, M., 2015. Optimal map conic projection—a case study for the geographic territory of Serbia. Tehnički Vjesnik 22 (2), 391−399.

Botley, F.V., 1951. A new use for the Plate Carree projection. Geogr. Review 41 (4), 640−644.

Buchroithner, M.F., Pfahlbusch, R., 2017. Geodetic grids in authoritative maps − new findings about the origin of the UTM Grid. Cartogr. Geogr. Information Sci. 44 (3), 186−200.

Buckingham, T., 2018. Review of Oxford Atlas of the World, twenty-fourth edition. Cartogr. Perspect. 90, 112−115.

Bush, C.N., 2015. Review of The National Geographic Atlas of the World (Deluxe 10th edition). Cartogr. Perspect. 81, 57−60.

Buterez, C.-I., Popa, A., Gava, R., Dumitru, R., Gruia, A.R., 2016. On the trail of a legend. The legacy of Lady Neaga seen through historical maps. e-Perimetron 11 (2), 77−89.

Deetz, C.H., 1918. The Lambert Conformal Conic Projection With Two Standard Parallels Including a Comparison of the Lambert Projection With the Bonne and Polyconic Projections. U.S. Department of Commerce, Coast and Geodetic Survey, Washington, D.C. Special Publication No. 47.

Dracup, J.F., 1977. Understanding the State Plane Coordinate Systems. U.S. Department of Commerce, National Oceanic and Atmospheric Administration, National Geodetic Survey, Rockville, MD.

Dracup, J.F., 1995. Geodetic surveying 1940−1990. U.S. Department of Commerce, National Oceanic and Atmospheric Administration, Washington, D.C.

Droissart, V., Dauby, G., Hardy, O.J., Deblauwe, V., Harris, D.J., Janssens, S., et al., 2018. Beyond trees: biogeographical regionalization of tropical Africa. J. Biogeogr. 45 (5), 1153−1167.

Dunn, S., 1774. A Map of the World, With the Latest Discoveries. Lionel Pincus and Princess Firyal Map Division of The New York Public Library, New York. <https://dp.la/item/539ad082c39861fbb3bdafa1b980d397> (accessed 04.05.19).

Edelman, R.R., Giri, S., Murphy, I.G., Flanagan, O., Speier, P., Koktzoglou, I., 2014. Ungated radial quiescent-inflow single-shot (UnQISS) magnetic resonance angiography using optimized azimuthal equidistant projections. Magn. Reson. Med. 72 (6), 1522–1529.

Edler, D., Bestgen, A.-K., Kuchinke, L., Dickmann, F., 2014. The effects of grid line separation in topographic maps for object location memory. Cartographica 49 (4), 207–217.

Ekblaw, W.E., 1928. The world on Goode's Homolo-sine projection, interrupted for the continents by J. Paul Goode. Econ. Geogr. 4 (1), 113–114.

Favretto, A., 2014. Coordinate questions in the Web environment. Cartographica 49 (3), 164–174.

Frye, A.E., 1895. Complete geography. Ginn & Company, Publishers, Boston, MA.

Gang, L., 2007. The Chinese inventor of bi-hemispherical world map. e-Perimetron 2 (3), 185–193.

Ghaderpour, E., 2016. Some equal-area, conformal and conventional map projections: a tutorial review. J. Appl. Geodesy 10 (3), 197–209.

Goode, J.P., 1905. A new method of representing the Earth's surface. J. Geogr. 4 (9), 369–373.

Goode, J.P., 1925. The Homolosine projection: a new device for portraying the Earth's surface entire. Ann. Assoc. Am. Geogr. 15 (3), 119–125.

Goward, S.N., Turner, S., Dye, D.G., Liang, S., 1994. The University of Maryland improved Global Vegetation Index product. Int. J. Remote Sensing 15 (17), 3365–3395.

Grafarend, E., Okeke, F., 2007. Transformation of Lambert conic conformal coordinates from a global datum to a local datum. Marine Geodesy 30 (4), 297–313.

Greenwood, D., 1964. Mapping. The University of Chicago Press, Chicago, IL.

Gudmundsson, G.A., Alerstam, T., 1998. Optimal map projections for analysing long-distance migration routes. J. Avian Biol. 29 (4), 597–605.

Guedes, T.B., Sawaya, R.J., Zizka, A., Laffan, S., Faurby, S., Pyron, R.A., et al., 2018. Patterns, biases and prospects in the distribution and diversity of Neotropical snakes. Global Ecol. Biogeogr. 27 (1), 14–21.

Helliwell, J., Layard, R., Sachs, J., De Neve, J.-E., Huang, H., Wang, S. (Eds.), 2017. World Happiness Report 2017. Sustainable Development Solutions Network, New York.

Ipbuker, C., Bildirici, I.O., 2005. Computer program for the inverse transformation of the Winkel projection. J. Surveying Eng. 131 (4), 125–129.

Jarcho, S., 1945. Equal-area projections and the azimuthal equidistant projection in maps of disease. Am. J. Public Health 35 (10), 1005–1013.

Jung, T., 2019. Compare map projections. <https://map-projections.net/index.php> (accessed 19.02.19).

Kessler, F.C., 2018. Map projection education in cartography textbooks: a content analysis, Cartogr. Perspect. 90, 6–30.

King-Hele, D., 1967. The shape of the Earth. Scientific American 217 (4), 67–76.

Kuniansky, E.L., 2017. Custom map projections for regional groundwater models. Groundwater 55 (2), 255–260.

Libecap, G.D., Lueck, D., 2011. The demarcation of land and the role of coordinating property institutions. J. Political Econ. 119 (3), 426–467.

Livieratos, E., 2008. The Anthimos Gazis world map in Kozani. e-Perimetron 3 (2), 95–100.

Loomis, F.B., 1961. Subsurfacegeology. In: Moody, G.B. (Ed.), Petroleum Exploration Handbook. McGraw-Hill Book Company, Inc, New York, pp. 13-1 to 13-74.

Morgan, R.S., Adams, C.W.M., 1974. A graticule for measuring atherosclerosis. Atherosclerosis 19 (2), 347–348.

National Aeronautics and Space Administration, 2000. Earth—The Blue Marble. National Aeronautics and Space Administration, Earth Observatory, Greenbelt, MD. <https://earthobservatory.nasa.gov/images/565/earth-the-blue-marble> (accessed 05.05.19).

National Geospatial-Intelligence Agency, 2014. Implementation Practice Web Mercator Map Projection, 2014-02-18, Version 1.0.0. National Geospatial-Intelligence Agency, Office of Geomatics, Springfield, VA. NGA. SIG.0011_1.0.0_WEBMERC.

Nyrtsov, M.V., Fleis, M.E., Borisov, M.M., Stooke, P.J., 2015. Equal-area projections of the triaxial ellipsoid: first time derivation and implementation of cylindrical and azimuthal projections for small solar system bodies. Cartogr. J. 52 (2), 114–124.

Olofsson, A.O.H., Torchinsky, S.A., Chemin, L., Barth, S., Bosse, S., Martin, J.-M., et al., 2009. Profiling the EMBRACE tile beam using GPS satellite carriers. In: Torchinsky, S.A., van Ardenne, A., van den Brink-Havinga, T., van Es, A.J.J., Faulkner, A.J. (Eds.), Widefield Science and Technology for the SKA, SKADS Conference 2009, pp. 253–257.

Parson, H.E., 1977. Settlement policy and land evaluation at the turn of the twentieth century in Quebec. Area 9 (4), 290–292.

Patterson, T., 2013. Mountains unseen: developing a relief map of the Hawaiian seafloor. Cartogr. Perspect. 76, 5–17.

Reeves, E.A., 1904. Van der Grinten's projection. Geogr. Record 6 (December), 670–672.

Robinson, A.H., 1974. A new map projection: its development and characteristics. In: Kirschbaum, G.A., Meine, K.-H. (Eds.), International Yearbook of Cartography, 14. Kirschbaum Verlag, Bonn-Bad Godeberg, Germany, pp. 145–155.

Šavrič, B., Jenny, B., White, D., Strebe, D.R., 2015. User preferences for world map projections. Cartogr. Geogr. Information Sci. 42 (5), 398–409.

Šavrič, B., Patterson, T., Jenny, B., 2019. The Equal Earth map projection. Int. J. Geogr. Information Sci. 33 (3), 454–465.

Schmunk, R.B., 2019. G. Projector, version 2.3.4. National Aeronautics and Space Administration, Goddard Institute for Space Studies, New York. <https://www.giss.nasa.gov/tools/gprojector/> (accessed 17.02.19).

Serenco Serviços de Engenharia Consultiva Ltda., 2017. Plano distrital de saneamento báscio e de gestão integrada de resíduos sólidos, Tomo II—Produto 2 (Diagnóstico situacional—caracterização distrital), Versão final. Serenco Serviços de Engenharia Consultiva Ltda, Curitiba, Brasil.

Sincovec, R.J., 2008. Working With Grid Coordinates. Edward-James Surveying, Inc., Colorado Springs, CO.

Skopeliti, A., Tsoulos, L., 2013. Choosing a suitable projection for navigation in the Arctic. Marine Geodesy 36 (2), 234–259.

Smith, D.B., Cannon, W.F., Woodruff, L.G., Solano, F., Ellefsen, K.J., 2014. Geochemical and mineralogical maps for soils of the conterminous United States. U.S. Department of the Interior, U.S. Geological Survey, Reston, VA. Open-File Report 2014-1082.

Snyder, J.P., 1978. Equidistant conic map projections. Ann. Assoc. Am. Geogr. 68 (3), 373–378.

Snyder, J.P., 1981. Space oblique Mercator projection mathematical development. U.S. Government Printing Office, Washington, D.C. Geological Survey Bulletin 1518.

Snyder, J.P., 1982. Map projections used by the U.S. Geological Survey. US Department of the Interior, US Geological Survey, Denver, CO. Professional Paper 1532.

Snyder, J.P., 1987. Map projections—a working manual. U.S. Government Printing Office, Washington, D.C. Geological Survey Professional Paper 1395.

Snyder, J.P., Voxland, P.M., 1989. An album of map projections. US Department of the Interior, Geological Survey, Denver, CO. Professional Paper 1453.

Sun, Y., Riva, R., Ditmar, P., Rietbroek, R., 2019. Using GRACE to explain variations in the Earth's oblateness. Geophys. Res. Lett. 46 (1), 158–168.

Taylor, J.W., 1951. Louisiana land survey systems. Southwestern Social Sci. Quart. 31 (4), 275–282.

Timár, G., Mugnier, C.J., 2010. Rectification of the Romanian 1:75 000 map series, prior to World War I. Acta Geodaetica et Geophysica Hungarica 45 (1), 89–96.

Tobler, W.R., 1962. A classification of map projections. Ann. Assoc. Am. Geogr. 52 (2), 167–175.

Tutić, D., 2010. Konformne projekcije za Hrvatsku s najmanjim apsolutnim linearnim deformacijama. Geodetski List 64 (3), 157–173.

U.S. Department of Commerce, National Oceanic and Atmospheric Administration, 2018. Final SPCS 83 (as of 2001). U.S. Department of Commerce, National Oceanic and Atmospheric Administration, National Geodetic Survey, Silver Spring, MD. <https://www.ngs.noaa.gov/SPCS/maps.shtml> (accessed 26.02.19.).

U.S. Department of State, 2017. Learn About Your Destination. U.S. Department of State, Bureau of Consular Affairs, Washington, D.C. <https://travel.state.gov/content/passports/en/country.html> (accessed 14.07.17.).

Washington State Department of Transportation, 2015. View and Print the State Highway Map. Washington State Department of Transportation, Olympia, WA. <https://www.wsdot.wa.gov/Publications/HighwayMap/view.htm> (accessed 22.02.19.).

White, C.A., 1983. A History of the Rectangular Survey System. U.S. Department of the Interior, Bureau of Land Management, Washington, D.C.

Williams, C.E., Ridd, M.K., 1960. Great circles and the Gnomonic projection. Professional Geogr. 12 (5), 14–16.

Williams, R., 1997. Gnomonic projection of the surface of an ellipsoid. J. Navigation 50 (2), 314–320.

Wycoff, W., 1986. Land subdivision on the Holland Purchase in western New York State, 1797-1820. J. Historical Geogr. 12 (2), 142–161.

Wyoming Department of Transportation, 2015. Survey Manual. Wyoming Department of Transportation, Photogrammetry & Surveys, Cheyenne, WY.

Yildirim, F., Kaya, A., 2008. Selecting map projections in minimizing area distortions in GIS applications. Sensors 8, 7809–7817.

Map Interpretation

Through the interpretation of maps, some insight about places and geography can be obtained without the need to physically visit the landscape. An effective map would assist a map user in understanding location, orientation, area, distance, and direction, and where appropriate, altitude (Kempf and Poock, 1969). Other aspects of maps (color, theme, etc.) can further enhance the message being conveyed, and help answer straightforward questions such as *What is located there?* and *How do I get from one location to another?* Close inspection of map features can address common questions such as *What is the tallest feature around here?* or *Where does this (road, river, etc.) go?* At some point, measurement of map features may be required to determine, *How far is it from here to there?* or *How large is this area?* Supportive information may also be required to understand, *What can be found here?* or *What lies between here and there?* Thematic mapping processes can also assist map users in understanding, *What happened here?* and *What do we plan to do here?*, or more simple questions, such as *What is mine?* and *What is yours?* or *Who owns that?* With this in mind, this chapter is meant to enhance the reader's map interpretation skills.

Scale

To be of value in providing a sense of direction, and generally describing land areas, maps are presented on reasonably-sized pieces of paper or on the screens of digital devices. Maps are thus drawn, rendered, or displayed as small models of the larger landscapes. *Scale*, as used in mapping endeavors, can refer to a proportion between two entities (i.e., a map and the real-life land or sea area it represents), to the act of measuring distance and area features through the use of a standard rate, or to the resizing of a map area to fit within the intended media. The former interpretation reflects the use of the word *scale* as a noun, while the latter two interpretations reflect the use of the word as a verb. For example, when used as a noun, one would indicate that the scale of a map might be 1 cm = 1 km (1 cm on the map equals 1 km in real life). In the latter three cases, one might attempt to understand the length of a particular road by *scaling it out*, or converting the length of it on the map to a real-life length (e.g., a road is 2.5 cm on the map, thus it must be 2.5 km long in real life). In this chapter, we refer to *map scale* as a subject or entity; therefore, the spatial relationship between a map and the land or system that it illustrates describes the *scale* of the map. However, it is not unreasonable to use the term in reference to an act, such as the need to *scale a map* to fit a desired piece of paper or digital display.

A map that has a 1 to 1 (1:1) scale relationship with the land or water body that it describes would be as large as that land or water body, which is impractical. Maps are therefore proportionally smaller than real-life systems, and this proportion is denoted as the scale. The scale chosen for a map affects the amount of detail that can be presented. A *large-scale map*, given a single output format size (e.g., a piece of paper), would contain a relatively small area and more detail than a *small-scale map*, which would contain relatively more land area and less detail. The format of a map (size of sheet of paper, size of computer monitor, size of portable device screen), along with the area to be mapped will both influence the scale selected. Further, the amount of detail desired within the map can influence the format and scale selected, and the area to be mapped (Greenwood, 1964).

Reflection 20
Consider the last small scale map that you remember using, that covered a very large geographic area. How much land or water body area did it cover (e.g., a county, a lake, an ocean, a continent)? What was its scale, theme, and purpose?

Imagine two maps drawn at significantly different scales (1:250,000 and 1:10,000). The larger scale map is the one that uses the 1:10,000 scale. This implies that one unit of distance on this map equals 10,000 of those same units in real life within the landscape being mapped. Therefore the model of the landscape being mapped would be larger than the model mapped using the 1:250,000 scale, where one unit of distance on a map equals 250,000 of those same units in real life. Of these two, the 1:10,000 scale map may contain less land area, but the opportunity to include more detail (e.g., roads, streams, and buildings) increases dramatically over the smaller scale map (Levin et al., 2010). This advantage can lead to the inclusion of more features on a map, as well as the ability to describe the same standard features with less generalization.

Often, scale is described by relating linear features (distances) on maps to real-life distances, but any feature, such as buildings or water bodies, can be used as the example to which a map scale is described (i.e., a lake on a map is 1/10,000th its real size). However, there are instances where common landscape features and other symbols may not be portrayed at the scale suggested on a map, creating a false impression of size or frequency of the resources described by the symbols. For example, in her paper investigating beer-trail maps, Feeney (2017) provides an example of a map where large symbols (pints of beer) are placed around landscape features (cities) to note where breweries may be located. However, the size of the symbols may give the map user a false impression that the State of Montana seems to have many very large breweries, when in fact it may have many small microbreweries. To further emphasize this point, imagine two maps, one using a 1:24,000 scale and the other using a 1:100,000 scale. On each map the same rural road may be drawn in such a way that it is 1/100 of an inch wide, given the drawing or printing technology employed. However, the road's implied real-life width would be different on each map. For example, on the 1:24,000 scale map, a 1/100th inch wide line would suggest that the road is 20 ft. wide in real life. To derive this, we need to remember that a 1:24,000 scale suggests that every 1 unit of measurement on the map represents 24,000 of those same units on the ground. So,

real-life width of road $= 0.01$ in. $\times$ 24,000
real-life width of road $= 240$ in.
real-life width of road $= 20$ ft.

Diversion 24
Let's close this discussion with a little work on your part. By simply converting the drawn width of the road to a ground distance, what is the suggested real-life width of the same road mentioned above on a 1:100,000 scale map?

Each map is presented to map users with a specific scale. The choice of scale to employ can be a very pragmatic one (fit the map to a specific size of paper). However, the choice of scale to employ may also be based on a desire to both present a sufficient amount of recognizable landscape character, along with some level of abstraction, to prompt in the mind of the map user an imaginative interpretation of the landscape (Kent, 2012). Although it is easy with digital devices to zoom into and out of a landscape, when the transition stops, the resulting map has a specific scale. Printed maps generally provide the scale graphically or numerically somewhere in the marginal area. Digital maps often only present a graphical version of a scale, if a scale is present at all. If a scale is known, measurements of and among features presented in a map can be made. A map scale therefore facilitates answering questions, such as *How far is it from one place to another?* and *How large is the area of interest?*

Representative Fraction Scale

Alternatively called a *natural scale*, a *ratio scale*, or a *fractional scale*, the representative fraction scales we have been using thus far are often considered the most natural representation of map scale. Expressed as *"1 : something"*, this map scale suggests that any one unit of measurement on the map (the left-hand side of the scale representation) is equal to some number of the *same units* on the ground or water body where the map features are situated. The representative fraction scale always begins with "1 :" (or, as some say, "*one to*") to indicate this relationship. When expressed orally, people often say "1 to 40,000", for example, for a map that employs a 1:40,000 scale. A map with this scale suggests that the mapped features (roads, properties, etc.) are 1/40,000 the size of the same features in real life. When measuring between two points (e.g., two road intersections) on a map that has a scale noted as 1:40,000, one unit of measurement on the map (e.g., centimeter, inch) is equivalent to 40,000 of those same horizontal units on the ground, between the endpoints of the places measured. In this respect, we arrive at two useful measurements:

Map distance: The distance measured on a map between two points
Ground distance: The actual horizontal distance between those same two points

These measures are related in that the result of a mathematical operation (*ground distance / map distance*) produces the right-hand side of the representative fraction scale, assuming, of course, that the two distances are provided in the same units of measurement. In other words, under the appropriate conditions, some rightly argue that

Map scale = (ground distance / map distance)

As we suggested earlier, a large-scale map would have a small right-hand side value in the representative fraction relationship (e.g., 1:1000), whereas a small-scale map would have a large right-hand side value (e.g., 1:500,000). As we also suggested earlier, given a consistent map size (a piece of paper or the screen of a digital device) a small-scale map would describe more land area than a large-scale map, since landscape features are proportionally smaller.

Translation 19
Imagine that you are discussing maps with some friends, and the subject of scale casually arose. Explain in general terms what a 1:10 scale implies and what a 1:1,000,000 scale implies. Of these two representative fraction scales, which represents a large-scale map, and which represents a small-scale map?

Reflection 21
In your opinion, in the development of a map, is it imperative that the representative fraction scale be a nice, round number (e.g., 1:40,000)? Or, do you think it is acceptable that a random scale (e.g., 1:42,143) might be used if the landscape features are adequately represented within the map? In either case, why?

As you may have observed, the representative fraction scale is *unitless*. As long as a map has not been reproduced in such a way as to modify the scale (enlarged, reduced), the representative fraction scale can be of value in understanding the absolute and relative sizes of features drawn on the map. Unfortunately, once enlargement or reduction of a printed map has occurred, the representative fraction scale, if unadjusted for these modifications to the map, may be of little value. This disadvantage is not unique to this type of scale, as you will soon learn.

In addition to assisting the map user in understanding distances, the representative fraction scale is also helpful in understanding area. To enable a relationship between true ground areas and areas drawn or represented on a map, we need to understand the *squared area equivalent*. For example, how much land area, in real life, would 1 cm^2 or 1 in.^{2} represent on a map? If a map scale is represented as 1:40,000 then

1 cm on the map = 40,000 cm on the ground
1 cm^2 on the map = 40,000 cm^2 on the ground
1 cm^2 on the map = 1,600,000,000 cm^2 on the ground
1 cm^2 on the map = 160,000 m^2 on the ground
1 cm^2 on the map = 16 hectares on the ground

In this metric (*System International*, or *SI*) unit example, the latter of these two relationships requires knowing that there are 10,000 cm^2 in a square meter (100 cm $\times$ 100 cm), and that there are 10,000 square meters in a hectare. What follows is an imperial system (English system of measurements) example along these same lines:

1 in. on the map = 40,000 in. on the ground
1 in.^{2} on the map = 40,000 in.^{2} on the ground
1 in.^{2} on the map = 1,600,000,000 in.^{2} on the ground
1 in.^{2} on the map = 11,111,111.1 ft.^{2} on the ground
1 in.^{2} on the map = 255.07 acres on the ground

In this imperial system example, the latter of these two relationships requires knowing that there are 144 in.^{2} in a square foot (12 in. $\times$ 12 in.), and that there are 43,560 ft.^{2} in 1 acre.

In aerial imagery interpretation applications (photogrammetry), the statement of a representative fraction scale can be shortened by simply referring to the right-hand side, or denominator, of the scale relationship. In these instances, it is called a *photo scale reciprocal (or PSR)* (Paine and Kiser, 2012). For example, an image or aerial photograph with a representative fraction scale of 1:40,000 has a PSR of 40,000. A large-scale image or aerial photograph would, therefore, have a smaller PSR value than a small-scale image or aerial photograph.

Equivalence Scale

In instances where the scale of the map is expressed in such a way that distance units are present on both sides of the equivalence sign (e.g., *1 in. = 1 mile*), the map user is presented with an *equivalence scale*. Some refer to this as a *verbal scale*, since it provides map users with a more familiar representation of the proportion of the map to the land area or water body it represents. Often, but not universally necessary, the units on either side of the scale relationship are different. The left-hand side of the relationship relates to a measurement made on the map between two landmarks; the right-hand side refers to the real-life equivalent horizontal (ground) distance between those two landmarks. For example, if one were to measure a distance of 2 cm between two landmarks on a map that was developed using a scale of *1 cm = 1000 m*, the real-life ground distance would be 2 $\times$ 1000 m, or 2000 m (each centimeter measured represents 1000 m in real life). An equivalence scale could be presented as *1 cm = 10,000 cm*, but we rarely encounter these types of examples in practice because the right side of the relationship should express some reasonable form of ground distance, usually in units different than the left side of the relationship. Also, we rarely would refer to, measure, or record real-life ground measurements in terms of small units of distance such as centimeters or inches, but rather in terms of feet, meters, kilometers, or miles. And further, through reduction, or removal of the common units (*centimeter* on both sides of the relationship), this type of scale would reduce to, and be better represented as a representative fraction scale (1:10,000). Thus, the two main distinctions between an equivalence scale and a representative fraction scale are (1) that an equivalence scale contains words that represent both map and ground distance units, and (2) most likely both sides of the scale relationship contain different distance units (e.g., centimeters on the left-hand side and meters on the right-hand side).

We can convert an equivalence scale to a representative fraction scale, and vice versa. For example, consider *1 cm = 1000 m*. If you know that there are 100 cm in each meter, this scale is the same as *1 cm = 100,000 cm*. Since the units on both sides of the scale relationship are now the same, we can drop the units (the words), and replace the equals sign with a colon, producing 1:100,000.

Similarly, we can convert a representative fraction scale to one or more equivalence scales. The following example takes a representative fraction scale (1:40,000) and converts it to three different equivalence scales, two of which contain different units on the left- and right-hand sides of the relationship:

1:40,000 is the same as
1 in. = 40,000 in. and
1 in. = 3,333.3 ft. and
1 in. = 0.63 m

Of course, to make these conversions one must need to know that with the imperial system, there are 12 in. in a foot and 5280 ft. in a mile.

Diversion 25

Using processes similar to what we have described above, develop metric and imperial system equivalence scales for the following representative fraction scales: 1:10,000, 1:20,000, and 1:100,000. Then, develop representative fraction scales that correspond to these equivalence scales: 1 in. = 1 m, and 1 cm = 50,000 m.

As with the representative fraction scale, the equivalence scale might suffer from misrepresentation if a printed map has been reproduced in such a way (enlarged, reduced) as to make the scale ineffective in describing landscape features. Once enlargement or reduction of a printed map has occurred, the equivalence and representative fraction scales, if unadjusted for modifications to the map, may be of little value.

Graphical Scale

A graphic in the form of a ruler, printed on a map or displayed on the screen of a digital device, acts as a *graphical scale* (Fig. 5.1). Graphical scales enhance the cartographic

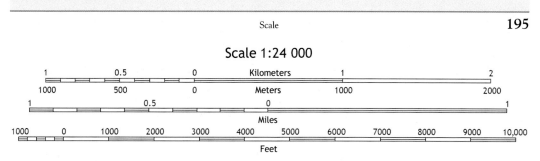

FIGURE 5.1 Graphical and representative fraction scales from the Midway, Georgia 7.5-minute U.S. Geological Survey quadrangle (USGSX24K29174).

quality of maps in that they may be more relatable from a cognitive perspective than the representative fraction and equivalence scales. Graphical scales explain map scale with an example of one or more physical units of distance, often displayed as bars or lines. When bars are employed in the graphic, they are called *bar scales* or *scale bars*. When lines are employed in the graphic, they are called *linear scales*. There is one main advantage to the use of a graphical scale over the previous ones described in this chapter: if a printed map has been reproduced (enlarged or reduced) in any way, the graphical scale is adjusted proportionately along with the landscape features and, therefore, retains its validity. When enlarged or reduced, a map that only includes a representative fraction or equivalence scale may be difficult to use, as these scales will no longer be accurate. The graphical scale, however, should still be of value. There is one disadvantage to the use of a graphical scale over the previous ones described in this chapter: the relationships suggested through the use of the representative fraction and equivalence scales are lacking. For instance, if you were comfortable knowing that 1 in. on a map represented 2000 ft. on the ground, in real life, a graphical scale may not easily provide this information.

When considering the use of a graphical scale, map developers might select one that provides a sufficient number of breaks in the lines or bars, so that map users can visualize subdivisions of distances. For example, a bar or line scale might indicate in total that the length of the bar represents a relatively long distance in real life (e.g., 10 miles), yet finer subdivisions of the bar also allow the map user to better understand shorter-distance relationships. For example, the three bars presented in Fig. 5.1 may allow the map user to understand multiples of imperial and metric system distances among the features within the map, yet the finer subdivisions of each can help the map user with more intimate spatial relationships.

Inspection 24
Access the 2015 map of Georgia (the former Soviet Republic), which can be found on this book's website (mapping-book.uga.edu). Roughly speaking, along a straight line, how far is the city of T'bilisi (the capital of Georgia) from the Black Sea? How far is T'bilisi from the city of Rustavi?

Other cartographic considerations that may increase the visual appeal of bar and line scales include placement, style, and size of the text used to describe the associated distance units (meters, kilometers, feet, miles, etc.). When developed by hand, the map user has

complete control over these aspects of a graphical scale. When using geographic information system (GIS) software, map developers may have the ability to control some of these
aspects. However, map developers may have no control over the finer details associated
with bar or line graphical scales when using some computer mapping programs or digital
devices. In order to address concerns of whether map users can understand scale, some
maps include both graphical and numeric scales (representative fraction or equivalence, or
both), as in Fig. 5.1.

Elevation

On maps, an *elevation* is generally expressed as the height of a place, usually the surface
of the ground (the *ground level*), relative to mean sea level. Often, people think of elevation
with respect to positive ground levels or, in essence, areas of dry ground above water.
However, elevation could also be expressed as the level of the sea bed below water, relative to mean sea level (Fig. 5.2). Thus mean sea level acts as a base level, or datum, to
which elevations are often referenced. While elevations are noted through the contour
lines drawn on maps in discrete value classes (e.g., 100 feet, 120 feet, 140 ft., etc.) elevations can also be expressed on maps as exact values that are precisely measured by surveyors. These marks on maps are considered *bench marks* (Fig. 5.3).

Expressing three-dimensional characteristics of landscapes on two-dimensional maps
can be challenging. One can imagine lines or paths, of land elevations or sea depths, where
the elevation or depth is constant. When drawn or otherwise presented on a map, these
lines connect places that are the same elevation above (or below) mean sea level, and are
therefore considered *contour lines*. Contour lines can generally indicate upward or downward progression of the land or bed of a water body. On dry land, they can be used to
interpret or represent slopes, tops of mountains, or depressions. As described in
Chapter 2, Map types, bathymetric contour lines are imaginary lines that describe the
depth of a submerged surface (river, lake, and ocean).

Reflection 22

For this exercise, imagine the recreational opportunities available at a national park,
such as the Grand Teton National Park in Wyoming. Why would the depth of water,
expressed as bathymetric contour intervals, be of value to a recreational map of this
park? Further, how can changes in the elevation of dry land be of value to a
recreational map of this park?

A *contour interval* represents the difference in elevation between successive or adjacent
contour lines. For example, when the contour interval is 10 m, contour lines will be presented on a map for each 10-meter difference in elevation (e.g., one line for 100 m above
sea level, another line for 110 m above sea level, another line for 120 m above sea level,
and so on). On a single map that uses a consistent contouring system, contour lines placed
very close together (Fig. 5.4) represent areas of steep slopes (the elevation changes quickly
over short distances) and contour lines placed farther apart represent areas of gentle
slopes (the elevation changes slowly or gradually over short distances). A densely

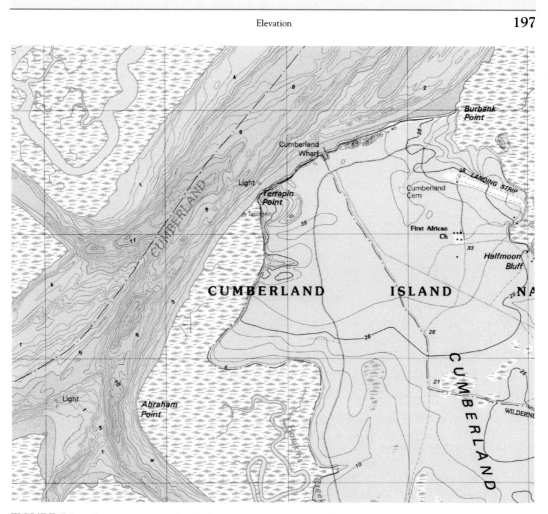

FIGURE 5.2   Elevation in general and elevation below water level [Cumberland Island North, Georgia 7.5-min-ute U.S. Geological Survey quadrangle (NIMA 4745 IV NW-Series V845)].

aggregated or packed set of contour lines can provide a perspective (through the resulting darker appearance on a map) of steep areas (Kennelly and Kimerling, 2006). As might be suggested, the slope of the land is 0 degree (or 0%) along the path where contour lines are drawn. Importantly, since these lines represent specific elevations, they never cross or merge, unless they describe an overhanging or vertical cliff.

Large contour intervals (expressing greater elevation changes) are used on maps where the elevation range among map features is significant (e.g., maps with significant coverage of steep lands). Smaller contour intervals are used on maps where the elevation range among map features is relatively insignificant (flatter landscapes). Although they often appear as lines on maps, on a complete map of a landscape, as we suggested, contour lines actually represent closed polygons (Fig. 5.5), yet occasionally contour lines are broken to enable labeling of the elevation value. Contour lines are drawn (or they *run*) roughly

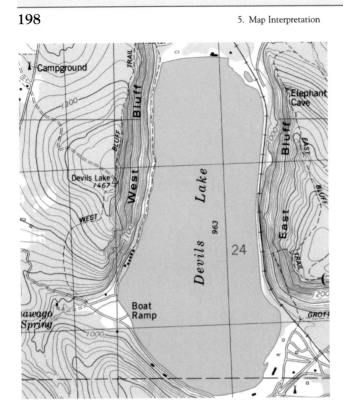

FIGURE 5.3 Devils Lake bench mark (1467 ft.), on top of the West Bluff, and an elevation of the surface of Devils Lake (963 ft. above sea level) [Baraboo, Wisconsin 7.5-minute U.S. Geological Survey quadrangle (NIMA 3070 I NW–Series V861)].

parallel to stream systems (Fig. 5.6); increases in elevation on either side of a stream and uphill within the stream system should produce a V-shaped set of lines pointing uphill (War Department, 1944). The base (or point) of each V indicates elevations are increasing in that direction. Often, a stream of some sort should be located in the middle of the V, if one were to connect the base of the V's. On ridges, contour lines may be viewed as a U-shaped set of lines (Fig. 5.7) that point toward areas of lower elevation (the base of the U pointing downhill), allowing one to understand the concept of *aspect* (north-facing slope and south-facing slope, etc.).

Translation 20
One of your friends is a grade school English teacher who wants to know more about the natural environment as expressed through cartography. She brings you a certain map that contains contour lines; the map indicates that the contour interval employed is 20 ft. Although you may be communicating orally with your friend, as an exercise, explain to your friend in writing what the contour interval, or the contour lines themselves, represent on this map.

Since contour lines should represent a closed polygon, when it represents elevations near the top of a mountain, it should appear as a small, closed polygon. It should be assumed that land areas inside the small polygon formed will be slightly higher in

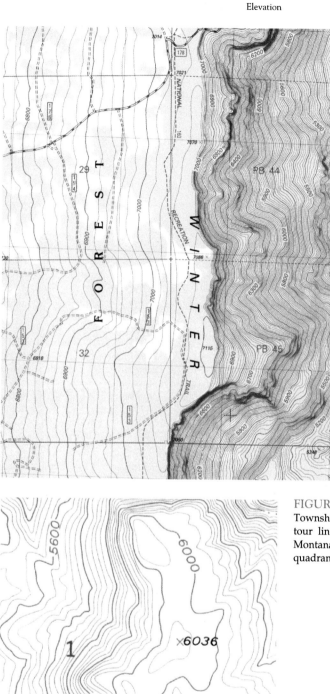

FIGURE 5.4 A transition from the gentle slopes of Winter Ridge (left) to the steep drop-off (right) that leads to Summer Lake in Central Oregon, where contour lines have a 20 ft. interval [Fremont Point, Oregon 7.5-minute U.S. Geological Survey quadrangle (42120-G7-TF-024)].

FIGURE 5.5 The peak of a hill in Section 1, Township 9 South, Range 10 West, where contour lines have a 20 ft. interval in the [Dalys, Montana 7.5-minute U.S. Geological Survey quadrangle (45112-A7-TF-024)].

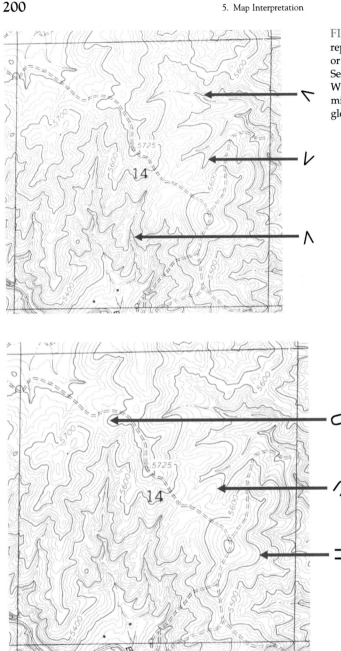

FIGURE 5.6 V-shaped contour lines representing potential stream channels or gullies in a mountainous area of Section 14, Township 7 South, Range 9 West of the Dillon West, Montana 7.5-minute U.S. Geological Survey quadrangle (N4507.5-W11237.5/7.5).

FIGURE 5.7 U-shaped contour lines representing the direction decreasing elevations in Section 14, Township 7 South, Range 9 West of the U.S. Geological Survey Dillon West, Montana 7.5-minute U.S. Geological Survey quadrangle (N4507.5-W11237.5/7.5).

elevation than the land areas under the contour line itself. Thus the peak of the hill or mountain should be found somewhere inside the polygon formed by the last contour line on the hill or mountain top. At times, a map may also contain a symbol that represents a surveyed bench mark indicating the elevation of the peak of the hill or mountain (Fig. 5.5).

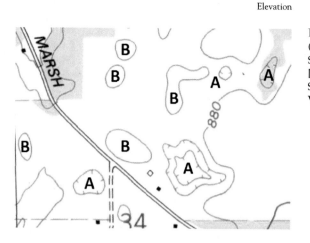

FIGURE 5.8 Depressions (A) and a hill tops (B) of sorts in south central Wisconsin, Section 34, Township 11 North, Range 7 East [Baraboo, Wisconsin 7.5-minute U.S. Geological Survey quadrangle (NIMA 3070 I NW—Series V861)].

In cases where a contour line represents a depression, it should be assumed that the land area inside the polygon formed by the contour line will be lower in elevation than the area under the line itself. Depressions, sinkholes, or dolines can consist of natural or man-made pits or dish-shaped closed areas formed by collapse of the underlying geologic structure, or perhaps by the process of water dissolution (Yang et al., 2015). The use of small marks, attached perpendicularly to a contour line, can indicate the presence of a depression (Fig. 5.8). These marks, or *hachures*, are symbols that complement the representation of relief on topographic maps. Hachures are short parallel lines primarily used to represent the direction of slopes, where gradient is emphasized (Magyari-Sáska, 2017). In contrast to the horizontal representation of consistent (zero) ground slope made possible through the use of contour lines, hachures are drawn parallel (upward or downward) to the gradient of the slope, or at right angles to the contour lines, to indicate the aspect of the slope. When used on a map, they may be more closely spaced in areas of steep slopes, or further apart in areas of gentle slopes (War Department, 1944). Hachure thickness or density can also be adjusted to indicate areas that are relatively steeper than others, or to provide the perspective of a landscape being illuminated by the sun from a specific aspect, producing the effect of shadows.

Finally, on large-scale maps and for areas where there may only be a few changes in discrete elevation classes, some have devised creative ways to illustrate the relative change in elevation. For example, in a manner similar to the hypsometric maps described in Chapter 2, Map types, the area between each set of contour interval lines can be filled with a distinct color to represent changes, or *steps* in elevation. In these cases, the user of a map would count the steps (changes in colors) and multiply the steps by the contour interval to understand in general how the elevation has changed between two places (Wessel and Widener, 2015).

Inspection 25

Examine the U.S. Geological Survey 1:24,000 scale Payment Quadrangle map in Michigan that can be accessed through the book's website (mapping-book.uga.edu). Alternatively, the map can be found near east of Sault Ste. Marie, Michigan, using the TopoView application developed by the U.S. Geological Survey National Geospatial Program (https://ngmdb.usgs.gov/maps/topoview/viewer/). When was the map originally created, and when was it revised? What are the contour intervals? Do they vary by country? Is the area around the north side of Sugar Island represented by more or less steep land than the area around Brassar? What is the highest elevation on this map, on Sugar Island? What is the elevation along Route 17, near the small town of French Bay?

Slope

Slope refers to the change in elevation between two points on a landscape over a given horizontal distance, and it is typically presented quantitatively as an angle, in degrees, or as a percentage change in elevation. If someone said there was 0% slope between two road junctures, for example, they are suggesting that there is no change in elevation between the road junctures, and more than likely the ground between these two points would be relatively flat. *Gradient* is a similar term for expressing changes in the surface of the Earth, yet the change in elevation between two points would be referenced to a distance (Jones and Jones, 2013). For example, a gradient might be expressed as a change in elevation of 10 ft./mile. Engineers and surveyors consider ground slope to be defined as *the rise over the run*. The *rise* represents the vertical change in elevation. The *run* is the horizontal distance over which the change in elevation is observed and measured. The term rise is a little bit of a misnomer, since it suggests one is looking upward (up a hill), when one may actually be looking downward (down a hill). Changes in elevation, therefore, suggest that the land is sloping upward or downward.

Reflection 23
Imagine that you are the land manager of a forested area located somewhere in the Rocky Mountains of the United States or Canada. Why would the slope of the land be important as it pertains to the development of roads? Why would it matter with respect to natural precipitation events? Why would one want to examine the slope of the land when considering the development of recreational facilities?

As we suggested, slope can be represented as the angle (in degrees) that is formed when viewing the land profile as a right triangle (Fig. 5.9), or as the percent change in elevation from one point to another. With respect to the latter,

Slope = (change in elevation between two points / horizontal distance between the points)

When expressed as a percent, slope requires that both the rise and the run are represented in the same units of distance (e.g., feet, meters). In using the equation above, the distance units cancel each other. The resulting slope is a decimal value that requires multiplication by 100 to reasonably call it a *percent slope*.

Percent slope = (change in elevation between two points / horizontal distance between the points) × 100

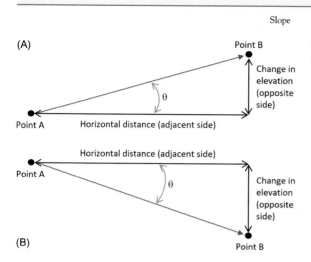

(A)

Point B

Change in elevation (opposite side)

θ

Point A   Horizontal distance (adjacent side)

Horizontal distance (adjacent side)

Point A

θ

Change in elevation (opposite side)

(B)   Point B

FIGURE 5.9   Right triangle model of ground slopes. (A) Uphill slope and (B) downhill slope.

Diversion 26

The elevation of a certain Point A is 560 ft. above mean sea level.
The elevation of a certain Point B is 730 ft. above mean sea level.
The horizontal reference distance between Points A and B is 820 ft.
What is the percent slope of the land between Points A and B?

Inspection 26

Examine the portion of the U.S. Geological Survey 1:24,000 scale map shown in Fig. 5.10. The contour interval is 20 ft. on this map. Featured on this map are humboldt Mountain, where you want to hike to, and a point marked X, where you will park your vehicle. The distance between the two places on the map is about 1.07 in. Along a straight line between these two points, what is the percent slope of the land?

Translation 21

You plan to embark on a lengthy hike with your friends, through a forested area near your current place of residence. You tell them through an e-mail message that the trails are well-maintained, and that the slope will be no more than 10% uphill. In their e-mail response, you find that they are not quite sure what this means. In a brief sentence or two, as if responding to their question, describe the concept of slope.

Slope can also be expressed as an angle and measured in degrees, or angles of incline (or decline). With reference to right triangle theory, the change in elevation between two points might be viewed as the vertical line representing the opposite side from an angle (Θ) in a right triangle model (Fig. 5.9). The horizontal distance between those same two points would be considered the horizontal line representing the adjacent side from angle Θ. Right triangle theory would suggest that angle Θ can be calculated as

Tangent (Θ) = (*opposite side length / adjacent side length*)

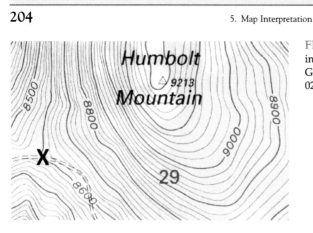

FIGURE 5.10 Inspection of slope for an area in Montana [Argenta, Montana 7.5-minute U.S. Geological Survey quadrangle (45112-C7-TF-024)].

This, of course, provides us with the tangent of angle Θ, and not the angle of the slope. Therefore, to determine the slope in degrees, the relationship becomes

Θ = Inverse tangent (*opposite side length / adjacent side length*)

When this is applied to ground slope calculations, the slope of land can be calculated as

Slope (Θ) = Inverse tangent (*change in elevation between two points / horizontal distance between the same two points*)

Diversion 27
The elevation of a certain Point A is 560 ft. above mean sea level.
The elevation of a certain Point B is 730 ft. above mean sea level.
The horizontal distance between Points A and B is 820 ft.
What is the degree of the slope of land between Points A and B?

In spreadsheet and computer programming applications, the inverse tangent function may be represented using an arc- prefix convention. For the inverse tangent computation, this may then require the use of an arctan or arctangent function.

Direction and Orientation

With respect to maps, *direction* often refers to the path between two places, using one of them as the starting point and the other as the ending point. A direction on a map can perhaps be expressed by an azimuth or a bearing. *Orientation* refers to the relative position of places. Suggesting that the city of Colorado Springs is located south of Denver is an example of the orientation of the two cities. When examining a map, understanding the orientation of landscape features helps map users position the landscape in their mind. In interpreting a map, we may simply want to answer the question, *Which way is north?* However, other inquiries (e.g., *Which direction is it from here to there?*) may require interpretation of direction and orientation. By tradition, north is located at the top of a printed map or digital screen, while south is located at the bottom, but this is not always the case. Therefore an explicit representation of direction or orientation is needed to avoid confusing the map user. As described in Chapter 3, Map components, one standard element of a map should be the north arrow. The north arrow can be as simple as an arrow, with the arrowhead pointing north, yet in Chapter 3, Map Components, we also described a compass rose, which explicitly represents the *cardinal directions* (north, south, east, and west) and at times finer divisions of direction.

Reflection 24

As a map user, is it important to you that a northward direction be located at the top of a map? Why, or why not?

When describing direction and orientation we either use *azimuths* or *bearings*. Azimuths are degrees, where north is represented as 0 degree, east is 90 degrees, south is 180 degrees, and west is 270 degrees (Fig. 5.11). Using this model, you will find that north is both 0 degree and 360 degrees. As an aside, when one changes directions entirely (either physically or conceptually), they may indicate that they have *done a 180*, meaning, for example, that north is the opposite of south, or any two azimuths directionally opposite of one another are 180 degrees different. To use the model in Fig. 5.11, you need to imagine that you are positioned in the center. And, as if looking outward to the rim of the diagram, you are considering travel in a certain direction. If you were to travel on a 120 degrees azimuth, you would be headed in a direction somewhere between east and south, thus traveling in a south-easterly direction.

Bearings divide the azimuthal model into quadrants, and each quadrant contains 90 degrees (no more). To describe a bearing, one needs to refer to the quadrant in two ways: where the degree increments of the quadrant begin (either north or south), and to which direction (east or west) they increase from 0 degrees to a maximum of 90 degrees. In the case of bearings, the place where the degree increments of the quadrant begin (either north or south) are considered to be 0 degrees. For example, one particular bearing may be

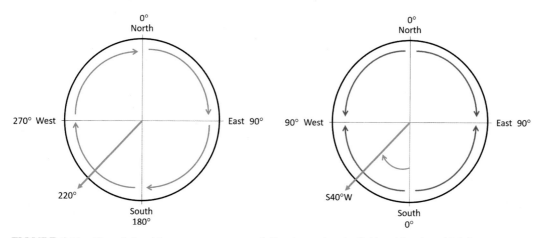

FIGURE 5.11   The relationships among measures of direction: azimuths (left) and bearings (right).

S40°W (Fig. 5.11). This bearing suggests that the direction of interest is located 40 degrees from south, towards the west.

Converting bearings to azimuths, or vice versa, requires some thought. In the case above (S40°W), since we know that the real azimuth that represents south is 180 degrees, the azimuth associated with S40°W is 220 degrees (begin at south (180 degrees) and add 40 degrees to it, since the direction of interest is toward the west of south). Care should be taken in converting bearings to azimuths, as one may need to visualize the conceptual models presented in Fig. 5.11 to confirm whether the bearing degree should be added to, or subtracted from, the place where the degree increments of the quadrant begin (either north or south).

Translation 22

Imagine you are working with an ecologist, examining their research study plots in some wild area of western Wyoming. From a road junction where you will park your vehicle, the directions to one study plot are provided as a bearing (N67°W). After a long day in the field, you find that the batteries in your GPS are drained and the GPS device no longer functions. You have a compass, but you need azimuths to use it. What is the azimuth associated with a bearing of N67°W?

Aspect

Aspect is a landscape feature that describes the direction that downhill slopes are facing. For example, we may say that one side of a ridge has a south or southeastern aspect. This suggests that if you were standing on the top of the ridge and began walking straight downhill, you would generally walk in a south (180 degrees) or southeasterly (135 degrees) direction. Precipitation from rainfall events, if sufficient enough to cause overland flow, would also travel in the direction of the aspect. We describe aspect in general terms since it may be obvious that at a local scale, there is a great amount of heterogeneity in the surface of a landscape. For example, in Fig. 5.12, we might describe the area denoted by the letter A as a north-facing slope, because while acknowledging the various localized quirks in the surface of the land, the slope generally faces north (water would flow downhill toward the stream denoted by the letter B). Aspects are often categorized into four (north, south, east, and west) or eight (north, northeast, east, southeast, south, southwest, west, and northwest) classes, although finer subdivisions (e.g., south-southwest, or about 202.5 degrees) might be used for specific purposes.

From a natural resource management perspective, understanding the aspect of a landscape is important because the presence of natural vegetation may differ based on it, often due to differences in solar input, and corresponding differences in air temperature and soil moisture availability. For example, in the southern Appalachian Mountains, north-facing slopes (north aspects) are generally cooler and wetter than south-facing slopes (south aspects). Therefore in natural forests one might expect to find certain deciduous trees (e.g., oaks) on north aspects and certain coniferous trees (e.g., pines) on south aspects.

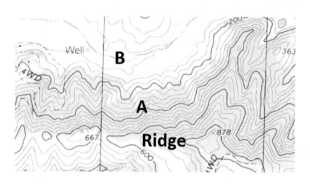

FIGURE 5.12 A north aspect slope (A) and stream (B) in an area south of San Luis Obispo, California [Pismo Beach, California 7.5-minute U. S. Geological Survey quadrangle (NIMA 1854 II NW−Series V895)].

Inspection 27
Examine the Homestake Reservoir, Colorado 7.5 minute quadrangle map provided on this book's website (mapping-book.uga.edu). In general, what is the aspect of the landscape if you were standing on top of Sugarloaf Mountain (approximately 106°25′ west, 39°16′ north) and were contemplating walking downhill on the shortest route to Turquoise Lake.

Relief

Within a given area, the *relief* of the landscape is described by the difference in elevation between the lowest point and the highest point (Jones and Jones, 2013). The indirect or conventional manner in which relief is represented on a map is through the use of contour lines (Pingel and Clarke, 2014). The difference in elevation between a contour line representing a higher elevation and a contour line representing a lower elevation is the relief of the area. Once a map user understands the concept of the contour interval, they should be able to discern areas on the map where there are steep slopes (higher density of contour intervals within a given area) as opposed to areas where there are gentle slopes (fewer contour intervals within a given area).

Inspection 28
Examine the Homestake Reservoir, Colorado 7.5 minute quadrangle map provided on this book's website (mapping-book.uga.edu). In general, given the density of the contour lines, describe and compare the relief around Homestake Peak, the North Fork of West Tennessee Creek, Hagerman Lake, and Virginia Lake. Please express the relative nuances of relief around each place by also referring to an aspect. For example, one might say that on the north, west, and south sides of Saint Kevin Lake, the relief is much more significant (steeper) than on the southeast side of the lake.

Rather than using contour lines to represent changes in elevation or gradations in slope of the land, other methods might be employed to address changes in relief. For example, the use of different colors to represent changes in relief or changes in elevation have been used on maps for at least 300 years (Bradbury, 1990). *Slope shading, shaded relief, or hill shading* effects have been suggested as ways to communicate the relief of land areas that may otherwise be difficult to interpret using contour lines. When developing a map that contains topographic relief, it may be beneficial to begin with the contour lines and shading effects (using slope and aspect), as these may influence where other map features may be located, and how they might be rendered (Kennelly and Kimerling, 2006). Care should be taken in the selection of a contour interval; one that is too small can dramatically increase the amount of information contained within a map, and may cloud or obscure the intended message of the map.

In applying the slope shading concept to a map, steeper slopes are drawn darker in a non photorealistic manner (Kennelly and Kimerling, 2006). Slope shading, in comparison with other methods for representing relief, may produce maps with greater white space, which may be beneficial for accommodating more map information (Pingel and Clarke, 2014). However, in areas where there is significant relief (Fig. 5.13), the dark areas can prevent the presentation of other pertinent information; thus, slope shading may require some creative adjustments to allow maps to have greater information carrying capacity. As another alternative, *layer tinting*, or the use of progressively different colors (or different tones of the same color) to represent perceptual changes in elevation (Fig. 5.14) on a map, might be used to express the relief of a landscape (Pingel and Clarke, 2014). With layer tinting, a graduated scheme of colors is selected to represent changes in elevation or altitude (or depth of water). Often, green is used for lower elevations, due to its association with agricultural and pasture uses of valley systems, while white is used to represent higher elevations. However, in arid climates the scheme should be adjusted to replace green with gray or olive colors, given that lower elevations may not contain actively growing vegetation (Bradbury, 1990). We discuss the use of color in Chapter 6, Map Colors, yet it is useful to have a *color wheel* that illustrates the range of colors that can easily be produced with a printer or digital device. Although the ability to determine elevation differences among places depicted on a map may be improved with the use of layer tinting, the ability to determine the coordinates of specific places on a map can be diminished in some cases (Kempf and Poock, 1969).

FIGURE 5.13 Shaded slope map of an area on the Talladega National Forest in Alabama.

FIGURE 5.14 Color relief map of an area on the Talladega National Forest in Alabama.

Distance

Generally, *distances* on maps are measured as the linear, horizontal path between two places. For example, the linear distance between the Washington Monument and the Lincoln Memorial in Washington, DC is about 4150 ft. (1265 m). However, with modern computer mapping technology, distances can be computed along irregular paths very quickly. The linear distance between the Washington Monument and the Lincoln Memorial in Washington, DC along a sidewalk path is about 4560 ft. (1390 m). This may be a difficult concept for some people to grasp, but the route from one place to another on a map is usually represented as a horizontal distance (Fig. 5.15), as if you were flying well above the landscape and looking down upon it. Thus, when a significant amount of slope exists, the walking distance between two points may be even further apart than the map suggests. Contour lines help us understand that while the horizontal distance between two points may be 1 km, for example, a person walking across the landscape may need to walk a longer distance uphill or downhill (or both) to successfully navigate between the points. Thus, the actual distance traversed on foot or by vehicle *on the slope* is usually longer than the horizontal distance depicted on a map, unless the slope of the land is negligible.

Translation 23
Imagine that you were examining a topographic map with your close friends, and you were trying to decide which of two trails to hike. One of your friends notes that the first trail seems shorter than the second. However, you notice that the relief is much more significant along the first trail (steeper) than the second trail. Explain to your friends how a map can provide you with both horizontal and slope distances.

In order to draw a map, field measurements of ground distances and ground slopes are necessary, and these need to be converted to horizontal equivalents. For example, a measured distance of 100 m across land that has a 15 degrees uphill slope results in a horizontal (mapped) distance of less than 100 m. To illustrate this point, we return to right

FIGURE 5.15 Two routes from the Washington Monument to the Lincoln Memorial in Washington, DC
Source: Image courtesy of (U.S. Department of Agriculture, Natural Resources Conservation Service, 2019).

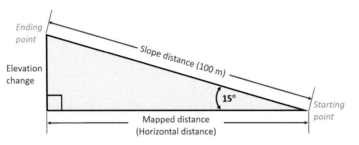

FIGURE 5.16 A specific right triangle model illustrating the difference between a slope (ground) distance and mapped (horizontal) distance when a significant slope exists.

triangle theory. The angle (Θ) of interest to us is 15 degrees in the model provided in Fig. 5.16. The hypotenuse of the model is 100 m, and we can see from the model that we moved some unknown distance uphill from the starting point to the ending point. The change in elevation can be determined using the methods described earlier, yet for this example, it is not necessary to know how much the elevation changed. We are interested, however, in the length of the adjacent side of the right triangle that is conceptually formed by moving 15 degrees uphill for 100 m. The adjacent side of this model is the horizontal distance equivalent of the movement, and the distance that would ultimately be drawn or represented on a map (i.e., the mapped distance). To determine this distance, we use the following equation:

Horizontal distance = (Cosine (θ) × measured slope distance)

In this case, the ground slope (Θ) is represented by a value denoted in degrees. If the incline were alternatively measured as a percent slope, we would need to first convert the percent slope measurement to a slope represented by degrees.

θ = Inverse tangent (*percent slope* / 100)

Returning to our example, the horizontal equivalent of 100 m traveled on a 15 degrees ground slope would be

Horizontal distance = Cosine (15 degrees) × 100 m
Horizontal distance = 96.6 m

Therefore, while we may have walked 100 m, we only traveled 96.6 m horizontally across the map.

Diversion 28
In an area of Colorado, a forest inventory is being conducted. The field personnel traveled a significant distance (500 m) between two sample plots, and they noted that the ground slope between these two plots was about 35%, on average. When drawn on a map, what would be the estimated horizontal ground distance between the two plots?

Area

Often, we find ourselves on a quest to understand the relative or absolute sizes of land areas. We may be interested in knowing how large one property may be compared to another. Perhaps we may want to communicate through the use of a map the size of a wildfire, the size of a wilderness, or the size of a city park. With respect to maps, the *area* of a place refers to the size of the surface of the land. Area, as with distance, is intimately related to the scale of a map. As with distance, area is calculated using horizontal or planar measurements. If one were to know the dimensions of a polygon, one could determine area by hand through rudimentary (scaled triangles, dot grids) or sophisticated (traverse survey) methods. Fortunately, mapping programs automate this function through mathematical calculations associated with the set of vertices that collectively represent the lines or borders of closed polygons.

A simple example of an area computation may be of value for map interpretation purposes. Imagine that an area on a map is formed as a rectangle, where there are roads on all four sides (Fig. 5.17). Corners A−D represent a simple rectangular area, ignoring the rounded corners, and the map area can be calculated as:

Map area = *(map width distance* × *map length distance)*

Approximate map measurements of the area in Fig. 5.17 suggest that the area is 2.0 in. (5.08 cm) long and 0.64 in. (1.63 cm) wide. Using imperial system units, the map area is then

Map area = 0.64 in. × 2.0 in.
Map area = 1.28 in.^{2}

Using metric units, the map area is

Map area = 1.63 centimeters × 5.08 centimeters
Map area = 8.28 centimeters^2

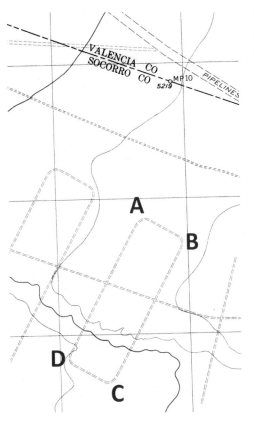

FIGURE 5.17   A rectangular area of land in the northern part of Socorro County, New Mexico [Becker, New Mexico 7.5-minute U.S. Geological Survey quadrangle (N3422.5-W10630/7.5)].

To determine the real-life *ground area*, we need to convert the map area to ground area using the appropriate conversion factor, which is based on the scale of the map.

Ground area = (map area × appropriate conversion factor)

If we were to assume that the scale of the map is 1:24,000, the appropriate conversion factor can be obtained in this manner:

1 in. on the map = 24,000 in. on the ground
1 in.^{2} on the map = 24,000 × 24,000 in. on the ground
1 in.^{2} on the map = 576,000,000 in.^{2} on the ground
1 in.^{2} on the map = 4,000,000 ft.^{2} on the ground
1 in.^{2} on the map = 91.827 acres on the ground

The appropriate conversion factor for a 1:24,000 scale map is thus 1 in.^{2} = 91.827 acres. Therefore if one measured 1.28 in.^{2} on a map, where each square inch equaled 91.827 acres, the ground area of this simple rectangular area is

Ground area = 1.28 in.^{2} × 91.827 acres/in.^{2}
Ground area = 117.54 acres

Diversion 29

Assume you are working in Belize, and the scale of the map you are using is 1:25,000. What is an appropriate conversion factor that allows one to measure an area on this map in metric units, and subsequently convert these measurements to hectares?

To determine the real-life ground area, we could have first converted the map measurements into ground distances, and then applied a formula similar to the one noted above for map area. The scale of the map is needed here as well to convert the map distances to the ground distances. Given this, we convert the map measurements to ground-based equivalents. For the imperial system,

$$2.0 \text{ in. on the map} = 2.0 \times 24,000 \text{ in. on the ground}$$
$$= 48,000 \text{ in. on the ground}$$
$$= 4000 \text{ ft. on the ground}$$

$$0.64 \text{ in. on the map} = 0.64 \times 24,000 \text{ in. on the ground}$$
$$= 15,360 \text{ in. on the ground}$$
$$= 1280 \text{ ft. on the ground}$$

To determine the real-life ground area for this rectangular piece of land, the following formula is used:

Ground area = (ground width × ground length)

Therefore, in our case,

$$\text{Ground area} = 4000 \text{ ft.} \times 1280 \text{ ft.}$$
$$= 5,120,000 \text{ ft.}^2$$
$$= 117.5 \text{ acres}$$

The conversion from square feet to acres in this example required an understanding that there are 43,560 ft.^{2} in 1 acre. The result is an approximate size of the land, subject to minor error in measuring the dimensions on a map using a ruler or engineer's scale. However, it provides perspective on the size of land features within the map.

Message

Technically, a *message* is a form of communication delivered from one person to another, through oral, written, or graphic means. Along these lines, the message(s) of a map can be transmitted through both written and graphic languages (Audije-Gil et al., 2017). With proper design, the messages that a map conveys can be clearly communicated to its intended customer through the use and presentation of map components, such as the map's title, legend, symbols, and geographic relations. With respect to the development of a map, this refers to the *symbology* of the map, or the specifications used for the placement and characteristics (line type, font type, color, etc.) of the words, symbols, and mapped features (Buckley and Frye, 2006). These graphic techniques can transcend verbal language barriers (Shirreffs, 1992). A map may, therefore, convey one or more of some very general map messages such as these:

- This is the place of interest
- This is mine (yours)
- This is where it happened
- This is the place we own or manage or control
- This is where we operate
- This is what we know about this land or water body
- This is the state of things
- This is a safe / not-so safe place
- These are the characteristics of the people found here
- This is the quality of the environment
- This is where there are hazards
- This is where problems have occurred
- This is the route
- This is our competitive advantage
- This is where we are promoting recreational activity
- This is where we plan to work
- This is the current / past weather
- These are our business locations

The interpretation of the message may require additional insight or knowledge of the landscape (or water body) on behalf of the map user, but the conveyance of the message can be facilitated through the use of various map components. Some digital maps delivered through personal computers, cell phones or other devices can rely mainly on colors and legends (representing a theme), as map titles may be absent. More difficult are cases where a map was created in a language other than the one(s) within the grasp of the map user. Here, the use of annotation on the part of the map developer may be of limited value. There may be occasions where the map developer used feature or color exaggeration to attract (or to distract) attention to (or from) areas of the map that relate (or not) to the message of the map. This use of artistic license can be of value in situations where precise representation of locations, areas, or distances is necessary for the normal course of business or for health and safety of humans and animals.

Inspection 29

A map of a wildfire can be found on the book's website (mapping-book.uga.edu). In examining this map, you might notice that the fire (red polygon) is located in the center of the map, between two sections (11 and 12) of a certain township in Idaho. What is the scale of the map? About how large was the Buck Fire on August 14, 2017? What message do you think the developer of the map intended to convey to the map user audience given the location of the fire, the area that is closed to public use, and roads and trails that are also closed?

As we noted earlier, for better or worse, some detail of the land or system to be mapped may be too small to be drawn true to the scale selected. If this detail needs to be included on a map, it may likely be exaggerated (Debenham, 1956). Others have commented on the ability of map developers to express falsehood with maps, through distortion and displacement of landscape features, and with regard to these topics, we refer readers to those publications (e.g., Monmonier, 1996).

Concluding Remarks

A map is a device for communicating a message to an audience; its development is a synthesis of knowledge and skill, of art and science (Shirreffs, 1992). In contrast to a table of data, to written prose, or to other graphics and charts, a map can give people a sense of distance, direction, area, relief, and composition of the landscape. Common tools for interpreting a map include the scale and indications of elevation and slope of land, which are related to aspect and relief. Direction and orientation can be understood through the use of a north-south graphic and an understanding of the relationship between north, south, east, and west. Actual distances and areas of land or water bodies are intimately associated with the scale of a map. The scale one uses in developing a map should be determined after considering the intended use of the map. Tourism maps and broad-area (county, Province, and state) road maps often are developed using a small scale to allow more information to be presented on a piece of paper or digital screen.

In discussing map interpretation concepts, we concentrated on maps made with modern mapping technology and techniques, and assumed that issues of the past (precomputer) are not necessarily inherent in the maps that can be produced today. However, there may be instances when historical maps need to be consulted to inform the human condition or to support management decisions. While we focus on what should be the standard methods for interpreting modern maps, readers should remember that map developers throughout history did not have the technology we enjoy today. Therefore in certain situations an appropriate level of concern should be placed on map development issues associated with historical maps, such as variations in scale that could occur across these maps as a result of distorted grids, unknown projections, potentially compressed higher elevations, potentially dilated depressions, and a general lack of accuracy (Jenny, 2006; Mastronunzio and Dai Prà, 2016).

References

Audije-Gil, J., Cambra-Moo, O., García Gil, O., González Martín, A., 2017. Cartografiando la anatomía. In: García Quintana, Á., Callapez Tonicher, P., Barroso-Barcenilla, F., (Eds.), Los mapas de la naturaleza, Memorias de la Real Sociedad Española de Historia Natural, Segunda época, Tomo XIV. Facultades de Biología y Geología, Universidad Complutense de Madrid, Madrid. pp. 343–362.

Bradbury, J.L., 1990. Specification of Cartographic Layer Tinting Schemes for Color Graphic Monitors. (Master of Science thesis). Department of Geosciences, Corvallis, Oregon State University, OR.

Buckley, A., Frye, C., 2006. A multi-scale, multipurpose GIS data model to add named features of the natural landscape to maps. Cartogr. Perspect. 55, 34–53.

Debenham, F., 1956. Map Making. Blackie & Son, Ltd, London.

Feeney, A.E., 2017. Beer-trail maps and the growth of experiential tourism. Cartogr. Perspect. 87, 9–28.

Greenwood, D., 1964. Mapping. The University of Chicago Press, Chicago.

Jenny, B., 2006. MapAnalyst – A digital tool for the analysis of the planimetric accuracy of historical maps. e-Perimetron 1 (3), 239–245.

Jones, C.E., Jones, N.W., 2013. Laboratory Manual for Physical Geology, Eighth ed. McGraw-Hill, Hoboken, NJ.

Kempf, R.P., Poock, G.K., 1969. Some effects of layer tinting of maps. Percept. Mot. Skills 29 (1), 279–281.

Kennelly, P.J., Kimerling, A.J., 2006. Non-photorealistic rendering and terrain representation. Cartogr. Perspect. 54, 35–54.

Kent, A.J., 2012. From a dry statement of facts to a thing of beauty: understanding aesthetics in the mapping and counter-mapping of place. Cartogr. Perspect. 73, 39–60.

Levin, N., Kark, R., Galilee, E., 2010. Maps and the settlement of southern Palestine, 1799–1948: an historical/GIS analysis. J. Hist. Geogr. 36 (1), 1–18.

Magyari-Sáska, Z., 2017. Automatic generation of hachure lines. Geographia Technica 12 (1), 75–81.

Mastronunzio, M., Dai Prà, E., 2016. Editing historical maps: comparative cartography using maps as tools. e-Perimetron 11 (4), 183–195.

Monmonier, M., 1996. How to Lie with Maps, second ed. The University of Chicago Press, Chicago, IL.

Paine, D.P., Kiser, J.D., 2012. Aerial Photography and Image Interpretation, third ed. John Wiley & Sons, Inc, Hoboken, NJ.

Pingel, T., Clarke, K., 2014. Perceptually shaded slope maps for the visualization of digital surface models. Cartographica 49 (4), 225–240.

Shirreffs, W.S., 1992. Maps as communicative graphics. Cartogr. J. 29 (1), 35–41.

U.S. Department of Agriculture, Natural Resources Conservation Service, 2019. Geospatial Data Gateway. U.S. Department of Agriculture, Natural Resources Conservation Service, Washington, DC <https://datagateway. nrcs.usda.gov/> (accessed 22.03.19.).

War Department, 1944. Advanced map and aerial photograph reading. War Department Field Manual FM 21-26. U.S. Government Printing Office, Washington, DC.

Wessel, N., Widener, M., 2015. Rethinking the urban bike map for the 21st Century. Cartogr. Perspect. 81, 6–22.

Yang, N., Wan, L., Zheng, G.-Z., Yang, J., 2015. Using hachures to construct a 3D doline model automatically. Cartographica 50 (2), 86–93.

The selection and use of colors on maps is perhaps the most challenging aspect of map development (Green and Horbach, 1998). The selection of colors to use in maps depends on both the technological aspects of the map production process and the reaction a map developer hopes to evoke among map users. The selection of colors is further complicated by the need to adhere to established conventions in cartography (Green and Horbach, 1998). Human perception of color is a visual sensation perceived when one is exposed to a certain range of electromagnetic radiation; our perception of color is a psychophysical phenomenon (Ohno, 2000). Color theory is in fact a complex topic, involving the fields of physics, psychology, and physiology (Smith, 1978). Color exists not only as a characteristic of the reflective or absorptive surface of an object, but also as a source of sensation that influences our experiences (Porteous, 1996; Whitfield and Whelton, 2015).

Formal conventions for the use of colors in maps have a long history. For example, Swedish regulations in the early 1600s instructed cartographers to follow specific protocols involving the use of color (green for forested areas, light blue for lakes, dark blue for streams, red for ownership boundaries, etc.), although later regulations allowed cartographers more freedom for allocating colors to various map features (Rystedt, 2016). Some map users may prefer viewing colored representations of landscape features over their panchromatic counterparts, and, therefore, the use of color may facilitate the achievement of greater accuracy in map interpretation (Brewer et al., 1997). Ultimately, map legibility, balance, and acceptance are influenced by the selection of colors (Green and Horbach, 1998).

Inspection 30
Examine the two maps provided in Fig. 6.1. Which of the two do you prefer? Why?

Historically, color was often employed to organize the display of information on maps, such as the geological foundations of Earth (Ksiazkiewicz, 2016) or the distribution of vegetation types within a specific country. In an interesting modern application, differences in color were used to help illustrate the relative location of people within an office building, based on the vertical location (floor) of their office (Nossum et al., 2013). Map users may need to be able to discriminate between numerous different colors in order to understand spatial pattern, landscape classifications, or other important information communicated by the map developer. Under normal circumstances, our eyes have the ability to effectively distinguish

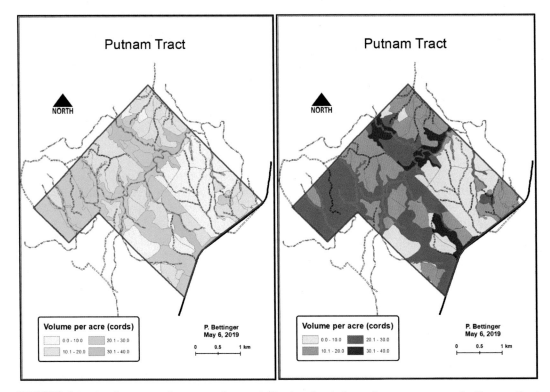

FIGURE 6.1   A color map and a gray scale map indicating the timber volume per acre on the Putnam Tract, central Georgia, United States.

differences in distance, light intensity, and color (Porteous, 1996). However, one issue we should remember during the development of a map is that certain individuals or groups of people may have trouble perceiving subtle changes in color (Brychtová and Çöltekin, 2017).

The colors that we perceive are within the range of possible types of energy bounded by, but not inclusive of, the ultraviolet and infrared regions of the *electromagnetic spectrum* (Fig. 6.2). Energy, or *electromagnetic radiation*, how energy is classified (Ball, 2007), can be considered either as a stream of photon particles or as a series of waves (Pollard et al., 2007). Using the latter conceptual approach, energy travels in a continuous sequence of peaks and valleys. The electromagnetic spectrum conventionally classifies energy based on wavelengths, or the physical distance between subsequent peaks or valleys. Energy can

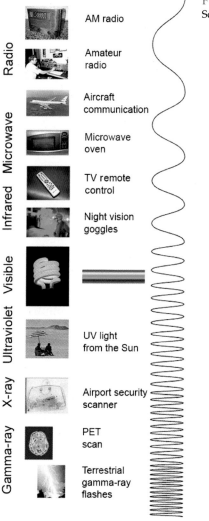

FIGURE 6.2 A conceptual model of the electromagnetic spectrum. Source: *National Aeronautics and Space Administration (2013).*

differ in the manner in which it interacts with matter (Pollard et al., 2007). Assuming the speed energy travels is constant, shorter wavelengths of energy pass a specific point during a fixed amount of time with greater frequency. Longer wavelengths of energy have a corresponding lower frequency. Thus, at one end of the electromagnetic spectrum are types of energy that have short wavelengths. These forms of energy are harmful to life on Earth, as they can cause changes to cellular structure when people, plants, and animals are exposed to them. These include gamma ray, X-ray, and ultraviolet short-wavelength energy. At the other end of the spectrum are types of energy that have long wavelengths. These include radio, cell phone, global positioning system (GPS), thermal, and other types of long-wavelength energy which are emitted all around us nearly all of the time in modern society. Visible light, facilitating our perception of color, lies between these two broad groups. Even though we often consider the Sun as the primary source, energy can be a product of both natural and anthropogenic sources. Energy should be viewed as a common resource that facilitates a wide number of modern conveniences (Herter, 1985) from every-day necessities (e.g., television and phone service) to medical practices (e.g., X-rays).

When we categorize energy into ranges of the electromagnetic spectrum, we create *bands* of electromagnetic energy. Within the range of energy that the normal human visual system can discern, the wavelength of light determines the perceived hue of color, and the frequency determines the intensity of the color (Joblove and Greenberg, 1978). The range of visible energy we perceive to be *light* has a wavelength between 400 and 700 nm, or 0.4 and 0.7 μm. By comparison, an average strand of human hair is about 50 μm in diameter. One nm is one-billionth of a meter in length, whereas one μm is one-millionth of a meter. Within the visible portion of the electromagnetic spectrum, we typically consider three bands of energy, expressed as the three primary colors of light:

Blue	400–500 nm	0.4–0.5 μm
Green	500–600 nm	0.5–0.6 μm
Red	600–700 nm	0.6–0.7 μm

The mixing of various intensities of the primary colors (blue, green, red) creates subtractive colors (yellow, magenta, cyan), and every imaginable combination of color from white to black. In contrast to the mixing of paint pigments, the mixing of high levels of red and green light (and very low or absent levels of blue light) creates the yellow color light (not brown).

	Intensity		
Color perceived	Blue	Green	Red
Yellow	–	High	High
Cyan	High	High	–
Magenta	High	–	High
White	High	High	High
Black	–	–	–

Reflection 25
Consider the last rather comprehensive map you examined, perhaps a U.S. Geological Survey topographic map. In examining the map, you might find that landscape features—points, lines, or polygons—are presented using a shade of blue. What types of landscape features might those have been? What types of features might you have found on the map that were presented using various shades of red? What types of features might you have found on the map that were presented using shades of green?

As is suggested in this discussion, light and color can be described as waves of electromagnetic energy; colors of visible light can be described as energy with wavelengths between 0.4 and 0.7 μm in length. Humans have evolved to capture visible light through photoreceptor cells within our eyes that respond to light through nerve impulses and subsequently transform these responses to images within our brains. In comparison, infrared energy is reflected or emitted naturally, or emitted from a device we have created, in wavelengths that are beyond (longer) the color red in the electromagnetic spectrum. Humans are not designed to see this energy. Since infrared energy is not necessarily a color we sense with our eyes, we have developed optical sensors to detect its presence and intensity. To put this further into perspective, GPS signals are emitted from a satellite using a wavelength of about 20 cm (nearly 1 foot), and amplitude modulation (AM) radio signals are emitted from radio towers using wavelengths that can be thousands of meters in length. While we cannot see this energy, human society has developed devices (radios, GPS receivers, etc.) that can emit, detect, and decode it.

Once landscape features are adequately displayed on a map, people can be further challenged in interpreting differences in color simply based on the environment in which they are using the map. For example, color perception can be affected by the intensity and composition of the illumination in the nearby environment of a map. In dark rooms, or at night, some colors on maps may be indistinguishable by humans, or lost within the range of color contrasts, as illumination conditions between the environment in which a map was developed and the environment in which the map is used have changed (Irigoyen and Herraez, 2003). Light reaching a specific map or a specific point on a map can be a result of a mixture of irradiation (direct and indirect) interacting with the map at a variety of angles, and modified by local occlusion and reflection (Foster, 2011). Therefore, viewing conditions will vary by time of day and position of the map with respect to various light sources. Previous knowledge of the landscape portrayed within a map and cognitive biases on the part of the map reader (perhaps due to extensive experience reading maps) may overcome some of these issues. For these types of reasons, it has been suggested that lighter colors might be used for filling polygons representing large places, and that darker colors should be used for smaller places (European Environment Agency, 2016).

Color Models

Colors and color combinations can help convey messages on maps and help facilitate the interpretation of information by users of those maps (Huang et al., 2007). For example, one might use a diverse color system to illustrate by political unit (e.g., country) differences in human condition (Fig. 6.3) or the status of interest in various sporting events (Fig. 6.4).

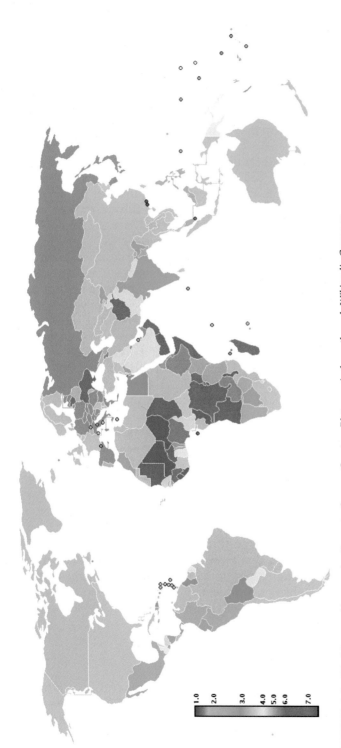

FIGURE 6.3   World map of human fertility rate. Source: *Courtesy PlatypeanArchcow through Wikimedia Commons.*

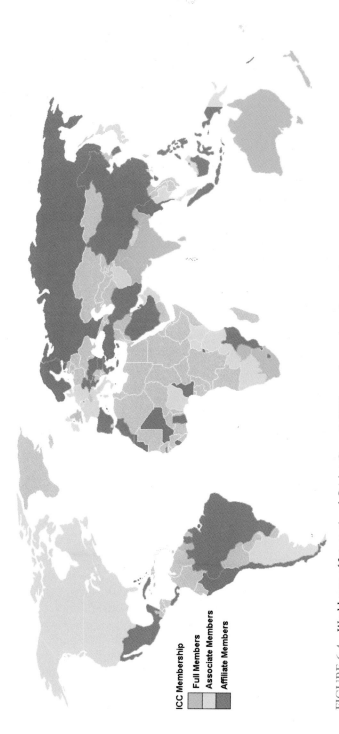

FIGURE 6.4 World map of International Cricket Council (ICC) member nations. Source: *Courtesy Fry1989 through Wikimedia Commons.*

FIGURE 6.5   Hypsometric tinting of Ireland. Source: *National Aeronautics and Space Administration (2005).*

As we suggested in earlier chapters, one might also employ *hypsometric tinting*, or elevation coloring, to convey differences in land elevations or sea depths (Fig. 6.5). *Cross-blended hypsometric tinting* has been suggested as a way to express lowland areas with colors that vary with elevation and environment, and thus may be more fitting to the natural environment of the region depicted in a map. For example, using a cross-blended hypsometric tinting approach, arid lowlands would be colored with a brown tint while polar lowlands of the same elevation would use a gray blue tint and humid lowlands might employ a green tint (Patterson and Jenny, 2013). It is interesting that different cultures around the world tend to encode their vocabularies with 11 or fewer basic categories of color: white, black, red, green, yellow, blue, brown, purple, pink, orange, and gray (Berlin and Kay, 1990), yet the color continuum is vast. Unless otherwise limited genetically, humans can discern around 2 million color hues; however, humans can probably only perceive less than 144 distinct colors (Zhang et al., 2017). Map developers are, therefore, often cautioned to carefully choose a color palette that includes a spectrum of hues discernable to most people (Levine, 2009). In this regard, several color models, color spaces, or color coordinate systems have been developed as a way to portray or describe color, and to organize the representation of color combinations (Joblove and Greenberg, 1978). *Colorimetry* encompasses the science and technology to quantify human perception of color (Ohno, 2000), and through these investigations, various models of the gamut of colors have been proposed. Interestingly, recent advances in colorimetry can in part be traced to the standards that were developed for the display of images on color television screens (Smith, 1978).

RGB Model

The most common color model is perhaps the *RGB model*, which portrays a particular color as containing certain levels of red, green, and blue light. The RGB color model has been called the *colorcube model*, a three-dimensional space with the dimensions of the edges represented by the primary colors (Smith, 1978). This color model describes how adding certain amounts of these primary colors of light will produce a specific shade of color on a map or the screen of a computer or other digital device, thus it is an additive model. The RGB model describes the inclusion of the three primary colors using discrete values that range from 0 to 255 (256 levels); this scale relates to an 8-bit depth of a computing device. Red, green, and blue each have a value in this model that ranges from 0 to 255 (Fig. 6.6). The lowest value (0) indicates no addition of a primary color to the development of a specific color. For example, when the red contribution to the RGB description of a particular

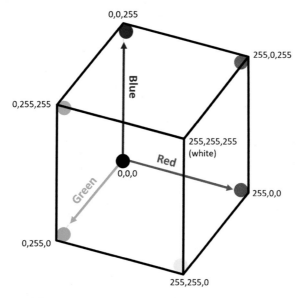

FIGURE 6.6 RGB color model.

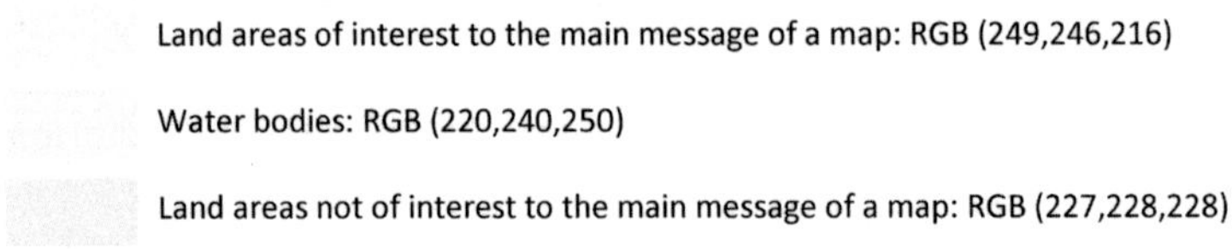

Land areas of interest to the main message of a map: RGB (249,246,216)

Water bodies: RGB (220,240,250)

Land areas not of interest to the main message of a map: RGB (227,228,228)

FIGURE 6.7   Suggested colors for basic land areas. Source: *European Environment Agency (2016)*.

color is 0, no red light is used in the development of the color. When the red contribution to a specific color is maximized, its value in the RGB model is 255. When red is at its highest level (255), and green and blue are at their lowest level (0), we create the pure color red. However, shades of red are not necessarily created when the red contribution to the RGB model is at its maximum, as red light may also be mixed with green and blue light to produce some other color. To connect these thoughts to our previous discussion on the creation of color, when all three primary colors of light are added together at their highest levels (255), the color white is produced. When red and blue are at their highest level (255), and green is at its lowest level, we describe magenta using the RGB model. When all three primary colors are at, or near their lowest level, we create black. The number of different color combinations that can be described using the RGB model is 256^3 (256 levels of red $\times$ 256 levels of green $\times$ 256 levels of blue), or 16,777,216.

Diversion 30
Using an Internet-based RGB conversion tool, what color is created with RGB values 200, 100, and 50? If you wanted to use a light blue color to represent water body areas on a map, what RGB values might you select for this color? If you wanted to use a dark, forest green color to represent wooded areas on a map, what RGB values might you select?

Some have proposed the use of the RGB model for standard communication of conditions of the natural environment, such as the mineral composition of water bodies (Nikolaev et al., 2018). Standards for the broader use of colors in maps have also been developed by various national and international agencies and organizations throughout the world. For example, the European Environment Agency (2016) suggests certain colors (Fig. 6.7) should be used to represent basic areas of land areas of interest to the main message of a map, water, and land areas outside of the interest to the main message of a map.

CMYK Model

The *CMYK model* portrays a color as containing levels of cyan, magenta, yellow, and black (key). These colors mask the reflectance of certain wavelengths of light and act as filters to produce the appropriate color. The CMYK model involves the subtractive colors of red (cyan), green (magenta), and blue (yellow) light; thus, in contrast to the additive RGB model, the CMYK model is considered a subtractive model. Black is included because while high concentrations of cyan, magenta, yellow theoretically can create the color black, they generally produce poor results (poor tone, highly saturated paper). With respect to printed maps, the CMYK model, therefore, reflects the amount of the subtractive colors (cyan, magenta, and yellow) and black mixed together to create a specific color shade. This is the method used in many modern plotters, the large printing devices made by Hewlett Packard and others for creating large format maps and posters.

For any specific color, the RGB value set from an additive process can be converted to the CMYK value set from a subtractive process. To convert a set of RGB values to CMYK values, we first divide the RGB values by their theoretical maximum value (255) to place them on a scale between 0 and 1.

red' = (original red value/255)
green' = (original green value/255)
blue' = (original blue value/255)

Then the CMYK values are calculated in the following manner, with black (K) being determined first, since it is necessary in determining the other three aspects of the CMYK color set.

a = the greatest value of the set of red', green', or blue' values
K = 1 − a
C = (1 − red' − K) / (1 − K)
M = (1 − green' − K) / (1 − K)
Y = (1 − blue' − K) / (1 − K)

As an example, assume that the color turquoise has the RGB value set (64, 224, 208). By dividing each of these values by the theoretical maximum value of 255, we can scale the red, green, and blue colors, and place them within the range of 0 and 1.

red' = (64/255) = 0.25098
green' = (224/255) = 0.87843
blue' = (208/255) = 0.81569

In viewing these scaled values, we can see that the greatest of these is 0.87843, therefore *a* = 0.87843. It then follows that the CMYK values, in decimal percentage form, are:

K = (1 − 0.87843) = 0.12157
C = (1 − 0.25098 − 0.12157) / (1 − 0.12157) = 0.71429
M = (1 − 0.87843 − 0.12157) / (1 − 0.12157) = 0
Y = (1 − 0.81569 − 0.12157) / (1 − 0.12157) = 0.07142

These values for C, M, Y, and K are then presented in the CMYK model as whole numbers. For example, while K = 0.12157 in decimal percentage form, it would be presented as 12% (0.12157 × 100). C would then be presented as 71%, M would be presented as 0%, and Y would be presented as 7%. Thus, while the color turquoise can be portrayed with an RGB value set of (64, 224, 208), it has a CMYK value set of (71, 0, 7, 12).

Diversion 31
The color of a lime (the fruit) might be described using the RGB value set (50, 205, 50). What is the CMYK equivalent of this color?

With respect to the production of maps, one organization (European Environment Agency, 2016) suggests using the CMYK model for printed maps, and the RGB model for the display of maps on computer screens and screens of other digital devices. Topographic maps developed by Geoscience Australia (2012) and competition quality orienteering maps (International Orienteering Federation, 2018) also utilize the CMYK color model.

HEX System

The HEX system is often used to represent colors in the HTML coding effort of Internet sites. In order to convey a desired RGB color in HEX form, two steps are necessary: (1) the RGB values need to be converted from decimal equivalents to an 8-bit value and (2) the values for red, green, and blue are then reduced to two characters each, and concatenated to form a HEX code.

In converting RGB values to an 8-bit value, the following fairly laborious process might be used:

1. Using one of the RGB values, determine (yes or no) whether, in succession, it (or its remainder after subtraction) is greater than or equal to the following series of numbers representing, in reverse order, the power of 2:

128, 64, 32, 16, 8, 4, 2, 1

As you can see, there are eight values in this list, and the answer (yes or no) is presented as a 1 (yes) or 0 (no). When the answer is yes (1), the number representing the power of 2 is subtracted from the current RGB value, and the remainder is used from that point forward. The following example illustrates how to convert a color with an RGB set of (50,205,50) to a HEX value. This color is, in fact, lime green.

RGB red = 50

Power of 2 value	Current RGB value	Answer[a]	Remainder
128	50	0	—
64	50	0	—
32	50	1	18
16	18	1	2
8	2	0	—
4	2	0	—
2	2	1	0
1	0	0	0

[a]1=current RGB value is greater than or equal to the associated power of 2 value, 0 otherwise.

As you may find in the above table, the first two rows (indicated by a *Power of 2 value* of 128 and 64) result in an *Answer** = 0 for each row. This implies that the *Current RGB* *value* associated with those rows (50) is lower than both 128 and 64. Once the *Current RBG value* is greater than the *Power of 2 value* on a row (e.g., row 3 of the table), the *Answer** = 1 for that row. The *Remainder* represents the outcome of (current RGB value − *Power of 2 value*) on rows where the *Answer** = 1. The 8-bit value for a situation where red = 50 is then 00110010. Here, each of the *Answers* above are transposed in order, and are then concatenated to form an 8-character (8-bit) string.

RGB green = 205

Power of 2 value	Current RGB value	Answer[a]	Remainder
128	205	1	77
64	77	1	13
32	13	0	–
16	13	0	–
8	13	1	5
4	5	1	1
2	1	0	0
1	1	1	1

[a]1=current RGB value is greater than the associated power of 2 value, 0 otherwise.

The 8-bit value for green = 205 is then 11001101. Since red and blue have the same RGB value, the 8-bit value for blue is the same as for red, or 00110010.

2. Once the 8-bit values for each piece of the RGB color are determined, they are split into two pieces (divided in half). The first half of the 8-bit red color from above (00110010) is 0011, the second half is 0010. Each half contains four elements. For the first half of the red color, the elements are 0, 0, 1, and 1, in that order. Each of the four elements is then multiplied by a specific constant (8 for the first element, 4 for the second, 2 for the third, and 1 for the last) to arrive at a value. If the resulting number for each half of the 8-bit color is less than 10, the actual number is used in the representation of color in the HEX system. If the resulting value is a number 10 or greater, it is then translated to a letter (A to F), where A = 10, B = 11, C = 12, and so on. The outcomes for each half are then concatenated to form a character set. For example,

RGB red = 50

8-bit red = 00110010
First half = 0011
Second half = 0010
Value for first half = $(8 \times 0) + (4 \times 0) + (2 \times 1) + (1 \times 1) = 3$
Value for second half = $(8 \times 0) + (4 \times 0) + (2 \times 1) + (1 \times 0) = 2$
Concatenate the two values: 32
Since neither the value for the first half nor the value for the second half are greater than or equal to 10, the actual numeric values (3 and 2) are used here to represent red.

RGB green = 205

8-bit green = 11001101
First half = 1100
Second half = 1101
Value for first half = (8 × 1) + (4 × 1) + (2 × 0) + (1 × 0) = 12, or C
Value for second half = (8 × 1) + (4 × 1) + (2 × 0) + (1 × 1) = 13, or D
Concatenate the two values: CD
Since both of the values for the first half and the second half were greater than or equal to 10, the actual values (12 and 13) are translated to letters (C = 12, D = 13).

RGB blue = 50

Since this value of blue has the same 8-bit value as red, it also has the same translated value (32).

3. The three 2-character pieces that represent red, green, and blue are then concatenated to create the HEX color code. However, this code is preceded by a hash mark (#).
Red = 32
Green = CD
Blue = 32
HEX code for lime green = #32CD32

> **Diversion 32**
> Admittedly, this is a challenge, but please give it your best attempt. The color of a banana (the fruit) might be described using the RGB value set (240, 255, 0). What is the HEX system equivalent of this color? Show your work.

As you might have noticed, higher values (those approaching 255) of red, green, or blue receive letters under this system. The highest letter that can be received for each half of a string related to red, green, or blue is F. Therefore, a HEX code of #FF0000 suggests that there are high amounts of red, and no (zero) amounts of blue or green, and therefore represents the brightest color of red. Similarly, #00FF00 represents the brightest green color, and #0000FF represents the brightest blue color. In addition, #FFFFFF (highest amounts of all three primary colors) represents the color white, and #000000 represents the color black.

HSV and HSL Models

For the two-color models presented in this section, a *hue* is a measure representing the pure color it resembles, as a position (often a degree) along a color wheel. *Lightness* is a measure referring to the reflective quality of a color, or the luminosity of the color. And *saturation* is a measure relating to the strength or intensity of the color in its purest form. It has a minimum and maximum value, and relates to the degree to which the color differs from neutral gray. At any constant lightness level, the saturation of a color can range from 100% (pure form of the color) to 0% (neutral gray). Finally, for one of these color models, a *value* relates to quantity of light reflected, or the darkness of a color, where when value = 0 a color is black.

The HSV (hue, saturation, and value) color model (or HSB color model, where B represents brightness) is a color coordinate system represented sometimes mapped to a cylinder or an inverted cone (Fig. 6.8). When represented as a cone, the hue value ranges from 0 to 360 degrees, and rotates around the axis of the cone (Stokking et al., 2001). As with the color wheel, red is positioned at 0 degree (or 360 degrees), green at 120 degrees, and blue at 240 degrees, and therefore the hue represents the angle around the color wheel (Fig. 6.9). An equidistant spacing of the three primary colors is inherent in the circular color space and other RBG-related color coordinate systems (Joblove and

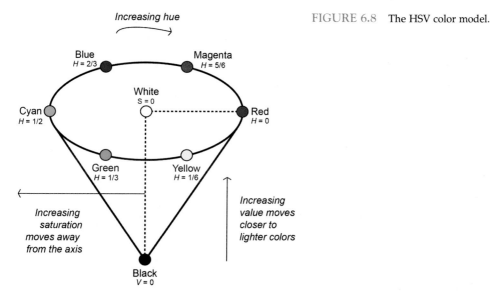

FIGURE 6.8   The HSV color model.

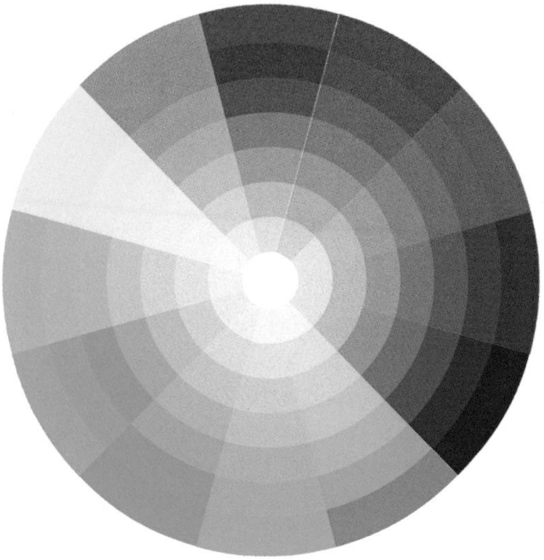

FIGURE 6.9   A color wheel, showing 12 main colors and the complements (on the opposite side), and where saturation of colors is greatest along the outside edge. Source: *Photo courtesy Good Free Photos (https://www.good-freephotos.com/).*

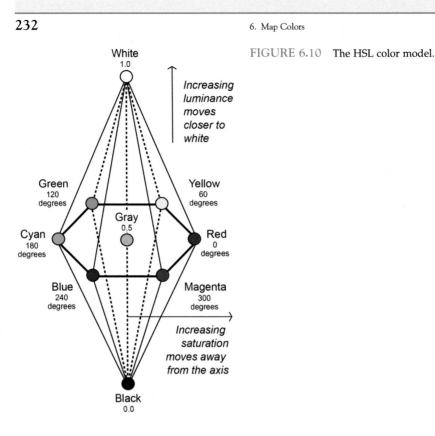

FIGURE 6.10   The HSL color model.

Greenberg, 1978). Along the outside edge of the cone is thus where the highest amount of the dominant hue will be found. In the HSV color model, *saturation* and *value* range from 0 to 1 (0% to 100%). Here, saturation measures the departure from an achromatic (gray or white) hue, or the distance from the axis of the conical model to the edge of the model. Any color positioned on the central axis of the model (along the gray scale from white to black) has a saturation = 0. In this model, *value* (an analog for lightness) measures the departure from the color black (Smith, 1978), where black (value = 0) is found at the tip of the cone, and white (value = 1) is found in the center of the base of the cone.

The HSL (hue, saturation, and lightness) color model can also be represented by a common cylindrical coordinate representation of color (Hou et al., 2017) or some other similar shape (Fig. 6.10). In the HSL color model, the hue also ranges from 0 to 360 degrees, and the saturation and lightness range from 0 to 1 (0% to 100%). The saturation is represented by the distance from the central axis of the model to the edge, and the lightness is represented by the distance from the bottom of the model (black) to the top (white). Dark colors are located near the bottom of the model, while lighter colors are located near the top. Within the HSL model, the lightness characteristic begins with the color black, then transitions through the color of interest to the color white. The lightness value in this model is meant to measure the energy in a color rather than its departure from the color black (Smith, 1978).

Diversion 33

Search the Internet for the w3schools.com colors tutorial, and select the *Color HSL* website. Using the HSL adjustments one can make below the large rectangle that represents the example color, what is the resulting color for any hue when the saturation value is 0? What is the resulting color for any hue when the lightness value is 0? What is the resulting color for any hue when the lightness value is 1?

Both of these models are of value in describing color for computer visualization systems. One of the main distinctions between the HSV and HSL color models is the manner in which pure hues are represented. In the HSV color model, there is a plane where all pure hues of colors are represented, yet this is absent from the HSL color model. Therefore the pure hues of red, white, and yellow (for example) may have the same *value* (zero blackness) and can be found on the same plane in an HSV color model, yet they have different lightness characteristics—red has one-third the lightness of white and one-half the lightness of yellow—and thus are not found on the same plane in an HSL color model (Smith, 1978).

The HSV and HSL color models have been used widely in society. Among many other uses, they have been used to objectively assist in quantification of the color of honey (Dominguez and Centurión, 2015), to map of changes in mineral composition of water bodies (Nikolaev et al., 2018), to detect the color of yarn (Kang et al., 1999), and to help understand the anatomical structure of the human brain (Stokking et al., 2001). When the aesthetic appeal of a map is being considered, adjustments to the brightness and saturation of hues can act to harmonize a message (Zhang et al., 2017).

It is possible to transform an RGB color set to HSV or HSL equivalents, yet the processes are more complex than transforming RGB values to HEX equivalents. If one is truly interested in these comprehensive processes, we encourage beginning the adventure with a visit to the paper *Color gamut transform pairs* by Smith (1978), and then perhaps examining the transformations presented in papers by Nagai et al. (2016) or Liu et al. (2007).

What You See is What You Get?

For some time, we have known that colors perceived by the human visual system may not be portrayed truly by digital or print graphic systems (McCamy et al., 1976). This holds for both viewing and capturing (or portraying) the landscape as well as viewing a map on a computer and displaying (or portraying) the map on different media or digital devices. When natural colors are necessary for display on a digital or printed map, we would hope that images produced through computer graphics, on screen or otherwise, are indistinguishable from the original photographs, or are considered *photorealistic* (Kennelly and Kimerling, 2006). When specific colors designed in a mapping program for a thematic map are desired, we would also hope that the resulting digital or printed maps contain features indistinguishable from those viewed in the mapping program. Therefore, one slight misstep in map preparation is the belief that *what you see* (on the screen) *is what you get* (on the mapping medium), often referred to as either the *WYSIWYG* or the *reality is what we observe* assumption (Marchal, 2015). In other words, the colors used in a map designed using RGB or CMYK (or other color scheme) values may not exactly match the colors presented within the printed or digitally rendered version of the map. This can be caused by differences in the color profiling method used by a computer (the device creating the map) and the color model conversion method used by the printer or digital device hosting the map (Katoh et al., 1999). Therefore, when preparing maps that will eventually be printed, *what you see may not be exactly what you get*. It is, therefore, recommended that map developers examine a proof (preprint) of the presentation of maps on the final display device or printed media to ensure the desired effects were achieved. Another option would be to create a color wheel for each plotter or printer, and each medium, so that the adjustments to colors can be considered as a map is being developed.

Of the nearly 17 million colors that can be described using the RGB system, the human sense of vision is unlikely to discern each independently of others that might be found in nearby regions of color space (Foster et al., 2006). The perceived matching of colors by humans, when the colors have different RGB (or CMYK, etc.) representations, is called a *metameric color match*. These *metamers*, or colors that seem the same even though they have different color measures, can be unique to an individual observer, and may change when illumination conditions change. In essence, one pair of colors that seem the same to one person may not seem the same to another person (Sluban et al., 2014). This phenomenon can be important when a color scheme (next section) for a choropleth or thematic map is designed. It is not easy to prove that a person may not be able to adequately judge the differences among similar colors. Metrics have been developed to describe metamerism, yet since the amount of ambient light available can influence one's impression that two different colors are the same, a reference illuminant would need to be stated when presenting results that describe the phenomenon (Berns, 2008). Within the field of color science, one important topic today involves efforts to decrease the amount of potential metamerism for applications that require high color fidelity, such as the archiving of digital images (Cao et al., 2016).

Color Schemes

Color schemes, or sets of colors displayed on thematic maps, can be arranged to promote effective interpretation by map users. They might be designed in such a way that different tones of different colors are employed (light to dark versions) to visually set each class of landscape feature apart. For example, the following color scheme (Fig. 6.11) might be employed to effectively represent three different classes of the same landscape features displayed on a map based on the density (amount) of some characteristic of interest

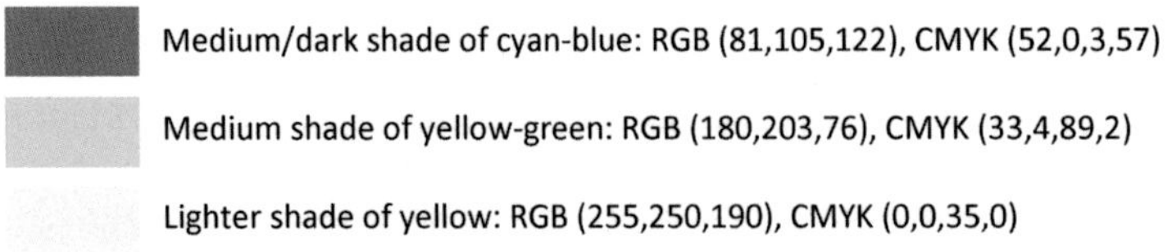

Medium/dark shade of cyan-blue: RGB (81,105,122), CMYK (52,0,3,57)

Medium shade of yellow-green: RGB (180,203,76), CMYK (33,4,89,2)

Lighter shade of yellow: RGB (255,250,190), CMYK (0,0,35,0)

FIGURE 6.11   A three-color density scale color scheme. Source: *European Environment Agency (2016).*

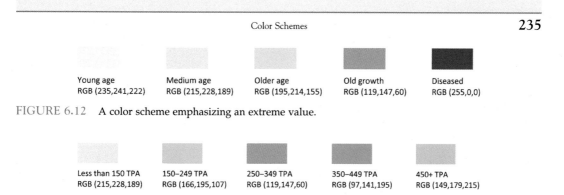

FIGURE 6.12 A color scheme emphasizing an extreme value.

FIGURE 6.13 A two-color mixed scale illustrating tree density.

Reflection 26

Think about a place (landscape or water body) for which you have great respect. Now imagine the perfect map of this place. What color schemes would be attractive to you as a user of this map? What color schemes, or individual colors, would repel your interest in this map? Why do you feel these ways?

Color schemes can also be designed to emphasize extreme values within a distribution of outcomes, or to emphasize values that are situated near the areas of greatest central tendency (mean or median) (Brewer et al., 1997). A scheme that emphasizes extreme values within five different classes of forested landscape features might involve different tones of a single color and one much different color (Fig. 6.12) that represents the extreme value. Alternatively, both variations in color and tone can be used to represent values closer to the mean or median of a classified set of data, as in the Fig. 6.13 example, where the mean number of trees per acre is 300.

A *color ramp* is a series of colors used to display information in a choropleth or thematic map. The colors chosen can be a result of various styles employed. For example, a *sequential scheme* can consist of ranges of lightness or brightness of a single hue. A light form of the hue might be used to represent lower values in a data range, and a dark form of the hue might be used to represent higher values. With respect to a forest management map, for example, a sequential scheme might be employed to illustrate the establishment year of forests (Fig. 6.14), or perhaps the amount of timber volume per unit area or some other quantitative value that varies across a landscape. This form of a color ramp may be of value when midpoint values within the data are not significant enough to emphasize (Light and Bartlein, 2004).

Diversion 34

From the book's website (http://mapping-book.uga.edu), acquire the GIS data associated with Compartment 220 of the Talladega National Forest. Using a GIS program, make a thematic map where the differences in forested stands (polygons) are illustrated through a set of colors that are based on the average age of the trees. In order to accomplish this task, use a color ramp based on a single base color.

A *gray scale scheme*, with colors ranging from white to black, might be employed as a panchromatic alternative to a color map (Fig. 6.15), or when (for example) publishing

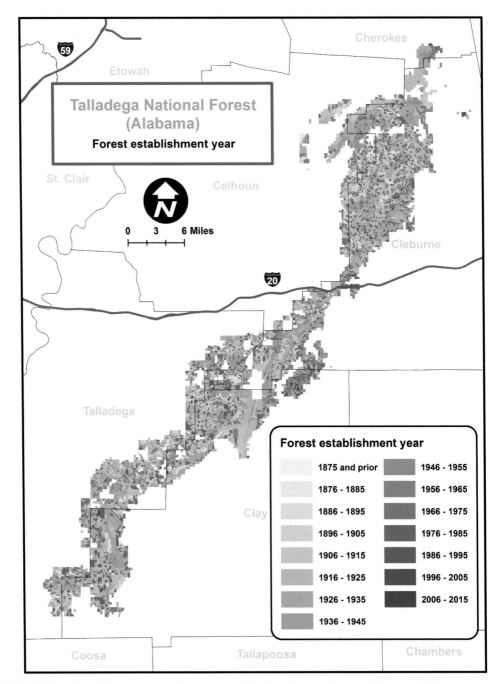

FIGURE 6.14   Establishment year of forests on the Talladega National Forest, Alabama, 2015.

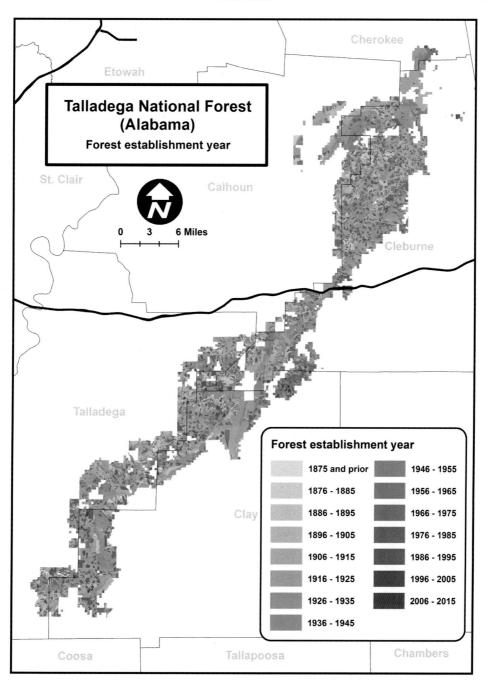

FIGURE 6.15   Gray scale version of the establishment year of forests on the Talladega National Forest, Alabama, 2015.

guidelines suggest color maps are to be avoided due to the reproduction costs. Maps might also need to be developed in panchromatic mode when the only alternative in printing a map involves the use of a basic black and white printer. Alternatively, a *spectral scheme* utilizes a series of colors one might consider acquired from a rainbow. The colors would include multiple saturated hues that are quite distinct from each other (Brewer, 1997). For example, a spectral scheme might be used to illustrate different opportunities for hunting within a managed wildlife refuge (Fig. 6.16). A *divergent scheme* has as its center a shared, or common color of two different hues. The ramp then departs to either extreme of the scale toward dark colors of the two different hues. For qualitative data, the shared color in the middle of the divergent scheme, perhaps the midpoint of the data distribution, is highlighted along with the extreme ends of the color spectrum employed. A divergent scheme can be created by combining two sequential schemes that share a common color (Light and Bartlein, 2004). A divergent scheme can also be designed with the critical point of interest in the center represented by a light color, and the low and high extremes of the data range represented by darker colors (Brewer, 1997). Fig. 6.17 illustrates the use of a divergent scheme to illustrate the forest establishment year of areas on the Talladega National Forest.

Diversion 35
Again, acquire Compartment 220 of the Talladega National Forest from this book's website (http://mapping-book.uga.edu). Using a GIS program, make two more thematic maps where the only differences among them is the manner in which the forested stands (polygons) are colored, based on the average age of the trees: (1) use a spectral color scheme based on a set of bright, distinctly different and vibrant colors and (2) use a divergent color scheme based on two colors. Which of the three maps (these two maps, and the map from Diversion 33) do you prefer in describing thematically the differences in average tree age in this area of Alabama?

Inspection 31
The Color Brewer *color adviser for cartography* tool (www.colorbrewer2.org) provides a map of county boundaries for US states, ranging from Nebraska to Pennsylvania in the north and Texas to Florida in the south (Brewer, 2013). How do the sequential, diverging, and qualitative color schemes differ when the counties on this map are colored?

   Gradients of colors, color palettes, color ramps, and color schemes may be necessary to effectively communicate complex ideas to map users. As was mentioned earlier, *hypsometric tinting* (relief shading) might be used as a color ramp to convey differences in land elevations or sea depths. In an application such as this, different elevation ranges could be represented by variations in color to illustrate graphically local changes in topography and underlying landforms (Leonowicz et al., 2009). Higher elevations, such as the tops of mountains, may be presented by a lighter color than the others. Perhaps the color white would be employed if snow or glaciers would be found there. Lower elevations are often represented by progressively darker green tints. Variations in the darkness of blue are often reserved to represent differences in water depth. Lighter blue tints would usually be

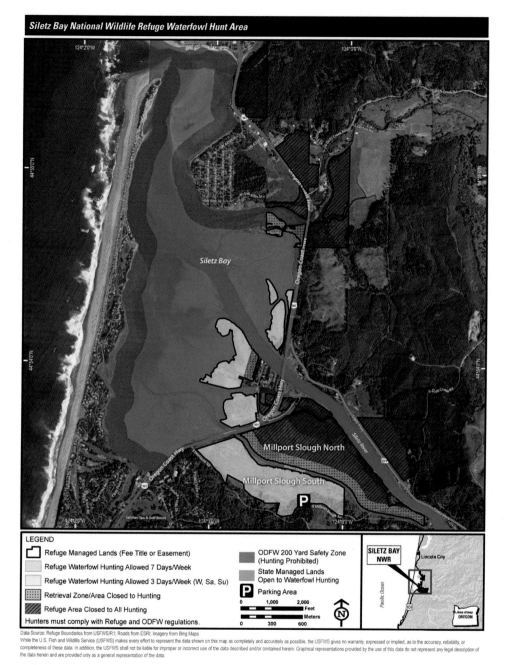

FIGURE 6.16   Waterfowl hunting opportunities in the Siletz Bay National Wildlife Refuge, Oregon, 2015. Source: *U.S. Department of the Interior, Fish and Wildlife Service, (2015).*

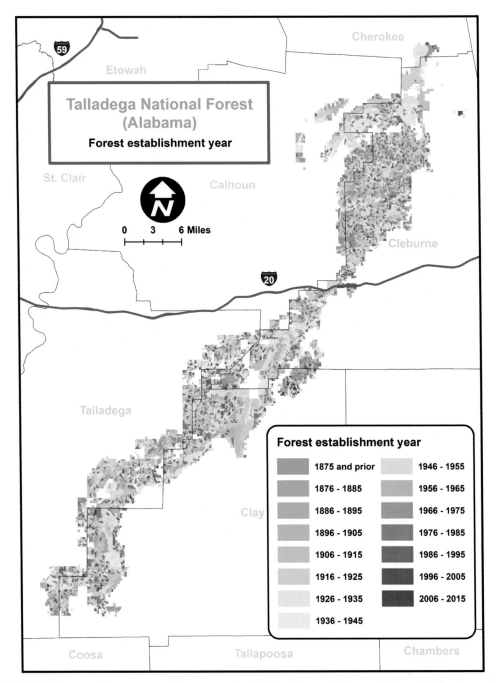

FIGURE 6.17   Establishment year of forests on the Talladega National Forest, Alabama, 2015, using a divergent color scheme.

employed to represent relatively shallow water, while darker blue tints would be employed to represent deeper water. A hypsometric tinting color ramp can be presented along with contour or bathymetric lines to enhance the set ideas communicated by a map.

Color Distance

The human visual system, through a mechanism called *visual attention*, reduces incoming information to a small, important subset of an entire scene; as this relates to maps, the distinctiveness of features presented on a map may be related to the color contrast among all of the features on a map since our visual attention is drawn to distinctively mapped regions (Lie et al., 2018). Color schemes for maps can be developed based on their ability to produce a certain amount of visual contrast among the colors selected, and a *color distance* can be used as a quantitative measure of the contrast among these colors. Choosing colors based on their associated color distance can be important for maximizing the readability of maps (Brychtová and Çöltekin, 2017). In its basic form, a color distance can represent the mathematical difference in three-dimensional Euclidean space between two sets of RGB values. A simple metric to describe the difference in two colors is:

$$Color\ distance = ((red_1 - red_2)^2 + (green_1 - green_2)^2 + (blue_1 - blue_2)^2)^{0.5}$$

Here, red_1, $green_1$, and $blue_1$ are the RGB values of one color, and red_2, $green_2$, and $blue_2$ are the RGB values of another color. Imagine two colors, violet (238, 130, 238) and orange (255, 165, 0) (Fig. 6.18). The color distance between these two is:

$$Color\ distance = ((238 - 255)^2 + (130 - 165)^2 + (238 - 0)^2)^{0.5}$$
$$Color\ distance = ((-17)^2 + (-35)^2 + (238)^2)^{0.5}$$
$$Color\ distance = (58,158)^{0.5}$$
$$Color\ distance = 241.16$$

Through this assessment, one might consider the color distance (241.16) between violet and orange to be quite large. In comparison, the color distance between violet (238, 130, 238) and lavender (230, 230, 250) is:

$$Color\ distance = ((238 - 230)^2 + (130 - 230)^2 + (238 - 250)^2)^{0.5}$$
$$Color\ distance = ((8)^2 + (-100)^2 + (-12)^2)^{0.5}$$
$$Color\ distance = (10,208)^{0.5}$$
$$Color\ distance = 101.03$$

FIGURE 6.18 Color differences between (A) violet and orange and (B) violet and lavender.

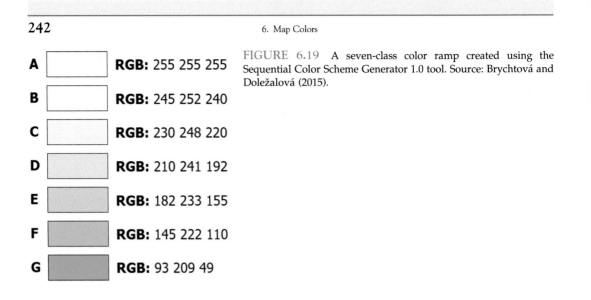

A
RGB: 255 255 255

B
RGB: 245 252 240

C
RGB: 230 248 220

D
RGB: 210 241 192

E
RGB: 182 233 155

F
RGB: 145 222 110

G
RGB: 93 209 49

FIGURE 6.19 A seven-class color ramp created using the Sequential Color Scheme Generator 1.0 tool. Source: Brychtová and Doležalová (2015).

This color distance (101.03) would be considered much shorter than the color distance between violet and orange, as violet and lavender are more similar colors than violet and orange.

Diversion 36

If the color distance can be used as a measure of contrast between two colors, what two vibrant colors, other than those mentioned in this section, might you select as having a large color distance? Given what you have selected, what are their RGB values, and what is the resulting color distance between them? Further, what two colors have the greatest color distance, and what is the resulting color distance between them?

Other mathematical relationships among colors are used in practice. Some that have been developed by the International Commission on Illumination (CIE) are device independent (as compared to RGB and HSV models), and are called the *CIE XYZ, CIELAB,* and *CIELUV* color space models (Bello-Cerezo et al., 2016). These and other similar color spaces are related to the HSV model, and their associated color distance values are used in certain Internet-based models for creating sets of color gradient schemes. One example program for developing color schemes is the *Sequential Color Scheme Generator* (Brychtová and Doležalová, 2015), which can automatically create a color ramp between two selected colors based on the color distance desired between each class in the ramp (Fig. 6.19).

Color Order Systems

The representation of color, through printing or presentation on digital devices, is an important consideration in the development of maps, and an accurate portrayal of one's ideal set of colors can be frustrating. The need for color order (matching, measurement) systems is, therefore, important for design and research purposes in that they provide a common language to describe colors and color combinations (Hård and Sivik, 1981). These *colorimetric representations*, or *color appearance models*, are often internationally recognized (Kasson and Plouffe, 1992), and are studied within the field of color science, or colorimetry (Vallari et al., 1994). Methods for describing the full spectrum of color are complex, but often couched in systems that some consider to be rational approaches for ordering and describing variations in colors. For the most part, three-dimensional conceptual models of color ordering systems we describe here attempt to make sense of the gamut of color through analytical or geometric approaches. Researchers continue to devise systems that can rationally order and describe color in *n*-dimensional conceptual space (Zhao, 2007).

As noted in the previous section, the CIE developed two-color space models in the early 1900s that are known as the *CIE 1931 RGB color space* and the *CIE 1931 XYZ color space*. While the *RGB model* was developed to uniquely describe the relative intensities of red, green, and blue primary color values and the resulting color stimulus (Vallari et al., 1994), the *XYZ model* (Fig. 6.20) attempts to emulate cone responses in the human eye, where X refers to a mixture of cone response curves, Y refers to luminance, and Z refers to blue stimulation. For any given level of luminance, the blue response and cone response curves contain every possible chromaticity (quality of color). It has been suggested that this model serves as the basis and international standard for calibration of all color management systems (Kasson and Plouffe, 1992; Ohno, 2000). The *RGB* and *XYZ models* are therefore related, as the *XYZ model* is a linear transformation of the *RGB model*, where *tristimulus values* (three integrated values to describe light stimuli of color) and color-matching

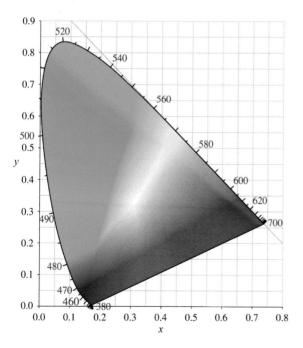

FIGURE 6.20   The *XYZ* color space model.

functions are all positive (Kasson and Plouffe, 1992). The *RGB model* can contain negative values of the primary colors when added to a monochromatic stimulus (Ohno, 2000), while the corresponding *XYZ model* only contains positive values. Further updates and adjustments to these models have been made since, to address the needs of computer-generated graphics. The *CIELUV model* is one example, created in 1976 to create a uniform color space along lightness scales, or different representations of white (Kasson and Plouffe, 1992). The *CIELAB model* is another, where the *XYZ* values have been transformed into coordinates (L^*, a^*, b^*) that further refine the range of colors perceivable within the visible human gamut of colors. The *CIECAM02 model* was recommended by the CIE in 2004 as a color appearance system that provides refinements on the shortcomings of previous models. Even though these are widely in practice, progress into representations of color appearance under various real-world conditions and within image reproduction systems continues (Luo and Pointer, 2018). As examples, other color system models have been developed for color-matching purposes of device-independent electronic media in efforts to replicate and reproduce real-world colors (Kasson and Plouffe, 1992) or to describe color for generally any sensing device (human or otherwise) based on the spectral sensitivity functions of those systems (Li and Lee, 2014).

The Munsell color order system was developed in the early part of the 20th century by artist Albert Munsell in an attempt to describe color by its hue, value (or lightness), and chroma (or difference from a gray of the same lightness) as perceived by an observer of the color (Judd, 1940; Nickerson, 1946; Evans and Swenholt, 1968). In the Munsell color system, lightness (value), hue, and intensity (chroma) define a three-dimensional model that depicts differences in colors (Smet et al., 2016). By the mid-20th century, the color system was used by many companies in efforts to standardize the color of various foods (e.g., tomato soup and mayonnaise) and to measure the color of products such as cotton, hay, milk, and butter (Nickerson, 1946; Nickerson, 1976). In the Munsell system, there are 40 basic hues. Depending on the hue, the conceptual planes of constant hue can be irregular (Shamey et al., 2011). The value scale of the Munsell system was intended to represent equal sensation steps within a psychophysical human system (Gibson and Nickerson, 1940). Even though it has a unique developmental history, it may also be possible to approximate the Munsell color system through transformations of others, such as the CIELAB model (Li and Lee, 2014).

The Pantone color-matching system is perhaps one of the most widely known for standardizing the communicating of color through a universal color language. The proprietary Pantone system is considered to be internationally recognized for its color identification codes (Engels et al., 2016). The system has been described as a 14-dimensional color space (Zhao, 2007) as compared to the three-dimensional RGB color model. This color system assists in the organization of color for fashion design, printing, and packaging design purposes, on the basis that the appearance of color is a function of the material on which it is placed (Pantone, 2018). Matching the colors of corporate and brand identities on media such as paper, paperboard, and corrugated board (cardboard) is critical, and printing technology often relies on this system to ensure colors are matched appropriately (Wu and El Asaleh, 2016). The color chip books, fan decks, color bridges, and swatches that communicate variations in color are common among this and other systems (Fig. 6.21). Pantone 3425 C, for example, is the green color that Starbucks coffee company uses for its logo

FIGURE 6.21   A Pantone color bridge.

(Lee, 2015). This shade of green also has the following color measures: RGB (0, 99, 65), CMYK (100, 0, 34, 61), HEX (006341).

Translation 24
Imagine for some important purpose of your dream job, you are in a situation where you need to develop the cover of a very important report. In your mind, you have an idea of the range of colors you would like to use. However, you need to work alongside a graphic designer and other colleagues to accomplish this task. You mention to your colleagues that you have already met with the graphic designer, and through their chip book, you selected Pantone 18-2120 (Honeysuckle) as the background color for the cover of your report. Some of your colleagues look perplexed at what you just said. Describe briefly for them the concept of a chip book.

Diversion 37
Through a search of the Internet, locate and report the RGB and CMYK equivalent values for Pantone 18-2120 (Honeysuckle).

The Swedish natural color system (NCS) was developed in an effort to describe colors and their associated relationships with other colors in a graphical and alphanumeric manner (Hård and Sivik, 1981). Within the NCS, colors are described in terms of their similarity to yellow, red, blue, green, white, and black (Shamey et al., 2015). The attributes of the NCS are the amount of white and black content, and the type of color that exhibits the greatest chromatic intensity. Within the NCS conceptual model (Fig. 6.22), chromatic intensity is considered relative, since the colors are positioned using a constant radial distance from a central gray scale (Shamey et al., 2011). With respect to both the Munsell and NCS

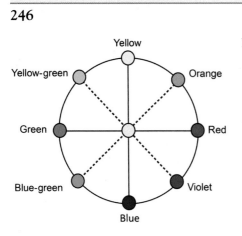

FIGURE 6.22   The Swedish NCS system.

■ Red: excitement, danger, power, passion, desire
▣ Orange: enthusiasm, fascination, determination, warmth
▫ Yellow: joy, happiness, energy, warmth, honor, loyalty, cheer
■ Green: health, harmony, freshness, safety, trust
■ Blue: peace, calmness, serenity, relaxation, intelligence, sadness
■ Purple: royalty, pureness, placidity, wealth, magic
■ Black: formality, power, sorrow, evilness, mystery, grief
□ White: goodness, innocence, purity, cleanliness
▨ Gray: impartiality, dullness, neutrality, maturity, boredom
▨ Pink: love, affection, femininity, optimism
▨ Brown: earthiness, naturalness

FIGURE 6.23 Basic colors and the emotions they may evoke.

color order systems, in some cases a constant hue in one system relates to a relatively wide range of hues in the other systems (Billmeyer and Bencuya, 1987).

Color Psychology

Even though reactions may vary by culture, much like music, colors can evoke emotions within people (Porteous, 1996). Variations in hue, saturation, and brightness of colors may influence human cognition and behavior (Elliot, 2015), and have been postulated to have an effect on the human body (Minguillon et al., 2017; Whitfield and Whelton, 2015). The reactions we have to the presentation of colors prompted the development of the alternative medicine field of *chromotherapy*. In any event, warm colors (red, orange, and yellow) have been suggested as being generally stimulating to people, while cool colors (green and blue) have been suggested as being generally soothing (Clarke and Costall, 2008). Some people prefer certain ranges of colors (e.g., pastels), yet people respond differently to sets of colors based on their age, gender, and cultural background. Eleven of the basic colors (Fig. 6.23) we can perceive with our eyes, of the millions of combinations possible, have been described as evoking various types of feelings in people.

Valentine Ridge Fire
Valentine Ridge Fire
Valentine Ridge Fire
Valentine Ridge Fire
Valentine Ridge Fire

FIGURE 6.24   Map titles presented with different colors.

As you may imagine, continuums among colors exist. For example, orange and yellow can both be considered as warm or happy colors, with orange offering a more relaxed state of happiness and yellow offering a more exuberant state of happiness (Clarke and Costall, 2008). The most preferred colors in industrialized countries seem to be shades of blues and purples (Porteous, 1996). In nature, colors are rarely found in isolation; the aesthetic experience one obtains from viewing scenery, or a map, is a function of all colors present, thus selection of color combinations for maps can affect a map user's preference for the map product (Schloss and Palmer, 2011). For example, examine the potential map titles in Fig. 6.24. Would they influence your impression of the professionalism or quality of a map?

Color schemes may, therefore, be selected as part of a marketing ploy, in a sense, on behalf of the map developer in an effort to engage map users and to elicit from them desired responses. A hierarchical structure of the categorization of landscape features (good to bad, high value to low value, etc.), along with psychologically perceived differences in colors, may both be considered by map developers in their desire to maximize the communicative ability of a map (Sun, 2015).

Inspection 32
Access and examine the malaria risk map of south and southeast Asia that is available on this book's website (mapping-book.uga.edu). Although the developers of the map may not have intentionally employed color psychology when making this map, describe how the colors selected for this map might have been used to illustrate the risk of contracting the disease in this part of the world.

Inspection 33
A map describing the path of Hurricane Mitch (1998) is shown in Fig. 6.25. Although the developers of the map may not have intentionally employed color psychology when making this map, let's assume that the color choice was not random. If we were to make this assumption, what emotions or feelings do you think that the map developer wanted to convey to people using the colors that they selected to represent the path of the hurricane?

Color Blindness

The color of objects is an important subject in commerce, and interestingly it may be the most advantageous trading metric for commodities such as honey (Dominguez and Centurión, 2015) and other products. However, color is not perceived equally by everyone

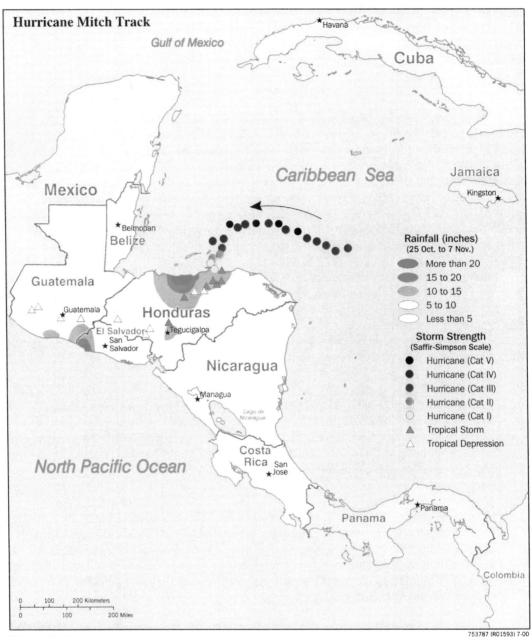

FIGURE 6.25 A map of the path representing the center of Hurricane Mitch in 1998. *Source: U.S. Central Intelligence Agency (2000).*

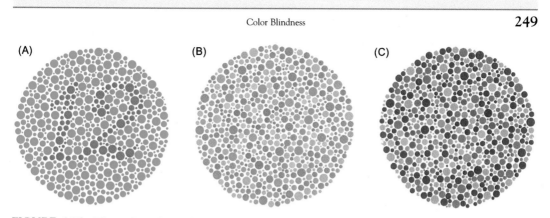

FIGURE 6.26 Three plates from the Ishihara (1972) tests for color blindness: (A) Plate 1, where both people with normal vision and color vision deficiencies should be able to recognize the number 12; (B) Plate 7, where people with normal vision will recognize the number 74, and those with red-green deficiencies may read it as 21; and (C) Plate 17, where people with normal vision will recognize the number 42, people with mild protanomaly may recognize only the number 2, and people with deuteranopia and strong deuteranomaly may only recognize the number 4.

(Tanaka et al., 2009). In the retina, the sensory membrane of our eyes, there are two types of light-sensitive bodies (the rods and the cones) that collect and send information to our brain via the optical nerve. In normal light conditions, the cones help us discriminate colors, thus these light-sensitive parts of our visual system receive red, green, and blue electromagnetic energy and allow us to perceive color (Joblove and Greenberg, 1978; Light and Bartlein, 2004). Interestingly, a small percentage of people with complete or partial loss of retina cone cell function have trouble distinguishing certain colors within the electromagnetic spectrum (Pridmore, 2014). Color perception tests, such as the Ishihara (1972) tests can be used to estimate color range deficiencies in people (Fig. 6.26). Three types of *color blindness*, as it is called, are *tritanopia*, *deuteranopia*, and *protanopia*. Each relate to the limitations within a person's ocular system, and all are often caused by congenital conditions. These limitations prevent a person from distinguishing certain colors in specific ranges of the electromagnetic spectrum. For example, tritanopia is a situation in both women and men (less than 1% of the population, however) where short-wavelength sensitive cones are lacking, preventing people from sensing some of the color in the blue range of the visible spectrum of light (the shorter wavelengths of visible light). People with tritanopia have difficulties discriminating between blue and yellow colors (Huang et al., 2007). Deuteranopia, on the other hand, is a situation where men (again, less than 1% of the population) may lack medium-wavelength sensitive cones, preventing them from sensing some of the color in the green and red range of the visible spectrum of light. People with deuteranopia may perceive some shades of green and red as gray colors. Protanopia similarly affects about the same amount of men in the human population, and like deuteranopia, green and red, along with orange and yellow, are less discernable and appear as gray colors. People with deuteranopia and protanopia conditions typically can only clearly see blue and yellow hues (Pridmore, 2014), and have difficulty discriminating between green and red (Huang et al., 2007). Milder versions of these conditions, where retinal cones are not lacking, but sensitivity to certain wavelengths may be diminished (or has been shifted), are *tritanomaly*, *deuteranomaly*, and *protanomaly*.

When developing maps, one might bear in mind that they should develop them in such a way as to make them *accessible* (capable of being understood) to people who have cognitive deficiencies, such as those with one or more forms of color blindness (Light and Bartlein, 2004). Therefore, with respect to color, when color schemes are used to portray different land or water resources, perhaps these should contrast sufficiently so that people with difficulties discerning different colors can understand the message of the map (Chisholm et al., 2000). Both spectral and divergent color ramp schemes can be designed so that people who have trouble separating red and green colors can be accommodated, by avoiding yellow-green colors in the set of colors chosen to represent data in a thematic map (Brewer, 1997). Similarly, this strategy might be employed to ensure that a map is useable when these are viewed on low-resolution digital screens. Methods have also been designed for digital devices to allow adaptive color selection based on results obtained through a color vision discrimination test performed by the map user (Kvitle et al., 2015). The printing or copying of maps in a gray scale (shades of gray between black and white) might provide a good test of information contrast for these types of situations. Also, the reassignment of colors to different spectral ranges of an image (or printed map) may also improve the visibility of certain color combinations (Huang et al., 2007; Tanaka et al., 2009) and allow people with red-green or blue-yellow color deficiencies to better understand the message of a map.

Reflection 27
Consider the last map that you developed. Whether purposefully or not, did you consider the accessibility of the map to a wide, potential audience? If you did, what aspects of the map and what alternatives were considered? If you did not, what could you have considered adjusting within the map to address or improve its potential accessibility?

An internationally accepted standard (World Wide Web Consortium, 2000) has been developed to provide advice on the use of color in graphs and maps delivered over the Internet; this advice may also be applicable to printed maps. This advice seeks to ensure that accessibility of information is improved for people who may be color blind, as it recommends certain color combinations that seem to provide a sufficient amount of contrast. The effort expended on making maps completely accessible to the entire human population should be thoughtfully explored, rather than ignored. While making color combination adjustments may effectively engage everyone, admittedly for widely distributed maps, specific adjustments to color schemes may not seem necessary, particularly when adjustments to a map color scheme may decrease the value of the map. However, in cases where important customers of a map (or segments of society) indeed have color discrimination limitations, a thoughtful consideration of the color scheme of a map would seem to be mandatory.

Concluding Remarks

For nearly every application in modern society, the choice of color is an essential aspect of the development of a map. Certainly, most map developers want to create maps that are effective in communicating messages. But deep down, map developers may also have a desire to create an attractive map, in a general sense, as opposed to one that is generally repulsive or unprofessional. The conscious selection of colors can assist in highlighting distinct differences among mapped features (e.g., water vs forest) as well as gradual differences within classes of mapped features (e.g., light blue for shallow water, dark blue for deeper water). In addition to the symbology and annotation a map may contain, color adds a dimension that may further increase the value of a map, as compared to its panchromatic counterpart. The choice of colors to use within a map may require thought on behalf of the map developer, especially when accessibility is a concern. The spatial distance between features of similar colors on a map can influence whether a map user is able to determine that the colors are similar or different (Brychtová and Çöltekin, 2017). Color choices for thematic maps may also influence the placement of a map's legend. Difficulties map users have in correlating colors presented in a map legend with colors displayed in a map have been noted for over 200 years (Ksiazkiewicz, 2016). Therefore, when color schemes involve similar hues, placement of a legend far from colored landscape features may result in map interpretation problems among map users.

References

Ball, D.W., 2007. The electromagnetic spectrum: a history. Spectroscopy 21 (3), 14−20.

Bello-Cerezo, R., Bianconi, F., Fernández, A., González, E., Di Maria, F., 2016. Experimental comparisons of color spaces for material classification. J. Electric. Eng. 25 (6), Article 061406.

Berlin, B., Kay, P., 1990. Basic Color Terms. Their Universality and Evolution. University of California Press, Berkeley, CA.

Berns, R.S., 2008. The proper use of indices of metamerism. Color Res. Appl. 33 (6), 509−511.

Billmeyer Jr., F.W., Bencuya, A.K., 1987. Interrelation of the natural color system and the Munsell color order system. Color Res. Appl. 12 (5), 243−255.

Brewer, C.A., 1997. Spectral schemes: controversial color use on maps. Cartogr. Geogr. Information Systems 24 (4), 203−220.

Brewer, C.A., 2013. Color Brewer 2.0 Color Adviser for Cartography. <http://www.ColorBrewer.org> (accessed 24.05.19.).

Brewer, C.A., MacEachren, A.M., Pickle, L.W., Hermann, D., 1997. Mapping mortality: evaluating color schemes for choropleth maps. Ann. Assoc. Am. Geogr. 87 (3), 411−438.

Brychtová, A., Çöltekin, A., 2017. The effect of spatial distance on the discriminability of colors in maps. Cartogr. Geogr. Information Sci. 44 (3), 229−245.

Brychtová, A., Doleželová, J., 2015. Sequential Color Scheme Generator 1.0. Palacký University, Olomouc, Czech Republic. <http://eyetracking.upol.cz/color/> (accessed 24.05.19.).

Cao, Q., Wan, X., Li, J., Liu, Q., Liang, J., Li, C., 2016. Spectral data compression using weighted principal component analysis with consideration of human visual system and light sources. Optical Review 23 (5), 753−764.

Chisholm, W., Vanderheiden, G., Jacobs, I., Eds., 2000. Core techniques for web content accessibility guidelines 1.0. World Wide Web Consortium. W3C Note 6.

Clarke, T., Costall, A., 2008. The emotional connotations of color: a qualitative investigation. Color Res. Appl. 33 (5), 406−410.

Dominguez, M.A., Centurión, M.E., 2015. Application of digital images to determine color in honey samples from Argentina. Microchem. J. 118, 110−114.

Elliot, A.J., 2015. Color and psychological functioning: a review of theoretical and empirical work. Front. Psychol. 6, Article 368.

Engels, G., Hofhuis, J., Lehr, C., 2016. The local color of colour marks. J. Intellect. Property Law Practice 11 (8), 628−639.

European Environment Agency, 2016. EEA Corporate Identity Manual 2015. European Environment Agency, Copenhagen, Denmark.

Evans, R.M., Swenholt, B.K., 1968. Chromatic strengths of colors, Part II. The Munsell System. J. Optical Soc. Am. 58 (4), 580–584.

Foster, D.H., 2011. Color constancy. Vision Res. 51 (7), 674–700.

Foster, D.H., Amano, K., Nascimento, S.M.C., Foster, M.J., 2006. Frequency of metamerism in natural scenes. J. Optical Soc. Am. A 23 (10), 2359–2372.

Geoscience Australia, 2012. Section 2 - National Topographic Map Series (NTMS) and General Reference Map Specifications. Commonwealth of Australia, Geoscience Australia, Canberra. <http://www.ga.gov.au/map-specs/topographic/v6/section2.html> (accessed 04.01.19.).

Gibson, K.S., Nickerson, D., 1940. An analysis of the Munsell color system based on measurements made in 1919 and 1926. J. Optical Soc. Am. 30 (12), 591–608.

Green, D.R., Horbach, S., 1998. Color – difficult to both choose and use in practice. Cartogr. J. 35 (2), 169–180.

Hård, A., Sivik, L., 1981. NCS—natural color system: a Swedish standard for coloer notation. Color Res. Appl. 6 (3), 129–138.

Herter Jr., C.A., 1985. The electromagnetic spectrum: a critical natural resource. Natural Res. J. 25 (3), 651–663.

Hou, J., Gao, T., Wang, P., 2017. Flow feature analysis and extraction for classifying axisymmetric vector field patterns. Multimedia Tools Appl. 76 (13), 14617–14634.

Huang, J.-B., Tseng, Y.-C., Wu, S.-I., Wang, S.-J., 2007. Information preserving color transformation for protanopia and deuteranopia. IEEE Signal Process. Lett. 14 (10), 711–714.

International Orienteering Federation, 2018. ISOM 2017 Appendix 1 - CMYK Printing and Colour Definitions, Version 1, 2018-06-20. International Orienteering Federation, Karlstad, Sweden.

Irigoyen, J., Herraez, J., 2003. Electromagnetic spectrum and color vision. In: Lončarić, S., Neri, A., Babić, H. (Eds.), Proceedings of the 3rd International Symposium on Image and Signal Processing and Analysis, Faculty of Electrical Engineering and Computing, University of Zagreb, Croatia, and Università degli Studi Roma Tre, Rome. IEEE Catalog Number 03EX651, 626-629.

Ishihara, S., 1972. The series of plates designed as a test for colour-blindness. Kanehara Shuppan Co., Ltd., Tokyo. 9 pp., 24 plates.

Joblove, G.H., Greenberg, D., 1978. Color spaces for computer graphics. In: Proceedings of the 5th Annual Conference on Computer Graphics and Interactive Techniques. Association for Computing Machinery, New York, 20-25.

Judd, D.B., 1940. The Munsell color system. J. Optical Soc. Am. 30 (12), 574.

Kang, T.J., Kim, C.H., Oh, K.W., 1999. Automatic recognition of fabric weave patterns by digital image analysis. Textile Res. J. 69 (2), 77–83.

Kasson, J.M., Plouffe, W., 1992. An analysis of selected computer interchange color spaces. ACM Transact. Graphics 11 (4), 373–405.

Katoh, N., Ito, M., Ohno, S., 1999. Three-dimensional gamut mapping using various color difference formulae and color spaces. J. Electron. Imaging 8 (4), 365–379.

Kennelly, P.J., Kimerling, A.J., 2006. Non-photorealistic rendering and terrain representation. Cartogr. Perspect. 54, 35–54.

Ksiazkiewicz, A., 2016. Unifying prospects: tinting geological maps in nineteenth-century Britain. Cartographica 51 (3), 159–174.

Kvitle, A.K., Green, P., Nussbaum, P., 2015. Adaptive colour rendering of maps for users with colour vision deficiencies. In: Proceedings of the SPIE-IS&T Electronic Imaging, Volume 9395, Color Imaging XX: Displaying, Processing, Hardcopy, and Applications. Article 939515.

Lee, K.S., 2015. A textual analysis of Pantone's color communication techniques through the application of Barthes' semiotic. Liberty University, School of Digital Media & Communication Arts, Lynchburg, VA. Master of Arts thesis. 104 pp.

Leonowicz, A.M., Jenny, B., Hurni, L., 2009. Automatic generation of hypsometric layers for small-scale maps. Comput. Geosci. 35 (10), 2074–2083.

Levine, T., 2009. Using colour in figures: let's agree to differ. Traffic 10 (3), 344–347.

Li, Y.-Y., Lee, H.-C., 2014. DLAB: a class of daylight-based uniform color space. J. Optical Soc. Am. A 31 (8), 1876–1885.

Lie, M.M.I., Borba, G.B., Neto, H.V., Gamba, H.R., 2018. Joint upsampling of random color distance maps for fast salient region detection. Pattern Recogn. Lett. 114, 22–30.

Light, A., Bartlein, P.J., 2004. The end of the rainbow? Color schemes for improved data graphics. EOS, Transact. Am. Geophys. Union 85 (40), 385–391.

Liu, Y.-S., Duh, D.-J., Chen, S.-Y., Liu, R.-S., Hsieh, J.-W., 2007. Scale and skew-invariant road sign recognition. Int. J. Imaging Syst. Technol. 17 (1), 28–39.

Luo, M.R., Pointer, M.R., 2018. CIE colour appearance models: a current perspective. Lighting Res. Technol. 50 (1), 129–140.

Marchal, B., 2015. The universal numbers. From biology to physics. Progr. Biophys. Molecular Biol. 119 (3), 368–381.

McCamy, C.S., Marcus, H., Davidson, J.G., 1976. A color-rendition chart. J. Appl. Photogr. Eng. 2 (3), 95–99.

Minguillon, J., Lopez-Gordo, M.A., Renedo-Criado, D.A., Sanchez-Carrion, M.J., Pelayo, F., 2017. Blue lighting accelerates post-stress relaxation: results of a preliminary study. PLoS ONE 12 (10), e0186399.

Nagai, S., Ichie, T., Yoneyama, A., Kobayashi, H., Inoue, T., Ishii, R., et al., 2016. Usability of time-lapse digital camera images to detect characteristics of tree phenology in a tropical rainforest. Ecol. Informatics 32, 91–106.

National Aeronautics and Space Administration, 2005. Ireland, Shaded Relief and Colored Height. National Aeronautics and Space Administration, Jet Propulsion Laboratory, Pasadena, CA. <https://www.jpl.nasa.gov/spaceimages/details.php?id=PIA06672> (accessed 06.05.19.).

National Aeronautics and Space Administration, 2013. The Electromagnetic Spectrum. National Aeronautics and Space Administration, Goddard Space Center, Imagine the Universe Program, Greenbelt, MD. <https://imagine.gsfc.nasa.gov/science/toolbox/emspectrum1.html> (accessed 05.05.19.).

Nickerson, D., 1946. Color measurements of standards for grades of cotton. Textile Res. J. 16 (9), 441–449.

Nickerson, D., 1976. History of the Munsell color system, company, and foundation III. Color Res. Appl. 1 (3), 121–130.

Nikolaev, V.F., Foss, L.E., Sulaiman, B.F., Agybay, A.B., Timirgalieva, A.Kh, et al., 2018. The unified scale of natural waters. Georesursy=Georesources 20 (2), 58–66.

Nossum, A.S., Li, H., Giudice, N.A., 2013. Vertical colour maps – a data-independent alternative to floor-plan maps. Cartographica 48 (3), 225–236.

Ohno, Y., 2000. CIE fundamentals for color measurements. In: Proceedings of the International Conference on Digital Printing Technologies (IS&T NIP16 Conference). Society for Image Science and Technology, Springfield, VA, 540–545.

Pantone L.L.C., 2018. Pantone color systems explained. <https://www.pantone.com/color-systems-intro> (accessed 08.04.19.).

Patterson, T., Jenny, B., 2013. Evaluating cross-blended hypsometric tints: a user study in the United States, Switzerland, and Germany. Cartogr. Perspect. 75, 5–16.

Pollard, A.M., Blatt, C.M., Stern, B., Young, S.M.M., 2007. Analytical Chemistry in Archaeology. Cambridge University Press, Cambridge, UK, pp. 275–279.

Porteous, J.D., 1996. Environmental Aesthetics. Ideas, Politics and Planning. Routledge, London.

Pridmore, R.W., 2014. Orthogonal relations and color constancy in dichromatic colorblindness. PLoS ONE 9 (9), e107035.

Rystedt, B., 2016. Swedish mapping in the Baltic countries. e-Perimetron 11 (4), 196–201.

Schloss, K.B., Palmer, S.E., 2011. Aesthetic response to color combinations: preference, harmony, and similarity. Attention Percept. Psychophys. 73 (2), 551–571.

Shamey, R., Shepherd, S., Abed, M., Chargualaf, M., Garner, N., Dippel, N., et al., 2011. How well are color components of samples of the natural color system estimated? J. Optical Soc. Am. A 28 (10), 1962–1969.

Shamey, R., Zubair, M., Cheema, H., 2015. Effect of field view size and lighting on unique-hue selection using natural color system object colors. Vision Res. 113 (Part A), 22–32.

Sluban, B., Peyvandi, S., Avval, S.C., Amirshahi, S.H., 2014. A general metric for the magnitude of observer metamerism. Coloration Technol. 130 (5), 368–375.

Smet, K.A.G., Webster, M.A., Whitehead, L.A., 2016. A simple principled approach for modeling and understanding uniform color metrics. J. Optical Soc. Am. A 33 (3), A319–A331.

Smith, A.R., 1978. Color gamut transform pairs. In: SIGGRAPH '78 Proceedings of the 5th Annual Conference on Computer Graphics and Interactive Techniques. Association for Computing Machinery, New York, pp. 12–19.

Stokking, R., Zuiderveld, K.J., Viergever, M.A., 2001. Integrated volume visualization of functional image data and anatomical surfaces using normal fusion. Human Brain Mapping 12, 203–218.

Sun, S., 2015. A perception-based color recommendation algorithm for hierarchical regions. Cartogr. Geogr. Information Sci. 42 (3), 259–270.

Tanaka, G., Suetake, N., Uchino, E., 2009. Visibility improvement of undiscriminatable colors by modification of yellow-blue components for protanopia and deuteranopia. In: 2009 International Symposium on Intelligent Signal Processing and Communication Systems, 562–565.

U.S. Central Intelligence Agency, 2000. Hurricane Mitch Track. U.S. Central Intelligence Agency, Cartography Center, Washington, D.C.

U.S. Department of the Interior, Fish and Wildlife Service, 2015. Siletz Bay National Wildlife Refuge Waterfowl Hunting Regulations. U.S. Department of the Interior, Fish and Wildlife Service, Oregon Coast National Wildlife Refuge Complex, Newport, OR.

Vallari, M., Chryssoulakis, Y., Chassery, J.M., 1994. Measurement of colour using a non-destructive method for the study of painted works of art. Measure. Sci. Technol. 5 (9), 1078–1088.

Whitfield, T.W.A., Whelton, J., 2015. The arcane roots of colour psychology, chromotherapy, and colour forecasting. Color Res. Appl. 40 (1), 99–106.

World Wide Web Consortium, 2000. Techniques for web content accessibility guidelines 1.0. W3C Note 6. <https://www.w3.org/TR/WCAG10-TECHS/> (accessed 02.03.19.).

Wu, Y.J., El Asaleh, R., 2016. Using wide format UV ink-jet printing for digital package prototyping. Int. Circular Graphic Educ. Res. 9, 37–46.

Zhang, Z., Qie, G., Wang, C., Jiang, S., Li, X., Li, M., 2017. Relationship between forest color characteristics and scenic beauty: case study analyzing pictures of mountainous forests at sloped positions in Jiuzhai Valley, China. Forests 8 (3), Article 63.

Zhao, X., 2007. Generic device color gamut description. Color Res. Appl. 32 (5), 394–408.

Map Development and Generalization

When a person makes a map, they convert spatial and textual information into a graphic that both communicates a message and provides people with a sense of place (Kraak, 1998). A certain amount of freedom in making maps is important, yet following appropriate cartographic practices may lead to the development of higher quality products (Green and Horbach, 1998). The development of a map is, therefore, an endeavor that involves a synthesis of ideas concerning the display of landscapes or water bodies on a flat medium such as paper or a digital screen (Schuit, 2011). In doing so, a balance must be struck between providing the minimum amount of relevant information, retaining the maximum amount of relevant information, and displaying this information as clearly as possible (Barbour, 2001).

Maps are developed in an attempt to frame a message in a context that will be effectively interpreted by the map user (Harley, 1989). Because maps are scaled down, highly simplified models of real systems, they are generalized descriptions of landscapes or water bodies. Contour lines, for example, can convey a considerable amount of information about a landscape, yet these lines may need to be generalized to improve the aesthetic quality of a map (Irigoyen et al., 2009). Perhaps this seems counterintuitive, but to be effective, every map excludes some information. The map developer acts as a filter, and the features contained in a map represent those that the map developer decides are important when crafting the story of the map (Metcalf, 2012). Through simplification, the map's message or purpose is often enhanced. Map generalization is also important for large organizations, such as national mapping agencies, that develop standard products at multiple scales. The detail provided on a large-scale map (1:10,000) would likely need to be generalized to produce a smaller scale map (1:50,000), since the amount of information contained in larger scale maps may be too great to illustrate on smaller scale maps (Stoter et al., 2014). This problem is also exemplified in the maps that are available through digital devices or computer mapping systems. As one zooms into, or out of, a portion of a landscape, the amount of landscape detail that is provided often changes. The features of a landscape are therefore reduced, simplified, combined, symbolized, omitted, or transformed during this process (Morrison, 1974). Perhaps for these reasons, maps have been described as the outcome of a negotiation between the inner and outer voices of a map maker (Wood, 2002). The internal negotiation process in the development of a map might include the following six general considerations:

1. Mapping of natural or human systems can be a rather subjective activity (Deng et al., 2016), so the first consideration in making a map is to *decide upon the purpose*. Without a purpose, there is likely very little use for a map (Harley, 1989). Along these lines, it would seem necessary to formally or informally assess the problems, opportunities, or conditions that the map needs to illustrate (Bi and Lin, 2009). Further, an understanding of the context within which the map will be used is important, and this may influence further map development decisions. For example, if a map is to be used only by one person, for a short period of time and for some personal reason, the map developer may decide to apply less effort to aesthetic quality and solely focus on its utility. A greater focus on aesthetic quality might apply in cases where a map might be inserted into a publication or report. In this latter case, the purpose of the map insinuates greater stature and potential use by a wider audience. The permanence of the map may also be greater, and therefore more attention to the aesthetic quality of the map may be necessary. In essence, the quality of a map should be reflective of its purpose.

   Translation 25
   Examine the map located on this book's website (mapping-book.uga.edu) that illustrates some aspects of the US states situated along the Gulf of Mexico and the Atlantic Ocean. Aside from the obvious information contained in the title of the map, what broader purpose do you think the map developer had in mind as this map was being created?

2. Along with determining the purpose of the map, the second consideration may be to *understand the potential audiences that may use the map*. At a minimum, the audience may be composed of only one person and used for only a short duration. As we suggested, at this far end of the audience spectrum a map developer may accept a lower level of aesthetic and cartographic quality that will be sufficient to simply meet the immediate needs of the audience (that one person). However, more than likely, someone else will use the map. This fact may instruct the map developer to employ several of the standard practices associated with mapping that we mentioned earlier in the book. The organization within which the map developer works may also require the insertion of

standard map elements, annotation, or other landscape features onto a map. Therefore, in considering the audience of a map, the map developer should think about what that audience expects to see in a map.

Inspection 34

Examine the map located on this book's website (mapping-book.uga.edu) of the vegetation types located on the Kiowa National Grassland in New Mexico (U.S. Department of Agriculture, Forest Service, 2011). Name 10 potential audiences of this map.

3. Another basic consideration in the development of a map is to *determine the area that the map should represent*. For example, in creating a thematic map of soil types within Section 6 of Washburn Township (T23N R1W) in Clark County, Wisconsin (Fig. 7.1), it would seem unreasonable to include the entire township, county, or state in the frame of the map, as Section 6 of Washburn Township encompasses but a small area of the

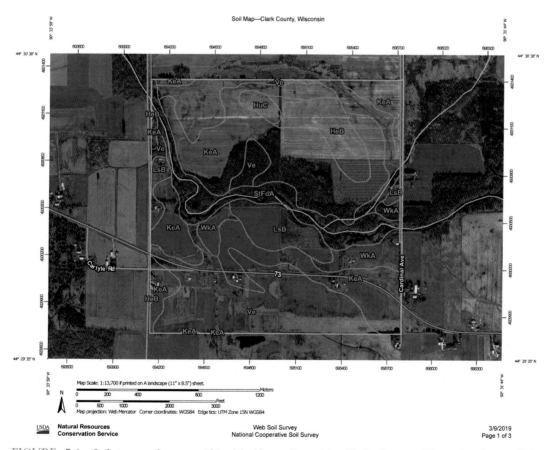

FIGURE 7.1 Soil type polygons within Washburn Township, Clark County, Wisconsin. *Source: U.S. Department of Agriculture, Natural Resources Conservation Service (2017).*

FIGURE 7.2   The position of Section 6 within Washburn Township. *Source: U.S. Department of Agriculture, Natural Resources Conservation Service (2017).*

township (Fig. 7.2). However, perhaps the county or state could be used within a locational inset to provide map users geographic context of Washburn Township's relative location. The aim is to adequately portray the geographic scope of the system to be mapped. The area of interest should be positioned appropriately in the main map frame, using a scale that neither diminishes the importance of the landscape of interest (a scale too small) nor results in omission of portions of the landscape of interest (a scale too large).

4. Once the basic considerations (purpose, audience, area) have been explored, *a list of the map elements that first seem necessary* might be compiled to communicate the map's message. In addition to the standard map elements (title, legend, orientation, scale, etc.), the thematic arrangement of landscape features and the appropriate color schemes (Fig. 7.3) might be considered, along with annotation that provides context (road names,

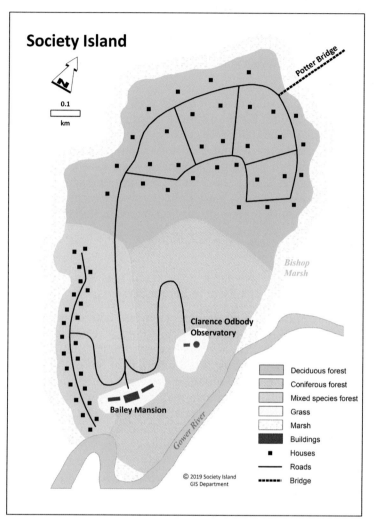

FIGURE 7.3 A map of a hypothetical location called Society Island, indicating the broad vegetation classes that might be found on the island.

contour lines, important points, etc.) for the user of the map. Concurrent with the data accumulation and acquisition process, documentation on the person or organization that developed the data, what point in time the data represent, how accurate the data seem to be, and whether there are limitations in using the data may be necessary. Through this compilation of desired map elements, the need to collect or develop new data for the map being created may become evident.

Diversion 38

Imagine that you need to develop a map of the urban trees in the downtown area of your hometown. In addition to the standard map elements (title, legend, orientation, scale, etc.), develop a list of other map elements or map features (e.g., databases of things found upon the landscape) that would be of benefit to this map. Now, rank these to identify the three main elements or features that should be added. What symbology (e.g., color scheme, font style, etc.) would you employ to illustrate these? If differences among or within each of these is necessary, what symbology (e.g., color scheme) would you then use?

5. A very important consideration in the development of a map is to *determine the media on (or in) which the map will be delivered*. Should the map simply be printed, the size of the paper (or other surface such as Mylar®) may limit what can adequately be displayed. For digital maps, the type of digital delivery system should be thoroughly considered. These products may range from digital maps stored as portable document format (PDF) files or images directly displayed on a cellular phone, tablet, or computer screen. The choice of software and hardware associated with these delivery systems may influence other aspects of the map, forcing a reconsideration of previous steps in the map development process.

Reflection 28
Try to remember the last map that you viewed. On what media was it presented? Could this map have been presented effectively on other media?

6. Finally, economics may play a role in the map development process. Therefore, an essential consideration in the map development process is to *estimate the time and cost to develop the map*. It is an inescapable truth that the process of developing a map requires time. Unless the map developer is altruistic, and is developing a map out of the goodness of their heart, someone (or some organization) is paying the map developer to create the map. The amount of time spent working on a map can therefore be translated to a cost. Further, there may be other expenditures associated with map development (data acquisition, media, software, hardware) that ultimately should contribute to the overall cost of a map. Unfortunately, the time a map developer spends working on a map may force them to not complete other work activities. This is important for those (e.g., such as engineers, foresters, and analysts) who have multiple roles in an organization. These time and cost considerations can influence the map development process and require a reconsideration of the previous steps described.

Admittedly, there are times when it would be reasonable for people to enter into the map development process with great speed and urgency. Maps of active wildfires or of storm damaged areas might need to be created relatively quickly. As we noted, there are a number of considerations and issues one should contemplate when developing a map. Some of these relate to the reduction (generalization) of a place of interest, and the presentation of this place of interest on some form of media (paper or digital screen). The level of confidence needed to create a quick and effective map may therefore be attained with an accumulation of mapping experiences.

In sum, logical considerations in the development of a map include:

- deciding upon the purpose of the map;
- understanding the potential audiences;

- determining the area that the map should represent;
- compiling a list of the necessary map elements;
- determining the media on (or in) which the map will be delivered; and
- estimating the time and cost to develop the map.

Drawing a Map by Hand

Although there are numerous computer programs that assist in map development, some of which are described in the next section. In this section, we describe how one might draw a map by hand, to a specific scale, and with reasonable arrangement of map elements and annotation to communicate a message to the map user. Here, the purpose of the map we will draw is ultimately to describe the vegetation conditions within a small property, yet through this discussion, we will only go as far as drawing the outline of a property, at a specific scale, on relatively normal writing paper (either A4 or 8.5 inches × 11 inches).

Reflection 29
The exercise that follows illustrates the development of a hand-drawn map of a small property. In addition to the purpose of making a map that describes the vegetation conditions on this property, what other purposes could be accommodated with a basic map of a property?

Assume that there is a US Bureau of Land Management property located in Gunnison County, Colorado (parcel 3431-000-00−002). The legal description of the parcel is W1/2 Section 7 Township 15 South Range 83 West within the US Public Land Survey System. From the description of the US Public Land Survey System in Chapter 4, Map Reference Systems, you may recognize that this represents one-half of a section of land, which theoretically suggests that the parcel size is 320 acres (129.5 ha) assuming it was surveyed and monumented perfectly on the ground. The property is a rectangle, but not a perfectly oriented one, for as you will see it leans slightly to the right. We will assume that the property boundary lines connecting the four corners are perfectly straight. The approximate Universal Transverse Mercator (UTM) coordinates (Zone 13) of the four corners are:

Lower-left corner	4,290,910 m north	350,600 m east
Upper-left corner	4,292,510 m north	350,613 m east
Upper-right corner	4,292,490 m north	351,558 m east
Lower-right corner	4,290,890 m north	351,527 m east

Translation 26
As you may recall, we discussed the UTM system in Chapter 4, Map Reference Systems. Imagine that a friend or family member is watching you draw this map of the US Bureau of Land Management property located in Gunnison County.

They see the coordinates and are curious what the alphanumeric characters represent. Describe for them very briefly the concept of a northing and easting, and what one set of these two values (4,290,910 m north, 350,600 m east), as an example, means in relation to this property.

It appears that the maximum length (or height) of the property when drawn on a sheet of paper, from a north-south perspective, is about 1620 m. This can be determined using the difference between the maximum and minimum northings. The upper-left corner of the property has the maximum northing (4,292,510 m) and the lower-right corner has the minimum northing (4,290,890 m).

maximum length of drawn property = maximum northing − minimum northing
maximum length of drawn property = 4,292,510 m − 4,290,890 m
maximum length of drawn property = 1620 m

It also appears that the maximum width of the property when drawn on a sheet of paper, from an east-west perspective, is about 958 m. This can be determined in this case by determining the difference between the maximum and minimum eastings:

maximum width of drawn property = maximum easting − minimum easting
maximum width of drawn property = 351,558 m − 350,600 m
maximum width of drawn property = 958 m

Given these dimensions, and the size of the media on which the map will be drawn, a map scale of about 1:10,000 seems reasonable. In this case, when drawn, the maximum length of the property on the map is:

maximum length of property on the map = maximum length of drawn property/map scale
maximum length of property on the map = 1620 m / 10,000
maximum length of property on the map = 0.162 m
maximum length of property on the map = 16.2 cm
maximum length of property on the map = 6.378 inches

The maximum width of the property on the map is:

maximum width of property on the map = maximum width of drawn property/map scale
maximum width of property on the map = 958 m / 10,000
maximum width of property on the map = 0.0958 m
maximum width of property on the map = 9.58 cm
maximum width of property on the map = 3.772 inches

If we assume we are using 8.5 inch × 11 inch paper on which to draw this map, we might begin by positioning the lower-left corner of the property on the paper. We will assume that the parcel itself does not need to be shifted aside to accommodate other map elements, and thus the parcel will be positioned directly in the middle of the page. The distance from the bottom of the page to the lower-left corner of the parcel involves knowing how much extra space we may have above and below the drawn parcel. If the page is 11 inches tall, and the *maximum length of property on the map* is 6.378 inches, this leaves about 4.6 inches above and below the parcel. If we divide this evenly, we would have

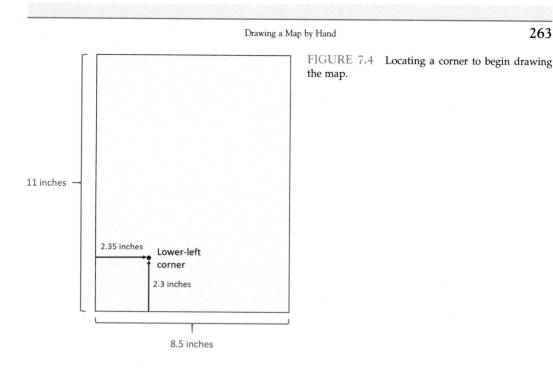

FIGURE 7.4   Locating a corner to begin drawing the map.

about 2.3 inches (about 5.8 cm) from the bottom of the page to the lower-left corner of the parcel, and about 2.3 inches from the top of the page to the top of the drawn parcel.

The distance from the left edge of the page to the lower-left corner of the parcel involves knowing how much extra space we may have to the left and right of the drawn parcel. If the page is 8.5 inches wide, and the *maximum width of property on the map* is 3.772 inches, this leaves us about 4.7 inches to the left and to the right of the parcel. If we divide this evenly, we would have about 2.35 inches (about 6 cm) from the left edge of the page to the lower-left corner of the parcel, and about 2.35 inches from the right edge of the page to the right side of the drawn parcel. With this information (2.3 inches from the bottom of the page, 2.35 inches from the left edge of the page), the location of the lower-left corner of the property can be drawn on the map (Fig. 7.4).

Once we position one of the corners on the map, when we continue to draw the parcel we might think in terms of the direction of travel on the ground. For example, after the lower-left corner has been defined on the map, we might draw the edge leading to the upper-left corner as if we were walking in that direction. The lower-left corner has UTM coordinates of 4,290,910 m north and 350,600 m east. The upper-left corner has UTM coordinates of 4,292,510 m north and 350,613 m east. This suggests that we moved 1600 m north and 13 m east when traveling from the lower-left to upper-left corners. With respect to the practice of surveying, this implies a change in *latitude* of +1600 m and a change in *departure* of +13 m. In other words, we moved north and east from the lower-left corner (Fig. 7.5A).

To determine the change in latitude:
Destination − upper-left corner northing: 4,292,510 m
Origin − lower-left corner northing: 4,290,910 m
Difference: +1600 m

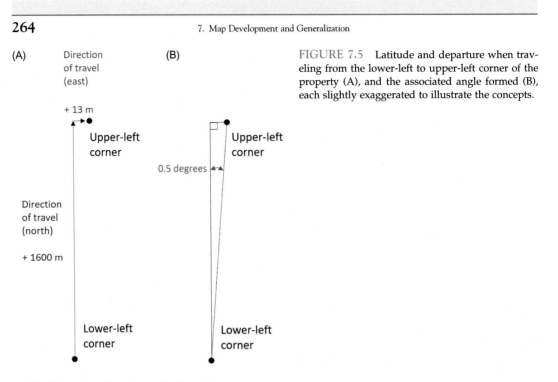

FIGURE 7.5 Latitude and departure when traveling from the lower-left to upper-left corner of the property (A), and the associated angle formed (B), each slightly exaggerated to illustrate the concepts.

To determine the change in departure:
Destination − upper-left corner easting: 350,613 m
Origin − lower-left corner easting: 350,600 m
Difference: +13 m

Reflection 30
When drawing a map such as this, changes in latitude are positive values (+) when traveling northward and are negative values (−) when traveling southward. Changes in departure are positive values (+) when traveling to the east and are negative values (−) when traveling to the west. The latitudes and departures (separately) can be summed as one draws a map around the property, beginning with movement away from the initial corner and ending with movement back to the initial corner (closing the polygon). Ideally, what numeric value should the sum of the latitudes have? How about the sum of the departures?

Using right triangle trigonometry, the angle traveled from the lower-left corner to the upper-left corner is about 0.5 degrees (Fig. 7.5B).

$\tan(\alpha)$ = (opposite side length/adjacent side length)
$\tan(\alpha)$ = 13 m / 1600 m
α = 0.4655

To draw this edge of the property, we would position a protractor with its center positioned on the lower-left corner and its representation of north pointing parallel to the edge of the page (assuming north is at the top of the page), then mark a location of approximately

0.5 degrees. This represents the direction we are headed from the lower-left corner to the upper-left corner of the property. The distance we are to draw involves using the changes in latitude and departure, the Pythagorean theorem, and the map's scale:

(Distance to draw on the map)^2 = (change in latitude)^2 + (change in departure)^2
(Distance to draw on the map)^2 = (1600 m)^2 + (13 m)^2
(Distance to draw on the map)^2 = 2,560,169 m^2
Distance to draw on the map = 1600.05 m

This is the hypotenuse of a right triangle formed using an angle α of 0.5 degrees, a hypotenuse of 1600 m, and an opposite side of 13 m. To draw this on the map, it needs to be scaled down:

Map distance = (distance to draw on the map/map scale)
Map distance = 1600.05 m / 10,000
Map distance = 0.160005 m
Map distance = 16.0005 cm
Map distance = 6.3 inches

Translation 27
Imagine again that a friend or family member is watching you draw a map of this US Bureau of Land Management property located in Gunnison County. While they enjoy your efforts to draw the boundary of the property, they recoil when they see you use mathematics and trigonometry. Describe for them very briefly why this effort is necessary.

We then draw a straight line that is 6.3 inches long, from the lower-left corner of the parcel in the direction of the upper-left corner of the parcel (Fig. 7.6), angling to the right by 0.5 degrees (azimuth = 0.5 degrees).

When moving from the upper-left corner to the upper-right corner, we travel south and east (Fig. 7.7A).

To determine the change in latitude:
Destination − upper-right corner northing: 4,292,490 m
Origin − upper-left corner northing: 4,292,510 m
Difference: −20 m

To determine the change in departure:
Destination − upper-right corner easting: 351,558 m
Origin − upper-left corner easting: 350,613 m
Difference: +945 m

Using right triangle trigonometry, the angle traveled from the upper-left corner to the upper-right corner is about 1.2 degrees (Fig. 7.7B).

tan(α) = opposite side length / adjacent side length
tan(α) = 20 m / 945 m
α = 1.2

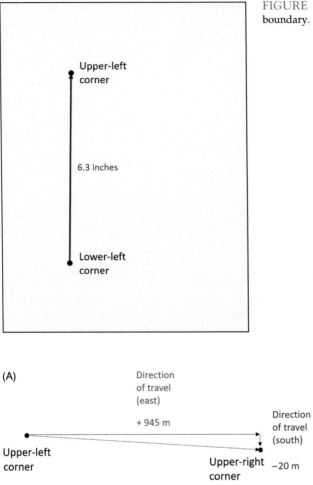

FIGURE 7.6   Drawing the first side of the property boundary.

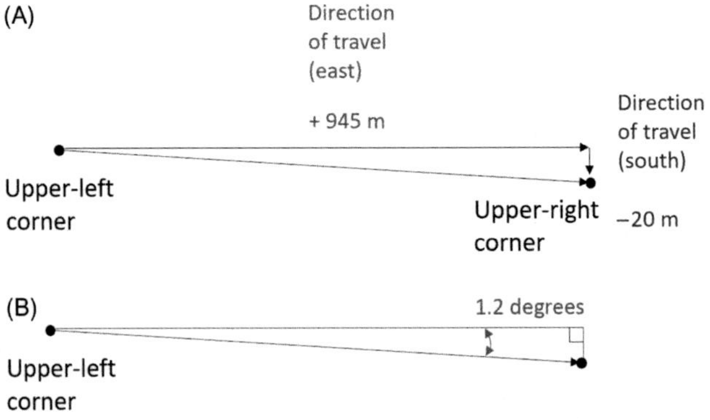

FIGURE 7.7   Latitude and departure when traveling from the upper-left to upper-right corner of the property (A), and the associated angle formed (B), each slightly exaggerated to illustrate the concepts.

This may be misleading without attention paid to the map. Given the model presented, the azimuth traveled is not 1.2 degrees, but rather 91.2 degrees since we are deviating 1.2 degrees from an easterly direction. Therefore, if you were to position a protractor with its center on the upper-left corner and north pointing parallel to the vertical edge of the page (assuming north is at the top of the page), and mark a location of 91.2 degrees, this would be the direction we are headed from the upper-left corner to the upper-right corner of the property. As before, this distance we are to draw involves using the changes in latitude and departure, the Pythagorean theorem, and the map's scale:

(Distance to draw on the map)^2 = (change in latitude)^2 + (change in departure)^2

(Distance to draw on the map)² = (945 m)² + (− 20 m)²
(Distance to draw on the map)² = 893,425 m²
Distance to draw on the map = 945.2 m

This is the hypotenuse of a right triangle formed using an angle α of 1.2 degrees, a hypotenuse of 945 m, and an opposite side of 20 m. To draw this ground distance on the map, the length of it (map distance) needs to be determined using the scale of the map:

Map distance = (distance to draw on the map/map scale)
Map distance = 945.2 m/10,000
Map distance = 0.09452 m
Map distance = 9.452 cm
Map distance = 3.72 inches

We then draw a straight line that is about 3.7 inches long, from the upper-left corner of the parcel in the direction of the upper-right corner of the parcel (Fig. 7.8), and angling 1.2 degrees to the south (azimuth = 91.2 degrees).

Diversion 39
For the other two sides of the map, using the four coordinates noted earlier, determine the changes in latitude and departure when moving from one corner to the other, and the length of each edge of the property (the map distance for each side). Then draw the additional two sides of the parcel on the map.

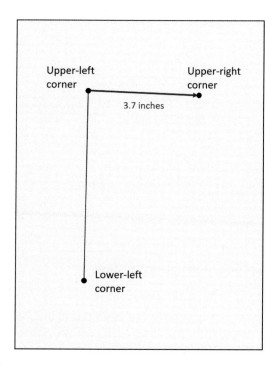

FIGURE 7.8   Drawing the second side of the property boundary.

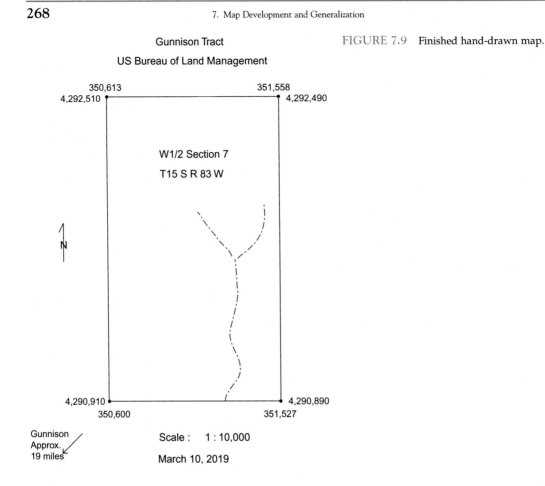

FIGURE 7.9   Finished hand-drawn map.

Once the parcel has been drawn to scale, you would then add standard map elements (title, north arrow, scale, etc.) to complete a professional, hand-drawn map (Fig. 7.9).

Diversion 40
Add to the map you have been creating some standard map elements. Further add to this map any annotation or symbology you feel would be of value in increasing the aesthetic appeal of the map.

Developing a Map in a Geographic Information System or Online Mapping System

Knowledge of and skill in using computer mapping programs are essential characteristics of new professionals beginning their careers in the management of natural resources (Bettinger and Merry, 2018). Perhaps the first issue to address is the difference between a geographic information system (GIS) and an online mapping program. Technically, a GIS encompasses the software, hardware, and infrastructure necessary to collect, manage, manipulate, store, and present geographic data. However, people often simply refer to the software as GIS. Some examples are ArcGIS (ESRI, 2018), SuperMap (SuperMap International Limited, 2018), and QGIS (QGIS Development Team, 2018). GIS facilitates the rapid integration of multiple types of data and knowledge and allows society to not only view phenomena of interest through maps, but also to conduct complex geographic analyses. With advances in computing technology, GIS is now migrating to *the cloud*, that mysterious place where computers and computing technology come together in a data center and allow people to develop maps and to conduct complex analyses without physically storing either the data or the GIS software on their personal computing device (Peterson, 2015). Online mapping technologies, including multiscale pan-able maps that assist us in navigating our daily routes, are of great benefit to society, but the analytical capabilities of these are often very limited. For example, within a GIS one can open a soils and a vegetation database, combine the two, and then ask (through a query) where the best places to plant a certain crop might be located. Geographic analysis along these lines is generally not yet possible in an online mapping program such as those provided by Google Earth (Google, 2018a), MapQuest (Verizon Communications, Inc., 2018), local taxing agencies, or government agencies. In addition to limitations in affordances (what online mapping technologies allow people to do), there are also challenges related to the interactivity of online mapping technologies (Poplin, 2015). However, online mapping technologies continue to evolve, and systems such as Google Earth Engine (Google, 2018b) may one day allow a crossover in functionality between traditional GIS and online mapping programs.

Inspection 35

Search the Internet for a tax assessor's office that provides online mapping or map search functions. In a memorandum, describe the capabilities of the system and make a list of the information that might be accessed through this online mapping system.

Whether one has access to GIS software or an online mapping program, the following steps will be beneficial in developing a map. The considerations we provided in the beginning of this Chapter will be of great value in the development of a map. What follows is a more mechanical description of the physical process of making a map.

1. *Select the area of interest.* Within a GIS, a map developer would open one or more geographic databases and depending on their extent (the amount of land or sea that they cover), zoom into an area of interest. Often the area of interest that is viewable is roughly the same area that will be presented in a printed or digital map. Within online mapping programs, a map developer will often be instructed to define the area of interest using a rectangle drawn with a computer mouse, or by supplying the online mapping system a geographic data file (e.g., a shapefile) with which the online system will understand the intended area of interest through an examination of the extent of the data.

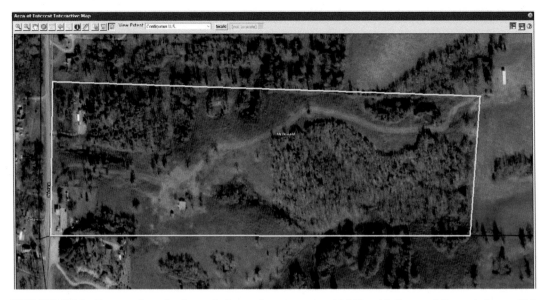

FIGURE 7.10   Farm and patch of woods in southwest corner of McDonald County, Missouri. *Source: U.S. Department of Agriculture, Natural Resources Conservation Service (2017).*

Diversion 41
The U.S. Department of Agriculture, Natural Resources Conservation Service (2017) supports access to nationwide soils databases through their Web Soil Survey. Locate this online mapping system and find the buttons that allow you to define an area of interest. Assume that your area of interest is McDonald County in the southwest corner in the state of Missouri. If you were to draw a rectangular area of interest that encompasses the entire county, what message would you receive? Assume now that your area of interest is the farm and patch of woods in the far southwest corner of McDonald County, just below (interestingly) Southwest City. If you were to draw a rectangular area of interest around this patch of farm and patch of woods (Fig. 7.10), how large is the area and what message might you have received from the Web Soil Survey system?

2. *Define the scale of the map.* The scale desired may need to be adjusted, depending on how much of a landscape or water body is included within a map. GIS programs allow users to define explicitly the scale of the map being developed. Online mapping systems may only present the user with one scale option, or perhaps the scale may be a function of the zoom level (or eye altitude above the landscape). Depending on the map scale selected, the area of interest may change.

Reflection 31
Imagine that you own a forestry consulting business, and that one of your clients owns a number of small forested parcels in the area where you live. Imagine further

that the landowner has asked you to develop a very comprehensive management plan for their collection of parcels. The scale of each parcel map might be different, simply due to differences in the size of the area of interest (the area around each parcel). Would these differences in map scale concern you as you develop maps for their management plan? Why?

3. *Add all of the features of concern to the map.* Most GIS programs allow the addition of user-defined geographic data to a map, while others (especially online mapping systems) force a map developer to select from a predefined set of data that usually cannot be edited. Some mapping programs also allow users to add symbols, features, and annotation that are not included in geographic databases. The specification of feature properties (font, color, style, etc.) and the placement of features on a map involves the *symbology* associated with mapping (Buckley and Frye, 2006), as we have mentioned in earlier chapters.

GIS programs allow map developers to both create and insert databases of landscape features into maps. For example, a biologist may want to create a database of the location of owl nesting sites. In doing this, they might create a database of point features, a basic vector data structure within GIS. Point features are simply single sets of *X,Y* coordinates that are independent of other nearby points, lines, or polygons. However, imagine that road or stream features are also of concern to a map, and that databases of these (and the nest sites) are unavailable. A map developer can create these features through the process of *digitizing,* an old term that we still use today to describe the creation of vector GIS data. Twenty years or more ago, a person would have used a special table (called a digitizing table) and a special computer mouse (often called a puck) to draw or *digitize* the landscape features. The person doing the digitizing would essentially trace the course of a road or stream using the digitizing puck, which would send a signal to the digitizing table (through the click of a button) to record *X,Y* coordinates of points (or the vertices of lines and polygons). Today, we essentially do the same thing with a computer mouse in conjunction with GIS software operating on a personal computer. For example, creating a point that represents an owl nest site would likely involve a single click of a computer mouse. For line features such as roads and streams, a person would use the computer mouse to draw the course of a road or stream, clicking a mouse button each time a new vertex is needed. When there are long, straight stretches in a road or stream, fewer vertices are needed to describe the course. When there are significant curves or bends in a road or stream, many more vertices may be needed to adequately describe the shape of the linear feature (Fig. 7.11). Lines, once drawn, have nodes (essentially special vertices) at each end. If other nearby lines are to be drawn, and these lines are to be connected, the nodes that represent where the lines connect should have the same geographic position or coordinates.

When one draws polygon features, they are creating enclosed areas that are significant enough in real life to be distinguished from other nearby areas. For example, in Fig. 7.12 the polygons reflect different ages of avocado plantations. Polygons are formed by connecting one or more lines to create an enclosed area. As with line features, more vertices may be needed to describe the shape of the edges of polygons where the edges are irregular, and less vertices are required where the edges are straight.

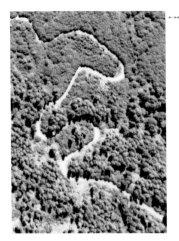 

FIGURE 7.11 A line feature, and its associated vertices, that describes a rugged mountain road in western Oregon.

FIGURE 7.12 Polygons, and the vertices of one, drawn around avocado plantations in La Habra, California, c.1973 (centered approximately 33.921°N latitude and 117.926°W longitude).

Diversion 42
From this book's website (mapping-book.uga.edu) acquire the GIS database related to the Gunnison BLM Tract. Using your preferred GIS mapping program, (1) digitize the streams that seem to flow through this property and (2) digitize the different vegetation patches (timber stands, management units, etc.).

4. *Add essential map elements to the map.* Where possible, essential map elements such as a north arrow (for orientation), a scale bar or scale text, a title, a legend, and the name and date of the map developer should be added to the map, in places that would seem most aesthetically pleasing to the intended audience. Other map elements we described in Chapter 3, Map Components, may also be necessary. For example, a warranty or caveat may be necessary when a governmental agency or a consultant develops a map for widespread public use.

Translation 28
Imagine again that you own a forestry consulting business, and that one of your clients owns a number of small forested parcels in the area where you live. The landowner has asked you to develop a very comprehensive management plan, and you are in the process of making the maps. You want to convey two things on these maps (1) that they are only for the landowner's use, and (2) you do not want to be held responsible for bad decisions that may result due to a map error caused. Very explicitly, how would you convey these concerns on the maps you develop?

5. *Print or develop the map and inspect its quality.* Unless you are in a great hurry to produce a map, it is always wise to create a draft and inspect it for potential revisions. Allowing a colleague or family member to view the map and offer comment may elicit some suggestions on content or aesthetics that were not evident to you during the map development process. Once the *final map* is printed or developed, it would also seem wise to view it again, for it may require additional work to meet its objective.

Diversion 43
Using the Gunnison BLM Tract GIS data available from this book's website (mapping-book.uga.edu), other data you may have developed for this tract, and any other information that may be available over the Internet, make a professional map of the area. Keep these issues in mind as you create the map: How far outside of the tract would you consider the area of interest to extend? What reasonable scales can you use in producing this map? What other landscape features would you add to the map? What map elements would seem necessary?

Generalizing Landscape Features

As we mentioned in the beginning of this chapter, *generalization*, within the guise of making a map, is the act of simplifying the features that you see by removing some of the detail that was once used to describe them. Nearly all maps that describe natural resource phenomena or human systems, such as maps of cities, national forests, and even paths of tornados (Deng et al., 2016), are simplifications of the real world or of real-world events. At certain map or viewing scales, the need for generalization of landscape features becomes important. In sum, generalization of landscape features is necessary in the development of a scaled model of a landscape (a map), and may require not only simplification in the shape of features, but also perhaps an offsetting (displacement) of features in order to maintain aesthetic clarity (Geoscience Australia, 2012).

With respect to vector data used within GIS, the generalization process may involve removing sharp points along the edge of a polygon or along the course of a line. Physically, this would involve the removal of vertices that describe the shape of these features, and perhaps the placement or adjustment of others in nearby locations that act to smooth the edges of the features. The density of points or vertices that describe lines or polygons within a vector GIS database can be controlled through automated processes such as *point density reduction*, where vertices describing lines are reduced or thinned. One purpose of generalization is to improve the aesthetic quality of a map produced at a certain scale. For a large-scale map (1:10,000 or less), the detail described by complex features might be retained. However, for a small-scale map (e.g., 1:50,000 or greater), complex detail may be indistinguishable and of no consequence. At smaller scales, it may also make sense from data storage, map printing, or map rendering perspectives to reduce the amount of detail necessary to communicate a message through a map. When these methods are applied to the data of a map, yet the map scale has not changed, they should be considered acts of *data reduction*; when the scale changes they might better be considered acts of *map generalization* (Li, 1993). For certain types of maps, such as orienteering maps, processes such as *selective generalization* and *graphical generalization* may be employed (International Orienteering Federation, 2017). Selective generalization involves deciding which features and how much detail should be incorporated into a map, based on the relative importance of features and their influence in the legibility of the map. Graphic generalization is more closely related to the forms of generalization we have previously discussed, and involves simplification, displacement or offsetting, and exaggeration of landscape features.

As an example of a generalization process, we will use a representation of the centerline of the Mississippi River, where it borders the western edge of the State of Mississippi (Fig. 7.13). The database that describes the river (and others) was developed as a source of reference information on the length of major rivers around the world (Basher, 2018). Prior to generalization there are 2163 vertices that are employed to describe the shape of the river, each ideally were created to illustrate a significant change in direction of river centerline. The bends and curvature of the river seem to be depicted fairly well with respect to imagery that is available through Google Earth.

Inspection 36
Although we provided a statement on the quality of the Mississippi River data, perhaps you should see for yourself. Acquire the Mississippi River KML file from the book's website (mapping-book.uga.edu) and open it in Google Earth. Examine the location of the centerline of the river and develop a brief summary of the quality of the data. Where you notice issues concerning the location of the river centerline, explain how these might have arisen.

FIGURE 7.13 Mississippi River along the west-
ern edge of the State of Mississippi. *Source: U.S.
Department of Agriculture, Natural Resources
Conservation Service (2017).*

A process to simplify the centerline of the Mississippi River was employed in ArcGIS (ESRI, 2018), retaining critical points that describe the centerline of the river yet using a simplification tolerance or minimum allowable spacing of 50 m (164 ft). Changes in the shape of the centerline are minimal when viewed at the scale of our map (Fig. 7.14) even though the number of vertices that represents this portion of the river is now only 719. Even employing simplification using a 100 m (328 ft) simplification tolerance provides no visual difference at the scale of the map presented, whereas the number of vertices has been reduced now to 464. However, when viewed more closely, one can clearly see differences in the representation of the river centerline (Fig. 7.15) among the original data and the two simplified databases. This form of generalization uses the Douglas–Peucker method for point (vertex) reduction (Douglas and Peucker, 1973). This algorithm has been used as a common data reduction (simplification) process that selects vertices (points) to thin or remain from a line based on how far they are from a straight line that connects other points already chosen to remain (Ebisch, 2002).

Reflection 32

Imagine, after spending several hours digitizing the intricate detail associated with the Mississippi River, your supervisor instructs you to now generalize or simplify

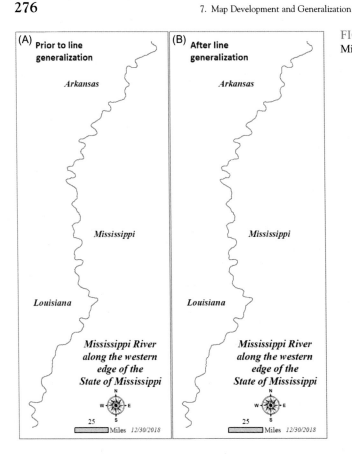

FIGURE 7.14 Generalization of the Mississippi River.

the data for public use. How would you feel about this type of data reduction process being applied to the database you very carefully created?

Another method for generalization of geographic data involves the retention or preservation of the overall structure of features, or critical bends within the course of a line or the shape of a polygon. This method is based on a line generalization process described by Wang and Müller (1998) that locates the bends within the shape of a line, and adjusts those deemed insignificant through elimination, combination, or exaggeration operators, while simultaneously adhering to the simplification tolerance. Other forms of map generalization may involve the aggregation of points, lines, or polygons that have similar characteristics with respect to the attributes that describe a landscape, or simply with respect to the fact that the points, lines, or polygons are geographically close to one another in the real world. Smoothing techniques can also be used to generalize lines or areas. These shift the positions of vertices in order to remove small perturbations (Irigoyen et al., 2009). The manner in which vertices are modified (or removed) is important, and a map developer needs to pay attention to unintended outcomes (e.g., overlapping polygons) that might occur when databases are generalized (or smoothed Shi and Charlton, 2013).

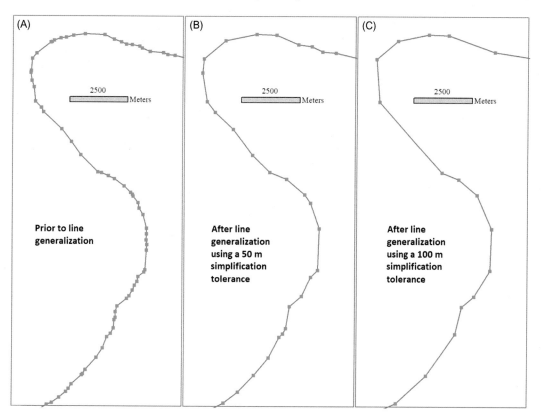

FIGURE 7.15   Closer view of the generalization of the Mississippi River.

Finally, it may be the case with advanced printing technology, such as 3D printers, that landscape features need to be generalized. For example, the rendering of tall, thin landscape features on a 3D map, such as cell phone towers or even the tops of some trees, may not be possible given the scale of the map and the physical ability of the printer (Kete, 2016). Therefore, the ability to illustrate very fine detail that might be available through aerial imagery may not yet be possible in a 3D map, and features such as these would be either generalized or absent from the map. Further, when printed three-dimensionally some landscape features that are suspended above ground may be sensitive to vibration and mechanical stress, given the materials used, and to prevent damage, these features may need to be reshaped or artificially connected to others to prevent map damage (Kete, 2016).

Rendering Digital Maps on Computers and Devices

Digital maps have several perceived advantages over printed maps: they can be updated and modified quickly, can be embedded into Internet-based content, can be easily transported and stored on handheld digital devices, and are widely used today (Kerski, 2019). In the digital world, creating an image of a real, 3D landscape on a two-dimensional (2D) plane is called *rendering*. Digital mapping involves a dynamically changing set of technologies. To be successful, a map developer may need to conduct a competitive analysis of the technology available, conduct an assessment of experiences and opinions of these technologies, and complete small case studies to assess their value when applied to a specific purpose (Roth et al., 2014). Realistic rendering of landscape features and terrain models on computers and other devices has been a concern for geographers and computer scientists for over three decades, as not only these processes are important for the development of maps, but also for applications such as flight simulators (Miller, 1986).

Static, analog maps, those presented both on flat media (e.g., paper) and on the screens of digital devices, undergo a map development process that requires consideration of the concepts presented in earlier chapters. These traditional maps engage the map user in a one-way interaction; the map user observes the features presented and gains insight or knowledge from the experience. Interactive maps are becoming more prevalent, particularly with the expansion of maps rendered on computers and digital devices. Although it may be debatable what constitutes an interactive map, Roth (2012) suggests that a two-way dialog engaging the map user with the map itself, often mediated by a computer, is necessary. As with traditional maps, interactive maps might be developed to promote visual thinking, to facilitate the development of geographic insight, and to assist users in synthesizing, exploring, and presenting messages posed within a geographic perspective (Roth, 2015). Speed, flexibility, and interface constraints may limit user satisfaction with interactive maps, but users of maps are more often becoming co-developers of those maps, whether they understand this or not, through interactive processes.

Reflection 33
After what you have learned so far about map design and development, society might take for granted the speed at which high-quality, interactive navigational maps are developed and delivered. Name another complex technology that you can reasonably compare to this type of societal reaction, and provide support for your opinion.

With respect to digital maps, generalization processes may be geared more toward compliance with user requests made through digital devices, rather than toward traditional legibility and perception issues associated with classical cartography (Lisitskii, 2016). However, color differences between the actual landscape feature and the mapped feature are a form of map generalization. Therefore, color rendering can be dependent on the technology available and other issues with digital devices that limit perfect color representation (Kete, 2016).

Adhering to a *responsive design* strategy is important for rapid updates of interactive maps delivered on digital devices. A responsive design strategy involves protocols and rules aimed at adjusting digital representations of maps according to display constraints of digital devices. For example, whether delivering a map to a personal computer monitor, tablet, or to a cellular phone screen, the system that enables the map to be viewed and used must be capable of displaying it in each of these various viewing environments to be considered responsive. To ensure user satisfaction with the digital delivery of maps, computer code is necessary to support the needs of digital rendering of maps on a variety of devices. A digital map, interactive or not, may be of limited value if it is only available through a few specific devices or web-browsing interfaces.

In the digital rendering process, points, lines, polygons are considered *primitives*, and various processes (clipping, projection, etc.) are applied to these before they are rasterized, or converted to pixels, and drawn (rendered) on a computer screen (Mustafa et al., 2006). The computer screens can include your desktop personal computer, laptop, tablet, or cell phone, and these are essentially 2D grids that accommodate a certain number of pixels (row potential $\times$ column potential). Pixels are assigned colors and a transparency value during the rendering process. For digital devices, there are also terrain rendering techniques that allow the placement of vector primitives on a 3D terrain surfaces (Schneider and Klein, 2007). When data are too extensive to support rapid display, the data are downsampled to improve visual quality at different scales of view (zoom extents) and to improve the speed at which data are displayed. *Pyramids* are these down-sampled datasets. Each successive smaller scale model pyramid dataset contains a resampled representation of more detailed data from the adjacent larger scale model. For example, a raster dataset may contain pixels with a spatial resolution of 15 m. This dataset can be downsampled to a set of pixels with a spatial resolution of 30 m by assigning to each 30 m pixel the dominant value from the four 15 m pixels it represents (or contains). The 30 m spatial resolution dataset can then be down-sampled to a set of pixels with a spatial resolution of 60 m by assigning to each 60 m pixel the dominant value from the four 30 m pixels it represents (or contains). With advances in 3D software products for small electronic devices, in the future we might expect that digital maps delivered to phones and tables will be augmented by 3D landscapes.

There are a number of technologies people can employ to integrate maps with computer applications or websites. These technologies may be designed to develop maps that produce products that are both visually suitable, illustrating important landscape features and landmarks, and also are subject to generalization processes (simplification, deformation, displacement) to promote recognition of some features and landowners and demote emphasis on others (Grabler et al., 2008). The *rendering engines* that are used are simply algorithms that are designed to convert vector primitives to raster images, combine these with other raster data, assign color schemes, and display the result in the frame that is available. A *graphics processing unit* (GPU), an electronic circuit within a computing device, may assist in efficiently rendering images for a digital display. At times, *tiles* are created to allow more efficient delivery of rendered data to computer devices. Tiles are pregenerated subsets of larger raster images that often serve as base maps in GIS or on the screens of other digital devices. They are presented to the user only when the user needs them, and they may consist of a set of pyramids that allow map users to view or develop maps very quickly.

Concluding Remarks

Map development is a form of storytelling. Effective map development requires a good understanding of the purpose of the map and the story the map attempts to convey. Items or features that are not important to the story should either be eliminated or generalized to avoid confusion. Advances in GIS and online mapping allow a map user to view maps at any scale that they desire. However, a map user should be aware of the scale that a map was prepared or rendered. In making a map available in a format (print or digital screen) and scale appropriate for the intended audience, it would seem evident that large, complex geographic databases may need to be generalized to save processing time, storage space, and rendering (or printing) effort. As a user of a map, we assume that the person who developed the map has engaged in an objective and sometimes scientific process to create and present knowledge (Harley, 1989). This is an assumption that should be softened in our minds. Certainly, experience is necessary to develop well-composed maps. Geographic information systems combine spatial and tabular data, and this facilitates a wide range of spatial analysis that would otherwise be nearly impossible without this integrative technology. By combining cloud computing and GIS, map developers can create dynamic, interesting maps that can be delivered to electronic computing devices (e.g., cellular phones, tablets, personal computers), but again we encourage caution and consideration for the quality of data that are drawn upon (no pun intended).

References

Barbour, D., 2001. Maps for fun and profit: one small company's experience. Cartogr. J. 38 (1), 81–85.

Basher, Z., 2018. Global high-resolution river centerlines. U.S. Department of the Interior, Geological Survey, Reston, VA. <https://www.sciencebase.gov/catalog/item/5a145fdde4b09fc93dcfd36c> (accessed 30.12.18).

Bettinger, P., Merry, K., 2018. Follow-up study of the importance of mapping technology knowledge and skills for entry-level forestry job postings, as deduced from recent job advertisements. Math. Comput. Forest. Natural-Res. Sci. 10 (1), 15–23.

Bi, H.B., Lin, D.K.J., 2009. RFID-enabled discovery of supply networks. IEEE Transact. Eng. Manage. 56 (1), 129–141.

Buckley, A., Frye, C., 2006. A multi-scale, multipurpose GIS data model to add named features of the natural landscape to maps. Cartogr. Perspect. 55, 34–53.

Deng, Y., Wallace, B., Maassen, D., Werner, J., 2016. A few GIS clarifications on tornado density mapping. J. Appl. Meteorol. Clim. 55 (2), 283–296.

Douglas, D.H., Peucker, T.K., 1973. Algorithms for the reduction of the number of points required to represent a digitized line or its caricature. Can. Cartogr. 10 (2), 112–122.

Ebisch, K., 2002. A correction to the Douglas-Peucker line generalization algorithm. Comput. Geosci. 28 (8), 995–997.

ESRI, 2018. About ArcGIS, The mapping and analytics platform. ESRI, Redlands, CA. <https://www.esri.com/en-us/arcgis/about-arcgis/overview> (accessed 02.02.19).

Geoscience Australia, 2012. Section 3 – National Topographic Database Production Information. Commonwealth of Australia, Geoscience Australia, Canberra. <http://www.ga.gov.au/mapspecs/topographic/v6/section3.html> (accessed 04.01.19).

Google LLC, 2018a. Google Earth. Google LLC, Mountain View, CA. <https://www.google.com/earth/> (accessed 29.12.18).

Google LLC, 2018b. Google Earth Engine. Google LLC, Mountain View, CA. <https://earthengine.google.com/> (accessed 29.12.18).

Grabler, F., Agrawala, M., Sumner, R.W., Pauly, M., 2008. Automatic generation of tourist maps. Proceedings of SIGGRAPH 2008. Association for Computing Machinery, New York.

Green, D.R., Horbach, S., 1998. Color – difficult to both choose and use in practice. Cartogr. J. 35 (2), 169–180.

Harley, J.B., 1989. Deconstructing the map. Cartographica 26 (2), 1–20.

International Orienteering Federation, 2017. ISOM 2017 International Specifications for Orienteering Maps. International Orienteering Federation, Karlstad, Sweden.

Irigoyen, J., Martin, M.T., Rodriguez, J., 2009. A smoothing algorithm for contour lines by means of triangulation. Cartogr. J. 46 (3), 262−267.

Kerski, J., 2019. Six ways to increase geoliteracy. ArcUser 22 (1), 64−65.

Kete, P., 2016. Physical 3D map of the Planica Nordic Center, Slovenia: cartographic principles and techniques used with 3D printing. Cartographica 51 (1), 1−11.

Kraak, M.-J., 1998. The cartographic visualization process: from presentation to exploration. Cartogr. J. 35 (1), 11−15.

Li, Z., 1993. Some observations on the issue of line generalisation. Cartogr. J. 30 (1), 68−71.

Lisitskii, D.V., 2016. Cartography in the era of informatization: new problems and possibilities. Geogr. Natural Resources 37 (4), 296−301.

Metcalf, A.C., 2012. Amerigo Vespucci and the Four Finger (Kuntsmann II) world map. e-Perimetron 7 (1), 36−44.

Miller, G.S.P., 1986. The definition and rendering of terrain maps. Proceedings of SIGGRAPH '86. Association for Computing Machinery, New York, pp. 39−48.

Morrison, J.L., 1974. A theoretical framework for cartographic generalization with emphasis on the process of symbolization. In: Kirschbaum, G.A., Meine, K.-H. (Eds.), International Yearbook of Cartography, 14. Kirschbaum Verlag, Bonn-Bad Godeberg, Germany, pp. 115−127.

Mustafa, N., Krishnan, S., Varadhan, G., Venkatasubramanian, S., 2006. Dynamic simplification and visualization of large maps. Int. J. Geogr. Information Syst. 20 (3), 273−302.

Peterson, M.P., 2015. Maps and the meaning of the cloud. Cartographica 50 (4), 238−247.

Poplin, A., 2015. How user-friendly are online interactive maps? Survey based on experiments with heterogeneous users. Cartogr. Geogr. Information Sci. 42 (4), 358−376.

QGIS Development Team, 2018. QGIS, A free and open source. Geographic Information system <https://qgis.org/en/site/> (accessed 29.12.18).

Roth, R.E., 2012. Cartographic interaction primitives: framework and synthesis. Cartogr. J. 49 (4), 376−395.

Roth, R.E., 2015. Interactivity and cartography: a contemporary perspective on user interface and user experience design from geospatial professionals. Cartographica 50 (2), 94−115.

Roth, R.E., Donohue, R.G., Wallace, T.R., Sack, C.M., Buckingham, T.M.A., 2014. A process for keeping pace with evolving web mapping technologies. Cartogr. Perspect. 78, 25−52.

Schneider, M., Klein, R., 2007. Efficient and accurate rendering of vector data on virtual landscapes. J. WSCG 15 (1-3), 59−66.

Schuit, W., 2011. A method for teaching topographic map interpretation. J. Geogr. 110 (5), 209−216.

Shi, S., Charlton, M., 2013. A new approach and procedure for generalising vector-based maps of real-world features. GISci. Remote Sens. 50 (4), 473−482.

Stoter, J., Post, M., van Altena, V., Nijhuis, R., Bruns, B., 2014. Fully automated generalization of a 1:50k map from 1:10k data. Cartogr. Geogr. Information Sci. 41 (1), 1−13.

SuperMap International Limited, 2018. SuperMap. Hong Kong. <https://www.supermap.com/en/> (accessed 29.12.18).

U.S. Department of Agriculture, Forest Service, 2011. Draft Environmental Impact Statement for the Kiowa, Rita Blanca, Black Kettle, and McClellan Creek National Grasslands Land and Resource Management Plan. U.S. Department of Agriculture, Forest Service, Southwestern Region, Albuquerque, NM. MB-R3-03-15.

U.S. Department of Agriculture, Natural Resources Conservation Service, 2017. Web Soil Survey. U.S. Department of Agriculture, Natural Resources Conservation Service, Washington, D.C. <https://websoilsurvey.sc.egov.usda.gov/App/HomePage.htm> (accessed 29.12.18).

Verizon Communications, Inc., 2018. MapQuest. Verizon Communications, Inc., Denver, CO. <https://www.mapquest.com/> (accessed 29.12.18).

Wang, Z., Müller, J.-C., 1998. Line generalization based on analysis of shape characteristics. Cartogr. Geogr. Information Syst. 25 (1), 3−15.

Wood, D., 2002. The map as a kind of talk: Brian Harley and the confabulation of the inner and outer voice. Visual Commun. 1 (2), 139−161.

8

Map Errors

Maps are simplistic representations of space, landscapes, and systems. There are no perfect databases of geographic features, as all are prone to error and inaccuracy due to generalization and measurement problems (Pascual, 2011). In fact, the most frequent complaint about today's digital maps may involve what seems to be mapping errors (Smith, 2008). Details drawn or displayed on maps can be important to the map developer in communicating their message, yet a map loses its meaning if it is inaccurate, out-of-date, or not legible (International Orienteering Federation, 2017). Map developers may impose unintentional distortions and omissions in their works (Wood, 2015), and certainly maps can contain some very distinct errors—blunders, positioning problems, typographical problems, and others. Many early 15th century European maps, for example, contained errors simply because many places were relatively unknown, and mathematics was not universally understood. Old world maps may have also been devised in such a way as to distort and falsify geography, perhaps to claim commercial rights to various resources, to the benefit of certain countries (Metcalf, 2017). Further, due to the complexity of the Earth, vegetation, water, and atmospheric resources, it can become necessary to approximate the descriptions of features (Ma et al., 2017); thus, when precise measurements of landscape features are difficult or impossible to capture, generalization and error may occur. Perhaps unintentional, perhaps not, scaled models of real systems will contain errors. And much to the dismay of map users, all maps contain some amount of uncertainty and error (Fisher and Tate, 2006).

Common Types of Map Errors

Errors can enter mapped information through the processes of measuring and recording the descriptions of landscape features (Arbia et al., 1998). Errors can further be propagated through both computer processing operations and human interaction in the map development process. It should come as no surprise that when a source map or database contains error, maps or databases created from this source will also contain some error. Within geographic information systems (GIS) error will not be alleviated simply by merging, buffering, adding, or overlaying one error-prone map with others. These processes can actually introduce issues and promulgate errors that need to be managed (Fig. 8.1). The attribute and positional locations of landscape features within a GIS database can each contain error, through the digitization process, global positioning system (GPS) data capture, and sampling or expert judgment on the characteristics of the features. One set of errors associated with attributes and positions of GIS database features are thus called *source errors*, and represent the discrepancies between the actual landscape and the landscape depicted in a map (Haining and Arbia, 1993).

In many cases in natural resource management, for example, aerial images are often used as base maps to draw and attribute thematic databases. These databases can include polygons representing similarly vegetated areas (stands of trees), lines representing courses of flowing water (streams), and points representing features of significance to a planning process (buildings, wildlife nest locations, etc.). Aerial images are of great value in obtaining a broader view of a landscape, yet they may be limited in their spectral resolution (number of bands of electromagnetic energy captured) and spatial resolution (size

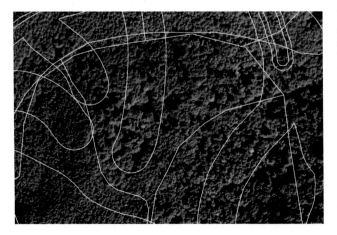

FIGURE 8.1 The creation of a number of small polygons through the intersecting of a riparian polygon database and a timber stand database in geographic information system.

of the grid cells in the image). And unless one is a well-trained aerial image interpreter with intimate knowledge of the land area being examined, a high degree of subjectivity may be used in creating new thematic databases from aerial images (Gergel et al., 2007). Therefore the development of mappable databases from very good (but not perfect) base maps can source errors into maps.

Translation 29
Imagine that you and a friend are viewing an old map. With today's perspective, roads have moved, buildings have been developed or demolished, and other land uses have changed. Your friend suggests, based on today's perspective, that the old map contained numerous errors. If one were expecting the map to represent current conditions, the differences between the landscape as depicted at the time that the map was created and the landscape as it exists today might be considered errors. But are they really? Explain to your friend how the old map might have contained less error than they think.

Omission and Commission Errors

A lack of obvious data or important information in a map is an *error of omission*, one common type of map error. Certainly, an omission of information may be necessary to reduce the amount of distractions one may find within a map. Omissions may also be intentional, in an attempt to hide or simplify certain landscape features that the map developer or the organization that employs them prefer not to reveal. This missing information could consist of features not being visible on a map, or features that are visible on a map, but not labeled. Often, archeological sites are omitted from maps to protect them from looting. Omissions may also have occurred simply because the map maker thought their inclusion was of no importance to the map, or that their inclusion made the map less effective at communicating its story. Further, through the processing of data collected *in situ* (in the field), it is not unreasonable to envision that omitted data were caused by problems within the data collection scheme. For example, features that are poorly measured or not measured at all may be omitted, or perhaps the data collected about the feature were lost, or perhaps the field measurement crew mistakenly failed to record the presence of the feature. For example, imagine that a field crew was using a GPS to record the locations of red-cockaded woodpecker (*Picoides borealis*) nest trees, with the purpose of making a map in support of an environmental assessment. The field crew may have (1) failed to recognize a nest tree, (2) lost the recorded GPS data point (or the file), or (3) recorded insufficient information about a nest tree, prompting the map developer to ignore it as the map was prepared.

Reflection 34
Think about the last map you made. Could any other landscape features have been included to improve the quality of the map? Were they not included due to data or time limitations, or due to other reasons that led to an omission of important information?

A rather subtle error of omission can be associated with the *minimum mapping unit* employed by some organizations. The minimum mapping unit is the smallest physical size, or quantity of resource, that would be recognized in a database. All features below this minimum size, or containing quantities below a minimum threshold, would be aggregated with nearby neighboring features under this rule. For example, with respect to the size of land areas, perhaps an organization employs a 1 acre (0.4 hectares) minimum mapping unit policy for the management units it recognizes in a GIS database. Under this policy, all areas less than 1 acre in size will be aggregated with nearby (usually adjacent) management units. Thus a small, natural opening contained within a coniferous forest that is less than 1 acre in size would likely not be identified and become part of the surrounding coniferous forest in this case. Other examples of these types of errors might include certain types of small scrub—shrub wetlands that are situated inside larger palustrine (nontidal, inland) forested wetlands, yet they are not present on a map since they are too small in size (Federal Geographic Data Committee, 2009). This type of database management policy inherently suggests that errors could be present in a map, when the policy simply prevents the display or recognition of features that are smaller than a minimum size to enhance the communicative ability of the map. The minimum mapping unit can be extended to line features; whereby, the rule may now be noted as a *minimum segment length*, based on a minimum physical distance or a minimum number of points (vertices) that describe landscape features.

Should a feature be included in a map, yet incorrectly labeled or categorized, this would be reflective of an *error of commission*. Commission errors are related to the misclassification of landscape features or to issues that arise due to limits imposed by the scale of a map (Federal Geographic Data Committee, 2009). Interestingly, a commission error might also be viewed as excess data contained in a geographic database (IDON Technologies, Inc., 2016). In essence, one form of a commission error may involve a feature displayed on a map, yet portrayed in character or quality as something other than its actual character or quality; in sum, it has been misclassified. For example, when developing a database of urban street trees for a city, an urban forester may record the position and size (diameter and height) of a certain tree along a certain city street. However, when recording this data, the tree may be assigned a species label (e.g., *Quercus alba*) that is inconsistent with the true species of the tree (e.g., *Acer rubrum*), which might be viewed as an error of commission.

In the field of remote sensing, errors of omission and commission are frequently examined through the use of an omission/commission matrix (Fig. 8.2) that is created during accuracy assessment of classified resources or landscape features. When using remotely sensed imagery, classification of the imagery into land uses or land or sea conditions is common. A number of mathematical processes can be employed to create thematic databases from the electromagnetic reflectance or emittances captured by satellite or aerial imagery. The classification of remotely sensed imagery results in a new map or a digital database, but in either case it represents a graphical description of a landscape. The accuracy of the new map or database is measured using locations on the ground acting as validation points (known conditions) that are compared against the land class on a map. When the true character or quality of a sample point is inconsistent with the classified map's suggestion of the character or quality of the resources found there (disagreements have been observed), an error of omission or commission has occurred. Errors of omission

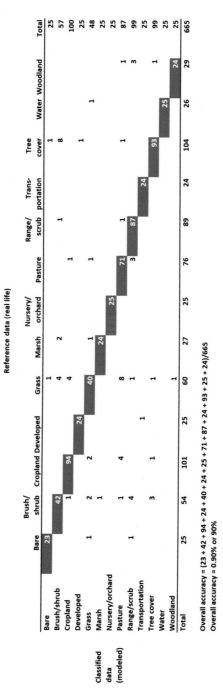

Reference data (real life)

Classified data (modeled)	Bare	Brush/shrub	Cropland	Developed	Grass	Marsh	Nursery/orchard	Pasture	Range/scrub	Transportation	Tree cover	Water	Woodland	Total
Bare	23										1		1	25
Brush/shrub		42	1		4	2					8			57
Cropland		1	94		4				1					100
Developed				24				1						25
Grass	1	2	2		40	1		1			1			48
Marsh						24						1		25
Nursery/orchard							25							25
Pasture	1	1	4		8			71	1		1			87
Range/scrub		4			2			3	87				3	99
Transportation				1						24				25
Tree cover		4			1						93		1	99
Water												25		25
Woodland					1								24	25
Total	25	54	101	25	60	27	25	76	89	24	104	26	29	665

Overall accuracy = (23 + 42 + 94 + 24 + 40 + 24 + 25 + 71 + 87 + 24 + 93 + 25 + 24)/665

Overall accuracy = 0.90% or 90%

FIGURE 8.2 An omission/commission matrix comparing modeled land classes to real-life land classes at select validation points.

and commission might be described as *measurement errors*. Unfortunately, no map developed using data classified in this manner is ever completely correct (Langford et al., 2006). In one example, satellite imagery was used in an attempt to delineate areas disturbed by fires over a 20-year period in Saskatchewan (Schroeder et al., 2011). In the land cover maps produced, much of the map error arose as a result of misclassifying persistent nonforested areas as being recently disturbed or as being persistent forests. Many of these persistent nonforested areas were wetlands that varied considerably in spectral value (as measured by the satellite sensor) from one year to the next due to differences in annual precipitation amounts, water levels, and vegetation conditions.

Gross, Random, and Systematic Errors

Another way to categorize map errors is to think about how the errors were propagated. A *gross error* may involve deliberate or unintentional placement or omission of a landscape feature on a map, or similarly involve deliberate or unintentional description (through symbology) of the landscape feature on a map (Fig. 8.3). For example, mistakenly recording the GPS location of a suspected nest tree of a red-cockaded woodpecker, when in fact it was not a nest tree, then subsequently illustrating this location on a map might be an example of a gross error. Gross errors can be the result of carelessness or mistakes that occur somewhere along the data collection to map development path. Often called *blunders*, they would ideally be detectable and controllable.

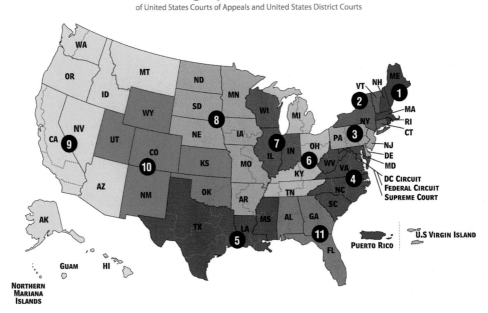

FIGURE 8.3   A map of the US federal courts circuit, with a small typographical error "US Virgin Island." Source: *U.S. Department of Justice (2016).*

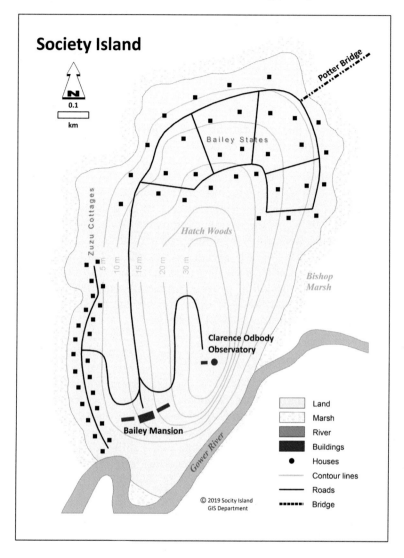

FIGURE 8.4 A map of Society Island containing several blunders.

Inspection 37

Examine the map provided in Fig. 8.4. In comparison to other maps of Society Island, can you identify all of the blunders?

A *random error* in the information contained in a map may arise due to uncontrollable fluctuations in variables that affect the data or information displayed. For example, perhaps while collecting data for an urban street tree inventory, a member of a field crew mistakenly classified a certain street tree in Memphis, Tennessee as *Quercus palustris* (pin oak) when in fact it was *Quercus nigra* (water oak) simply because they were distracted by some external event while recording the information. Random errors such as these may be undetectable through an inspection of the data collected prior to the development of a map. Perhaps the display of data on a map (in a different format than a data table) might help locate some of these. For some types of data and information, random error can be described through the development of a probability distribution for the resource of interest, and inspections of various measures of shape and spread of the distribution (Haining and Arbia, 1993).

Reflection 35
As a user of a map, or as the supervisor of a person who is developing maps, what processes might you employ to locate random errors in maps?

A *systematic error* in a map may be caused by problems related to the protocols and methods employed to describe the character and quality of resources, by inherent problems with the instruments used to collect data and information, or by consistent human-caused problems in the measurement or recording of data. In the latter case, perhaps while collecting data for an urban street tree inventory, a member of a field crew consistently labeled certain street trees in Dayton, Ohio as *Acer rubrum* (red maple) when in fact they were *Acer saccharum* (sugar maple). Systematic errors might be detectable through a close inspection of the data collection and manipulation processes that are employed prior to the development of a map, and technically, systematic error can be described by various measures of bias in the underlying data (Haining and Arbia, 1993).

Quantity and Location Errors

From a technical perspective, a map may contain a *quantity error*, where the total amount of some resource (e.g., area of a land class) is inconsistent with the total area of the same land class on a different map that represents the same area at the same point in time (Schaffer et al., 2016). Imagine two different organizations developed a map of the peanut farms in Pulaski County, Georgia, and each map contained a different total amount of land currently being used to grow peanuts. This might be considered an example of a quantity error associated with one (or both) of the maps.

Translation 30
Examine the two maps in Fig. 8.5, indicating the amount of volume per acre for timber stands in 2019. These maps may have been created by the forest management organization within which you work. Explain to your coworker the reasons why there seems to be a quantity error in one of the two maps.

A *location error* might be observed in a map when a feature is found in a different place than where it is located in real life (Schaffer et al., 2016). These may be caused by database management processes, deliberate movement of annotation to avoid overlapping labels, or simply by blunders. Fig. 8.6 provides an example of a wildfire ignition point that was very precisely positioned, to 6 decimal places of a degree of latitude and longitude. If the positional data were to have been entered or stored in the database using 2 decimal places of a degree, the coordinates that define the ignition point may not exactly represent the starting point of the fire. The difference between the two ignition points is location error, which

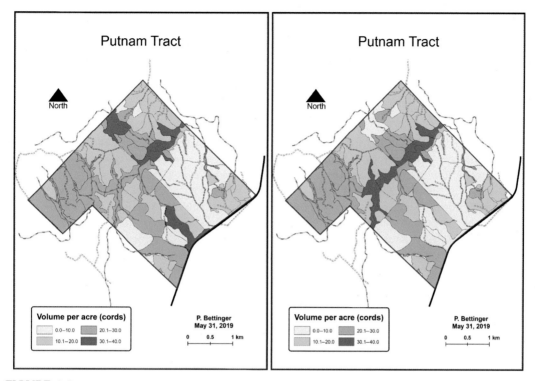

FIGURE 8.5   Volume per acre on the Putnam Tract (left) and a similar map with potential quantity errors (right).

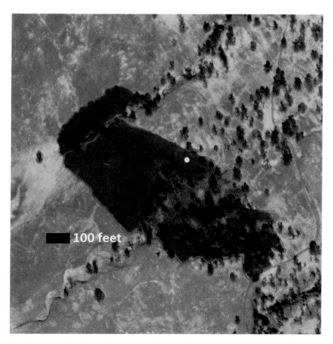

FIGURE 8.6   Precisely positioned wildfire ignition point (*red dot*, 40.128743 degrees latitude, −122.058885 degrees longitude), and less precisely positioned wildfire ignition point (*yellow dot*, 40.13 degrees latitude, −122.06 degrees longitude) for a fire that occurred in 2017 approximately 3.2 miles east of Dairyville, CA.

may (or may not) be of concern to the natural resource managers or landowners associated with this land.

Scale and Generalization Errors

Maps can easily be developed using a scale that is inappropriate for the purpose of the map. Further, map error may be propagated by map generalization processes as well. It might seem obvious that reductions in content and resolution can occur as map scale decreases (e.g., from 1:24,000 to 1:100,000). With respect to a digital map, one might envision a reduction in content as they zoom out (increase the eye altitude) of a landscape. In small-scale printed maps (e.g., 1:100,000), as compared to large-scale printed maps (e.g., 1:10,000), one may find that roads are drawn smoother (a simpler shape with fewer curves) and that fewer landscape features are present given the limited printable or displayable space. With respect to linear features or edges of closed areas (polygons), map errors can arise from error propagated by the generalization and positioning of the vertices that represent changes in direction along these lines (Liu and Shi, 2016), as we described in Chapter 7, Map Development and Generalization.

> **Diversion 44**
> Prior to the delivery of geographic information through digital means, an organization may have already generalized landscape data to ensure that the computer processing time and effort are reasonable for users of their services. For example, line simplification (reduction in vertices that describe lines) can result in lines that change in direction more abruptly than you might expect in a carefully created geographic database. As a subtle example of this, using Google Earth Pro, navigate to US Forest Service road FS 323 on the De Soto National Forest in Mississippi (31.078127 degrees north latitude, 88.932214 degrees west longitude). Within the *layers* sidebar window, enable the viewing of roads. Given the slight changes in direction of the road found here, about how far apart are the vertices that describe the road, on average? Assuming the road was generalized, what issues may arise if you were to use this data on a map?

Generalization and simplification of geographic features and their attributes (as features get combined) often are applied to features we assumed were well-defined before these processes were employed. However, map error may propagate through the use of poorly defined landscape features, those that were initially vague and ambiguous in character and quality before a map was developed. Thus, we might characterize map error in this sense as errors associated with well-defined features, and errors associated with poorly defined features (Fisher, 1999). Well-defined features might be thought of as those with sharp edges, such as roads, buildings, or other clearly defined features. The polygon and line representations of orchards and roads described in Fig. 8.7 may contain errors even though they are very clearly identifiable on the aerial image. The error might be associated with incorrect georeferencing or drawing of the feature, or with incorrect attribution (assigning the feature the wrong label or class). Poorly defined features might be

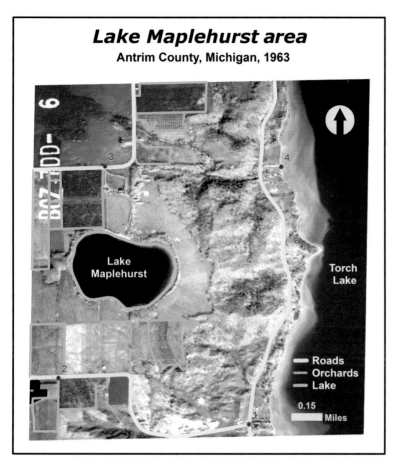

FIGURE 8.7   Roads, orchards, and a lake in Antrim County, Michigan.

considered those with fuzzy boundaries, such as the transition between soil types or natural forests. The shape of the polygons representing soil management units (MUTYPE) in Fig. 8.8 likely contain some error because a change from one soil type to another is usually not as evident as the location of a centerline of a road or the edge of an orchard. Vagueness or ambiguity among landscape features being represented in a map can lead to error when the boundaries of these features are fuzzy. Each field in the study of science may classify nature in a different manner, and these differences in perception can cause ambiguity in map interpretation (Schaffer et al., 2016). In addition, natural features may not have clearly defined land area boundaries; some exceptions perhaps being the intersection of water bodies and land, or certain rock formations (e.g., cliffs).

Consistency and Completeness Errors

Map consistency problems can be viewed from two perspectives: consistencies within a single map and consistencies across similar maps. *Logical consistency* of mapped

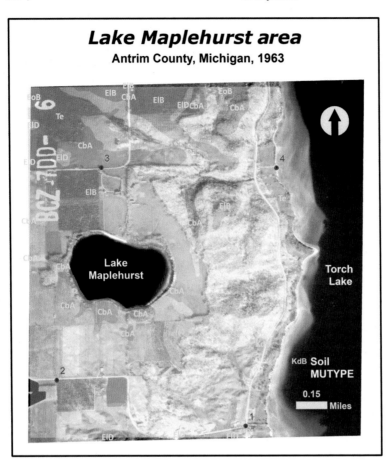

FIGURE 8.8 Soil management units of an area in Antrim County, Michigan. Source: *U.S. Department of Agriculture, Natural Resources Conservation Service (2017).*

information refers to an adherence of the data to conceptual, logical, or physical data structure rules. Within a single map, one might expect that information assigned to landscape features (the attributes of features) should be presented in a consistent manner. This may require adherence to data collection and processing standards by all involved in the map development process (Fürst, 2002). An example of an inconsistency would be to record the dominant species of one forest as *Pinus taeda*, and the dominant species of another forest in the same map as *P. taeda*. The use of valid values for the region in which one works would also be of concern here, as *Pinus hyacinthum* (blue pine) would be considered an invalid tree species if recorded on a map of southern US forest types, since it technically is not consistent with the taxonomy of tree species found in this region of the world.

Topological problems in geographic data (streams flowing uphill, slivers between polygons, etc.) might be considered as sources of inconsistencies in maps. Across multiple maps, consistency may refer to the methods used and the relationships observed (Fürst, 2002). Within GIS databases that are used to develop maps, several topological issues

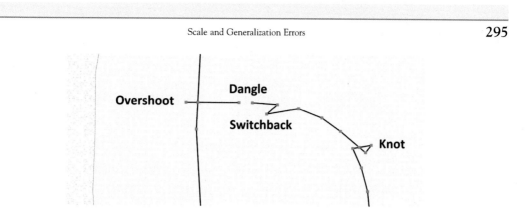

FIGURE 8.9   Line feature positional errors.

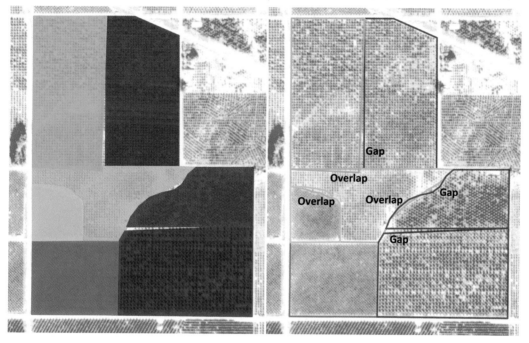

FIGURE 8.10   Polygon feature positional errors.

might arise that need to be addressed. Linear features may contain dangles, overshoots (a line extending beyond an intersection), switchbacks, or knots (loops) that would not be present in the real-life landscape (Fig. 8.9). Point features may simply be located in the wrong place. Polygon features could include gaps and overlaps among a set of polygons (Fig. 8.10), or slivers (small polygons) that result from overlay (clip, erase, intersect, etc.) processes among two or more databases. These issues are created during the digitizing process by the person that created the database or by some other mathematical

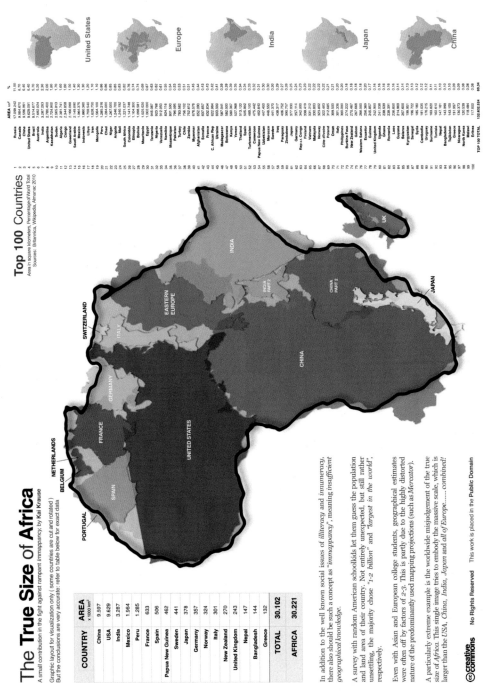

The True Size of Africa

A small contribution in the fight against rampant *immappancy*: by Kai Krause

Graphic layout for visualization only (some countries are cut and rotated)
But the conclusions are very accurate: refer to table below for exact data

COUNTRY	AREA x 1000 km²
China	9.597
USA	9.629
India	3.287
Mexico	1.964
Peru	1.285
France	633
Spain	506
Papua New Guinea	462
Sweden	441
Japan	378
Germany	357
Norway	324
Italy	301
New Zealand	270
United Kingdom	243
Nepal	147
Bangladesh	144
Greece	132
TOTAL	**30.102**
AFRICA	**30.221**

In addition to the well known social issues of *illiteracy* and *innumeracy*, there also should be such a concept as "*immappancy*", meaning *insufficient geographical knowledge.*

A survey with random American schoolkids let them guess the population and land area of their country. Not entirely unexpected, but still rather unsettling, the majority chose "*1-2 billion*" and "*largest in the world*", respectively.

Even with Asian and European college students, geographical estimates were often off by factors of 2-3. This is partly due to the highly distorted nature of the predominantly used mapping projections (such as *Mercator*).

A particularly extreme example is the worldwide misjudgement of the true size of Africa. This single image tries to embody the massive scale, which is larger than the *USA, China, India, Japan and all of Europe……. combined!*

Top 100 Countries

Area in square kilometers, Percentage of World Total
Sources: Britannica, Wikipedia, Almanac 2010

		AREA km²	%
1	Russia	17,098,242	11.50
2	Canada	9,984,670	6.70
3	China	9,596,961	6.40
4	United States	9,629,091	6.40
5	Brazil	8,514,877	5.70
6	Australia	7,692,024	5.20
7	India	3,287,263	2.20
8	Argentina	2,780,400	2.00
9	Kazakhstan	2,724,900	1.80
10	Sudan	2,505,813	1.70
11	Algeria	2,381,741	1.60
12	Congo	2,344,858	1.60
13	Greenland	2,166,086	1.50
14	Saudi Arabia	2,149,690	1.40
15	Mexico	1,964,375	1.30
16	Indonesia	1,860,360	1.30
17	Libya	1,759,540	1.20
18	Iran	1,628,750	1.10
19	Mongolia	1,564,100	1.10
20	Peru	1,285,216	0.86
21	Chad	1,284,000	0.86
22	Niger	1,267,000	0.85
23	Angola	1,246,700	0.83
24	Mali	1,240,192	0.83
25	South Africa	1,221,037	0.82
26	Colombia	1,141,748	0.76
27	Ethiopia	1,104,300	0.74
28	Bolivia	1,098,581	0.74
29	Mauritania	1,030,700	0.69
30	Egypt	1,002,000	0.67
31	Tanzania	945,087	0.63
32	Nigeria	923,768	0.62
33	Venezuela	912,050	0.61
34	Namibia	824,116	0.55
35	Mozambique	801,590	0.53
36	Pakistan	796,095	0.53
37	Turkey	783,562	0.53
38	Chile	756,102	0.51
39	Zambia	752,612	0.51
40	Myanmar	676,578	0.45
41	Afghanistan	652,090	0.44
42	Somalia	637,657	0.43
43	France	632,834	0.42
44	C. African Rep.	622,984	0.42
45	Ukraine	603,500	0.41
46	Madagascar	587,041	0.39
47	Botswana	582,000	0.39
48	Kenya	580,367	0.39
49	Yemen	527,968	0.35
50	Thailand	513,120	0.34
51	Spain	505,992	0.34
52	Turkmenistan	488,100	0.33
53	Cameroon	475,442	0.32
54	Papua New Guinea	462,840	0.31
55	Uzbekistan	447,400	0.30
56	Morocco	446,550	0.30
57	Sweden	441,370	0.30
58	Iraq	438,317	0.29
59	Paraguay	406,752	0.27
60	Zimbabwe	390,757	0.26
61	Japan	377,930	0.25
62	Germany	357,114	0.24
63	Rep.o.t. Congo	342,000	0.23
64	Finland	338,145	0.23
65	Vietnam	331,212	0.22
66	Malaysia	330,803	0.22
67	Norway	324,220	0.22
68	Côte d'Ivoire	322,463	0.21
69	Poland	312,685	0.21
70	Italy	301,336	0.20
71	Philippines	300,000	0.20
72	Burkina Faso	274,222	0.18
73	New Zealand	270,467	0.18
74	Gabon	267,668	0.18
75	Western Sahara	266,000	0.18
76	Ecuador	256,369	0.17
77	Guinea	245,857	0.16
78	United Kingdom	242,900	0.16
79	Ghana	238,539	0.16
80	Romania	238,391	0.16
81	Uganda	241,038	0.16
82	Laos	236,800	0.16
83	Guyana	214,969	0.14
84	Belarus	207,600	0.14
85	Kyrgyzstan	199,951	0.13
86	Senegal	196,722	0.13
87	Syria	185,180	0.12
88	Cambodia	181,035	0.12
89	Uruguay	176,215	0.12
90	Suriname	163,820	0.11
91	Tunisia	163,610	0.11
92	Nepal	147,181	0.10
93	Bangladesh	143,998	0.10
94	Tajikistan	143,100	0.10
95	Greece	131,957	0.09
96	Nicaragua	130,373	0.09
97	North Korea	120,538	0.08
98	Malawi	118,484	0.08
99	Eritrea	117,600	0.08
100			
	TOP 100 TOTAL	**133,632,524**	**89.34**

United States

Europe

India

Japan

China

FIGURE 8.11   The true size of Africa. Source: *Krause (2011)*.

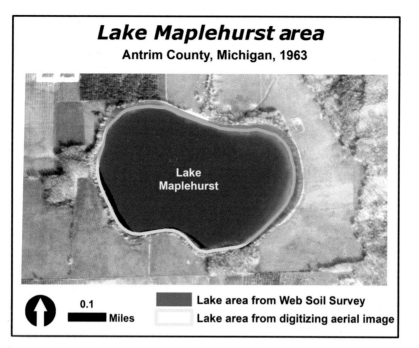

FIGURE 8.12 Minor map inconsistencies in two representations of Lake Maplehurst, Michigan.

manipulation that may have been applied to the GIS database. Often these issues need to be corrected by hand, through close inspection of the database.

To be consistent across multiple maps, one might also use the same coordinate and projection systems (e.g., Universal Transverse Mercator), and the same political (country, state, province, county, and city) and administrative (ownership) boundaries. Certain map projections, as we noted in Chapter 4, Map Reference Systems, can introduce errors in orientation (direction), distance, area, and angular relationships among mapped features. Some map projections, such as the Mercator, may mislead map users about the true size of certain land areas the farther they are from the equator (Fig. 8.11). Another example of an inconsistency across or among maps might involve two thematic maps obtained from different sources; an overlay of these may reveal a difference in the shapes of features that should share common edges. For example, one set of features thematic map might have been developed through the interpretation and subsequent digitizing of aerial imagery, while another may have been developed through the digitizing of maps (Fig. 8.12). To be considered highly consistent, maps should use similar labeling styles for landscape features, and similar processes for assigning labels to features. Finally, if one is able to view a map on a digital device, or change scale within a computer mapping program, the information presented at different scales should be consistent (Fürst, 2002). For example, a small-scale map of precipitation levels covering a large area (such as a country), should be reflective of, and consistent with, a localized map of precipitation levels.

Reflection 36

The presentation (through text or logos) of the entity that developed a map, a person or an organization, can add a sense of legitimacy to the map itself and tacitly suggest a higher quality and thus lower chance of map error. Maps produced by reputable organizations likely undergo multiple levels of quality control prior to being released for general public sale or use. However, maps developed by individuals may have a limited amount of associated review prior to release. In your opinion, which types of maps should contain information describing the source of data and the developer's information, and which types should not?

As we described in Chapter 4, Map Reference Systems, a number of maps contain disclaimers and warranties that warn map users of the potential problems related to the completeness of maps. *Completeness* relates to the presence or absence of features, and whether they are described correctly. The issues of both consistency and completeness of maps require verification or auditing of the information presented in the map product. In Chapter 4, Map Reference Systems, we also described various ways in which a warranty or set of caveats can be inserted onto maps to protect the developer of the map from subsequent litigation and to warn the map user of potential map problems that may arise from incomplete or inconsistent information. *Warranties* are designed to protect the map developer against legal claims for losses based on incomplete or inconsistent information contained in maps. *Caveats* simply inform users of maps about the limitations and specific conditions under which the information should be used. Some maps and digital map products contain these warranties and caveats, some maps do not contain these. When provided, they might alert the map user to potential map errors.

Map Accuracy

Map error and map accuracy are very closely related. An assessment of map accuracy may result in a quantification of the errors in information (thematic accuracy) contained within a map, or the errors in accuracy of the data used (positional accuracy) within a map (Geoscience Australia, 2012). In the latter case, mapped features are often compared to the coordinates or characteristics of *well-defined points*, or those places on a map that can be accurately identified and described. These may include road intersections, or other linear features such as railroads that intersect with roads or each other. A well-defined point more technically may refer to a place where, with a high degree of accuracy, the horizontal position is known with respect to the datum used (Federal Geographic Data Committee, 1998). An assessment of map accuracy can be made in a quantitative fashion such as this or in qualitative manner, describing more explicitly (for example) the completeness of a mappable database.

Positional Accuracy

Positional accuracy refers to the proximity or closeness of a measurement to some standard or true measurement (American Society for Photogrammetry and Remote Sensing,

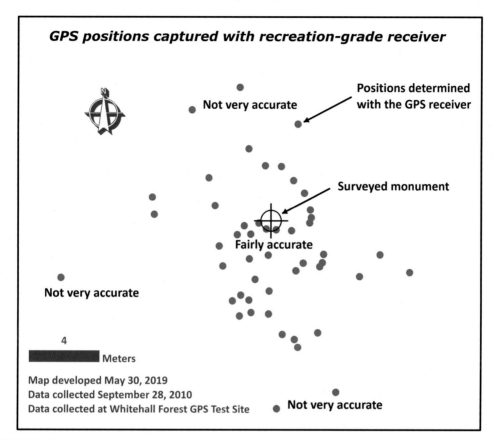

FIGURE 8.13   GPS positional accuracy map based on points collected with a recreation-grade receiver.

2014). In other words, positional accuracy describes the spatial correspondence of mapped data with the real-world data that they are meant to represent. For example, if one were to collect a GPS position while standing over a surveyed monument (Fig. 8.13), and the GPS receiver recorded a position very close to the position of the surveyed monument, the measurement might be judged *fairly accurate*. However, if the GPS receiver recorded a position that was far from the surveyed position, perhaps 100 ft. (about 30 m) away, the measurement might be judged *not very accurate*. Accuracy can be affected by the processes we noted earlier, or the quality of the measurement tool. Positional accuracy can be affected by, and suffer from omission (complex shapes that are too generalized), blunders (poorly drawn features, incorrectly placed features), random and systematic errors, consistency (some features positioned well, others not so well), and completeness error (some features from certain groups are missing from the map).

The *precision* of measurements refers to the repeatability of the measurement device or measurement process to develop data that agree, even though bias may exist. While they may seem similar, accuracy and precision are often described in classic graphics using point features (Fig. 8.14). Returning to the GPS data collection example, while standing

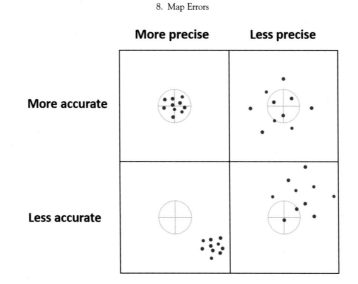

FIGURE 8.14   A conceptual model illustrating the difference between accuracy and precision.

over a surveyed monument and recording multiple positions with a GPS receiver, ideally, one would want the collection to consist of similar geographic positions. If the positions (measurements) are very similar, we would judge the process or tool used as being *fairly precise*. When the positions are not very similar (spaced far apart), we might judge the process or tool used as *not very precise*.

Reflection 37
Imagine, in an effort to develop a database of the city's infrastructure, you are using a recreation-grade GPS receiver to determine the position of a fire hydrant. You record this position several times with the GPS receiver. Why wouldn't this collection of GPS positions, even when the GPS receiver is stationed firmly over the hydrant, have the same northing and easting coordinates? In other words, why might the recorded positions not all be the same?

Positional errors can be assessed in a horizontal (northing and easting) manner or in a vertical (elevation) manner (Federal Geographic Data Committee, 1998). Often, a derivation of the root mean squared error (RMSE) in point observations is performed to assess the accuracy of GPS-determined positions, which are compared to well-known (surveyed) positions. The error (Table 8.1) in both horizontal aspects of each GPS observation of the position (northing and easting) is determined to be the difference between well-known (surveyed) position and the positions determined by the GPS device. Using the Pythagorean theorem, the horizontal error for each observation can then be determined.

As an example of determining the horizontal error using the Pythagorean theorem, consider observation 1:

TABLE 8.1 Example error in data points collected using a GPS device.

Observation	Northing error (m)	Easting error (m)	Horizontal error (m)
1	0.51	− 0.25	0.568
2	− 0.34	0.64	0.725
3	− 0.68	0.49	0.838
4	− 0.07	− 0.58	0.584
5	0.19	− 0.71	0.735

$$(\text{Horizontal error})^2 = (\text{Northing error})^2 + (\text{Easting error})^2 \quad \text{or}$$

$$\text{Horizontal error} = \left((\text{Northing error})^2 + (\text{Easting error})^2\right)^{0.5} \quad \text{thus}$$

$$\text{Horizontal error} = \left((0.51)^2 + (-0.25)^2\right)^{0.5}$$

$$\text{Horizontal error} = 0.568$$

Once the horizontal errors are determined for each observation, they are squared, then summed, and then applied the square root function. For this data,

$$\text{RMSE} = \sqrt{\frac{(0.568)^2 + (0.725)^2 + (0.838)^2 + (0.584)^2 + (0.735)^2}{5}}$$

To visualize the concept of "RMSE," think of it in reverse order:

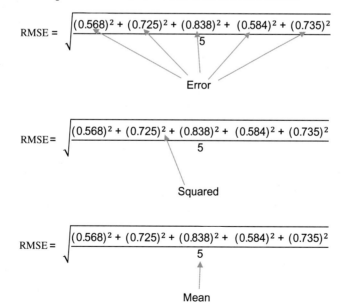

TABLE 8.2   UTM Zone 11 coordinates for six urban trees from GPS and from Google Earth.

Tree	GPS northing (m)	GPS easting (m)	Google Earth northing (m)	Google Earth easting (m)
1	5,193,620.25	728,260.45	5,193,613.09	728,263.48
2	5,193,572.45	728,225.68	5,193,579.04	728,231.87
3	5,193,551.85	728,270.58	5,193,553.35	728,272.95
4	5,193,587.32	728,307.81	5,193,584.31	728,314.53
5	5,193,532.89	728,196.93	5,193,538.58	728,193.89
6	5,193,540.84	728,249.34	5,193,543.99	728,245.09

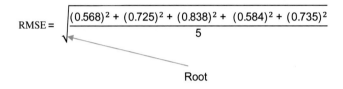

$$RMSE = \sqrt{\frac{(0.568)^2 + (0.725)^2 + (0.838)^2 + (0.584)^2 + (0.735)^2}{5}}$$

Root

In this case,

$$RMSE = \sqrt{\frac{(0.3226) + (0.5252) + (0.7025) + (0.3413) + (0.5402)}{5}}$$

$$RMSE = \sqrt{\frac{2.4318}{5}}$$

$$RMSE = \sqrt{0.48636}$$

$$RMSE = 0.6974 \; m$$

Diversion 45

Imagine that you were given positional data (points) describing the locations of street trees in the city within which you work as an urban forester. Upon viewing these in Google Earth, you notice that the points do not seem to fall directly on top of the center of each tree, as you had hoped. Perhaps, this is an outcome of collecting the data with a GPS device, or due to some other reason concerning the data collection process (e.g., data collector not standing close enough to the tree). Although you know there may be some minor positional error in Google Earth data, you use it as the reference information in determining the positional accuracy of the tree location data. Using the data provided in Table 8.2 for a sample of some of these trees, describe your estimate of the root mean squared error of the tree location data.

The result of computing positional accuracy of points in this manner is the raw RMSE for the data that was collected. Often, this is what is presented in publications or statements of positional accuracy of static point data collected by GPS devices under various environmental conditions (e.g., see Bettinger and Fei, 2010).

Thematic Accuracy

Thematic accuracy relates to the labels or attributes assigned to features on a map (Congalton and Green, 2009). Attributes, from a geographic data point of view, are those pieces of information that describe geographic features such as points (individual trees), lines (roads), or polygons (land parcels). In GIS programs, an *attribute table* (much like a spreadsheet) would contain this information, which would be related directly to the geographic features so that queries and other operations can be performed based on what the table contains. Therefore, *attribute accuracy* refers to the quality of the information contained in an attribute table, and should be of great interest to those who use it to make maps. As with positional accuracy, attribute accuracy can be affected by, and suffer from omission error (some attributes missing), blunders (incorrect attributes assigned to features), random and systematic errors, consistency (more than one method is used to illustrate an attribute), and completeness error (some features have no attributes).

In developing a geographic database, the attributes that describe geographic features are either created by a person or generated automatically through mathematical operations. If a person was charged with entering the attribute information into the table, the opportunity for data processing error certainly exists. Should attribute information be created through a mathematical process, the opportunity for unreasonable values may exist if the necessary controls are not employed. For example, a regression model may be used to predict the average height of trees within a forest based on the average age of those trees and the quality of the site. However, the regression model may only be valid for certain ranges of average tree ages and site qualities, and therefore unreasonable average tree heights can be created when predictor values are outside of a desired range. Without quality control processes, errors may be present in the attribute tables of GIS data, and further propagate into other mapped databases developed from the primary data.

A very nice geospatial database of land ownership parcels in the State of Florida was developed by the University of Florida GeoPlan Center (2018). This was a massive undertaking, given the 67 counties and 10,174,752 parcels involved. The database contains ownership information (name, address of owner, etc.), property size, land use designation, legal description, and many other pieces of information used by property assessors in each county. A total of 67 attributes could have potentially been assigned to each land ownership parcel, thus there are nearly 700 million cells in the attribute table. Certainly, correcting minor problems would likely entail a considerable amount of effort.

Inspection 38
Within a GIS program, open the Seminole County data available on the book's website. Conduct a thorough examination of the owner information and land use description in the attribute table. What issues might you attempt to correct if you had the time and resources necessary to visit or otherwise inspect (perhaps through remote sensing) these parcels?

Omission errors involve attributes that are missing for one reason or another. Within the Florida statewide parcel database, there are 86,758 parcels that lack a general description of the property (i.e., have a *parcels with no values, parcels with no values (row)*, or *no data available* attribute within the DESCRIPT attribute field). This type of assessment may also be used in determining the completeness of the database. Blunders in attributes would be represented by obviously incorrect data. Where landowners live outside of the United States, the country associated with their mailing address is placed in the OSTATE field along with state names. In this database, there are several parcels that indicate that the owners live in either *Belgioum, Belgique, Belgium,* or *Belguim*, according to the OSTATE attribute field. Certainly, at least two of these are correct, depending on a person's perspective of the country's name, but *Belgioum* and *Belguim* might be considered blunders or errors in attribute consistency. Whether these are random or systematic problems is difficult to ascertain. Issues with the consistency of attribute information may be further observed if one were to conduct a query of the land use descriptions. For example, to understand the extent of multifamily housing units across the state, one could conduct a query, but would need to use both *multifamily* and *multifamily less than 10 units* attributes within the DESCRIPT attribute field. In the OSTATE field one may also find *AL* and *ALABAMA* to represent the state where this landowner is located. These are examples of how standards for data through the database design can help avoid map error. These issues aside, as we noted, the database contains over 700 million cells, which may require a significant effort to address attribute accuracy problems. For smaller maps and GIS databases, these types of attribute issues can be addressed through careful examination of the information.

Map Standards

Many governmental organizations, and a few larger private land management organizations, have developed mapping standards, often posed as desired (or required) specifications for maps that are published by the organization. In some respects, these might be viewed as procedures to employ when developing a map to ensure the quality of the end product. The reach of these standards can be as far as the specification of a coordinate and projection system, a datum, and the symbology (color and style) of features displayed within a map. Standards and rules regarding the selection of place names, their placement with respect to a landscape (or water) feature, and their type style (font), size, color, and case (upper and lower) might all be defined for standard map products produced by an organization (e.g., Geoscience Australia, 2012). Organizations that are intimately associated with the use of maps, such as the International Orienteering Federation, have also developed standards for the development of maps used for specific purposes, such as orienteering competitions. In this case, these standards describe levels of data accuracy, the type style, size, color, and case of annotation and labels, as well as standard symbols for landscape features and their precise representation on maps (International Orienteering Federation, 2017).

Conformance to positional location standards is assessed for various forms of coordinate reference systems (e.g., geodetic, projected, and derived), depending on the circumstance, and *well-known text* examples have been provided to describe these (Open Geospatial Consortium, 2015). Standards have also been devised for the definitions of features used in geographic databases that describe land and facilities. For example, the Open Geospatial Consortium (2016) describes a *property unit* as

> *A Property Unit is a unit of ownership in land. . . . In some jurisdictions, a land parcel is a single, contiguous area of land, possibly including water. In other jurisdictions, a single land parcel may contain several disjoint areas of land. . . .*

Verification tests may subsequently be applied to inspect and ensure that the requirements are met, and that data conform to the standards. For reasons of tactical and operational importance, accuracy requirements of point positions, in both horizontal and vertical directions, have also been published by military agencies in efforts to standardize definitions employed and to document the processes that might be used to understand positional accuracy (U.S. Department of Defense, Defense Mapping Agency, 1990).

Standards for data capture accuracy are also often used by governmental and larger land management companies, which in effect indicate that data collected will not be used unless it meets all accuracy and precision thresholds. This is a difficult policy to accept, particularly long after data have been collected, since nonconforming data are essentially discarded. Standards may also be developed to manage the *topology* of data, or the spatial relationships among landscape features. These types of rules might involve ensuring that lines which should not intersect, such as contour intervals, do not intersect in the spatial database. They might involve ensuring that polygons which should not overlap, such as water bodies, do not overlap in the spatial database. As might have been suggested in the discussion of a minimum mapping unit and a minimum segment length, standards may be employed to reduce or adjust a GIS database to conform to these issues. In addition, for line features, standards may be employed to limit the distance between two consecutive vertices that describe a line, or to limit the total length of a line segment (distance between the two end nodes).

Translation 31

You are developing a map of wetlands for the county or province within which you work. Your supervisor suggests that on this map, all wetlands should be displayed. But, you know that the database of wetland polygons you are using employed a minimum mapping unit of 1 hectare. Explain to your supervisor what this means, and why the map you are developing may not illustrate all of the wetlands in the county or province.

With nearly every aspect of geographic data used in maps, as well as the process of compiling a map, there may have been developed a standard. In the United States, the National Geospatial Program and the Federal Geographic Data Committee have compiled many different standards for geospatial data that enable the sharing of data, and for which may be followed in the development of maps. For example, standards for design of geologic map features have been developed to assist in the development of maps by the U.S. federal government or federally funded contractors (Federal Geographic Data Committee, 2006). These standards attempt to ensure that geologic maps are consistent in appearance and quality; thus, the maps will facilitate effective communication and comprehension of information by map users by requiring map developers to use familiar map symbols. Examples of standards for common transportation features (roads and railroads) are illustrated in Fig. 8.15, along with their associated cartographic specifications. Standards for the development of topographical maps have also recently been developed (Davis et al., 2019) that detail how map data should be collected, compiled, stored, and displayed. The standards may not only address the design of the symbols, but also the colors and patterns associated with the symbols. However, standards may not be rigid, as they may suggest that modifications to the sizes, colors, or line weights of symbols are permitted as long as the modified appearance is not similar to other symbols contained on the map.

The United States *National Standard for Spatial Data Accuracy* (Federal Geographic Data Committee, 1998) is another standard that pertains to the positional accuracy of data presented in maps and other geospatial databases, to inform users of data quality and accuracy. Interestingly, the standard does not suggest the threshold values that would insinuate a map was of high or low quality, and suggests that organizations develop these for their own product and application purposes.

Inspection 39

Locate the International Orienteering Federation (2017) *International Specification for Orienteering Maps*. In your own words, through a short memorandum, briefly describe their thoughts on map accuracy and legibility.

Another example of a national standard is one for vegetation classification (Federal Geographic Data Committee, 2008), which represents a process by which U.S. federal organizations who manage local and regional vegetation inventories can consistently classify and code these resources so that they can easily be aggregated upward to facilitate assessments of national vegetation conditions. These are essential procedures an organization would follow to adhere to the hierarchical nature of vegetation classification, through the phases of data collection, analysis, and presentation. Further, standards on the naming conventions and characteristics of soil survey maps have been developed for organizations who collaboratively are developing the soil survey of the United States (Federal Geographic Data Committee, 1997). This standard pertains mainly to the attribute data (name, data type, value ranges, precision, etc.) that describe soil map units (polygons), including the physical and chemical properties of soils and other aspects of soils that are important in interpreting the quality of soils. Further, mapping standards have been devised to support mapping and classification of wetlands in the United States, and provide specifications for minimum data quality components for potential inclusion of mapped wetlands into a national database (Federal Geographic Data Committee, 2009).

The American Society for Photogrammetry and Remote Sensing (ASPRS) has also published a number of standards, one of which describes the accuracy thresholds for common mapping applications. In the *positional accuracy standards for digital geospatial data* (American Society for Photogrammetry and Remote Sensing, 2014) horizontal and vertical accuracy standards for digital data are described using the RMSE as the basic metric of quality for satellite or aerial imagery of varying characteristics (pixel size, etc.).

Ref no	Description	Symbol	Cartographic specifications*
28.1	Highway (generic)		*lineweight .325 mm; line color 70% black*
28.2	Road or street (generic)		*lineweight .25 mm; line color 50% black*
28.3	Primary highway, undivided (Class 1)		outlines: lineweight .125 mm in 100% black .5 mm fill: lineweight .5 mm; line color 100% red
28.4	Primary highway, divided by centerline (Class 1)		.5 mm — .5 mm
28.5	Primary highway, divided by median strip (Class 1)		.5 mm — spacing may vary
28.6	Secondary highway, undivided (Class 2)		fill: dash length 3.0 mm; space 3.0 mm
28.7	Secondary highway, divided by centerline (Class 2)		.5 mm — .5 mm
28.8	Secondary highway, divided by median strip (Class 2)		.5 mm — spacing may vary
28.9	Light-duty road, paved (Class 3)		outlines: lineweight .125 mm in 100% black .5 mm fill: lineweight .5 mm; line color 50% black
28.10	Light-duty road, gravel (Class 3)		.5 mm fill: dash length 3.0 mm; space 1.5 mm
28.11	Light-duty road, dirt (Class 3)		.5 mm fill: dash length 1.5 mm; space 3.0 mm
28.12	Street in urban area; light-duty road, composition unspecified (Class 3)		lineweights .125 mm .5 mm
28.13	Unimproved road (Class 4)	============	lineweights .125 mm .5 mm ============ dash length 1.25 mm; space .5 mm
28.14	Four-wheel-drive road (Class 5)	4WD ============	lineweights .125 mm .5 mm 4WD ======= HI-5 dash length 1.25 mm; space .5 mm
28.15	Trail	- - - - - - - -	lineweight .15 mm - - - - - - - - dash length 1.25 mm; space .5 mm
28.16	Interstate route marker	(70)	H-6 (100% red) draft as shown (70) lineweight .2 mm; line color 100% red
28.17	US route marker	(25)	H-6 (100% red) draft as shown (25) lineweight .2 mm; line color 100% red
28.18	State route marker	(36)	H-6 (100% red) circle diameter 4.375 mm (36) lineweight .2 mm; line color 100% red
28.19	Railroad (single track)	—+—+—+—	all lineweights .125 mm 5.0 mm 1.0 mm
28.20	Railroad (more than one track)—showing number of tracks	4 TRACKS	all lineweights .125 mm 5.0 mm 1.325 mm 4 TRACKS .5 mm HI-5

FIGURE 8.15   Standards for the display of common transportation features. Source: *Federal Geographic Data Committee (2006).*

International standards have also been developed for maps. The International Organization for Standardization (ISO) is a federation of national standards organizations, essentially a clearinghouse for standards used throughout the world (ISO Technical Committee 211, 2009). The geographic information—data quality standard (ISO 19157:2013) describes procedures for assessing data quality in geographic databases, including the components to describe data quality, and the principles for reporting data quality. As with the United States *National Standard for Spatial Data Accuracy* this ISO standard does not define minimum acceptable levels of data quality. Another international standard (ISO/TS 19158:2012) provides a framework for the development of systems used to assess the quality of geographic databases as they are being developed. Data quality assurance standards that are specific to certain countries, such as Australia and New Zealand, are also consistent with this ISO standard (Joint Technical Committee IT-004, 2013).

Translation 32
The development of standards seems like a lot of tedious work, one member of your team commented over dinner one night, during a debate about the quality of GPS positions. Develop a succinct and clear argument why (1) an organization would want standards and (2) why an employee of an organization would want to follow those standards.

Some mapping standards go as far as stating who will be responsible for issues concerning the accuracy and completeness of maps. For example, the process of transferring the ownership of water rights in the State of Utah from one person to another requires identifying the professional who prepared the report of conveyance; this is the person responsible for the completeness of the associated map. Many other required map components, such as a north arrow and a scale, are also noted in these rules (Utah Office of Administrative Rules, 2019).

Assessment of Mapping Skills

Certainly, we all have opinions about art; mapping is a form of art even when it is a rote requirement of one's employment. A well-designed map that is visually pleasing may be able to communicate a message more easily than one that is not well-designed. These opinions we either express outwardly through oral or written feedback, or we hold internally, to be exposed only at our discretion. The concepts of scale, direction, area, location, color, annotation, legends, and other spatial relations all influence our interpretation and appreciation for a map. As we have suggested throughout this book, maps should be designed to communicate effectively a message to the intended audience or customer. Maps can be very creatively crafted using highly honed cartographic skills. They can also represent a very simple interpretation of a resource using basic mapping skills. Given a wide range of perspectives on map models, how does one assess the skills of the map developer? Alternatively, some might ask *should* one assess the skills of the map developer?

Reflection 38

Imagine that you manage an office where your employees frequently create maps to accompany contracts for proposed urban developments or recreation area management actions. Every one of the employees in your office has a responsibility to create the maps that accompany these contracts. If you had to assess the quality of your employees' maps, how would you do so?

At some point, we may want to assess the knowledge, ability, and even motivation of a map developer for making effective and aesthetically pleasing map products. Certainly, for jobs that require a considerable amount of mapping effort, a set of qualifications (or a test) may be employed to determine whether a person meets a minimum standard in terms of knowledge and skill of mapping processes (Estes et al., 2016). Whether a person is attempting to secure a job or whether they currently are in the act of making maps for professional or personal reasons, a rubric might be constructed to rank or classify the apparent level of core competency or foundational skills (those essential to the development of effective maps). This rubric might include assessments of past performance (the aesthetic quality and completeness of maps developed), the communicative ability of maps developed (the message being delivered clearly), and the ability to create a map in a timely manner, among other characteristics of a preferred map developer. Other metrics may be employed to assess the quality of certain types of maps through expert opinion or panel assessments (Rossiter et al., 2017). One could also envision an extension of cost-efficiency measures (Hengl et al., 2013) applied to general mapping expertise. An assessment of mapping skills can therefore become quite challenging if it is needed to evaluate quality or performance of map developers.

Concluding Remarks

Maps are simplifications (models) of real-world human and natural systems, and as a result, they can contain errors introduced accidentally or purposefully by the map developer. In these respects, the management of geographic data used in maps is important from a technological point of view (mapping systems working effectively), a legal point of view (e.g., liability and privacy concerns), and a data quality point of view. In general, map errors have been attributed to issues related to skill in map development, knowledge of map development processes, and adherence to standard mapping rules (Rodriguez and Plaia, 2008). One might need a set of formal or informal processes (e.g., omission/commission validation) to judge the quality of maps, or perhaps a process to assess the potential adverse consequences (risks) of making decisions when error or uncertainty is evident in the map (Agumya and Hunter, 1999). The fitness of a map for use in specific applications may be a function of the amount of error and uncertainty that the map developer imputes on the product versus the needs of the map user. However, when there are no alternative maps to refer to, assessing the fitness of a map may be a moot subject. Ultimately, as users of maps, we should be concerned about their ability to help us navigate the landscape, and help us understand what resources might be found across a landscape.

For better or worse, requirements for mapping accuracy may be reduced in today's postindustrial information era, where it is necessary to produce maps quickly using a variety of available source data with uncertain quality, and present them on digital devices (Lisitskii, 2016). Information for digital online maps is obtained from a number of sources, therefore problems may arise, for example, in attempts to pinpoint addresses due to feature location (and shape) generalization and the need to ensure consistency among feature naming conventions (Smith, 2008). Similar to static, printed maps, digital maps can contain errors—incorrectly labeled places and newly developed features (e.g., roads) that are missing. Fortunately, there may exist methods for society to propose changes to online mapping databases such as Google Maps (Google, 2019). Through services such as these, anyone can suggest changes or edits to the mappable databases; therefore mapping services employ people to verify the suggestions. Even though improvements in data using crowd-sourced concepts such as this may be realized, the quality of the data is difficult to verify (Estes et al., 2016). However, the use of these processes should result in improvements to the positional and thematic character (name, status, address, and hours of operation) of mapped features in digital mapping services (Nield, 2018).

References

Agumya, A., Hunter, G.J., 1999. Assessing "fitness for use" of geographic information: what risks are we prepared to accept in our decisions? In: Lowell, K., Jaton, A. (Eds.), Spatial Accuracy Assessment, Land Information Uncertainty in Natural Resources. Ann Arbor Press, Chelsea, MI, pp. 35—43.

American Society for Photogrammetry and Remote Sensing, 2014. ASPRS positional accuracy standards for digital geospatial data (edition 1, version 1.0. - November 2014). Photogramm. Eng. Remote Sensing 81 (3), A1—A26.

Arbia, G., Griffith, D., Haining, R., 1998. Error propagation modelling in raster GIS: overlay operations. Int. J. Geogr. Inf. Sci. 12 (2), 145—167.

Bettinger, P., Fei, S., 2010. One year's experience with a recreation-grade GPS receiver. Math. Comput. For. Nat.-Resour. Sci. 2 (2), 153—160.

Congalton, R.G., Green, K., 2009. Assessing the Accuracy of Remotely Sensed Data, Principles and Practices, second ed. CRC Press, Boca Raton, FL.

Davis, L.R., Fishburn, K.A., Lestinsky, H., Moore, L.R., Walter, J.L., 2019. US topo product standard, Chapter 2 of Section B, U.S. Geological Survey standards, Book 11, Collection and Delineation of Spatial Data. U.S. Department of the Interior, Geological Survey, Reston, VA. Techniques and methods 11-B2.

Estes, L.D., McRitchie, D., Choi, J., Debats, S., Evans, T., Guthe, W., et al., 2016. A platform for crowdsourcing the creation of representative, accurate landcover maps. Environ. Model. Software 80, 41—53.

Federal Geographic Data Committee, 1997. Soil Geographic Data Standard. U.S. Geological Survey, Reston, VA, Federal Geographic Data Committee Document Number FGDC-STD-006.

Federal Geographic Data Committee, 1998. Geospatial Positioning Accuracy Standards, Part 3: National Standard for Spatial Data Accuracy. U.S. Geological Survey, Reston, VA, Federal Geographic Data Committee Document Number FGDC-STD-007.3-1998.

Federal Geographic Data Committee, 2006. FGDC Digital Cartographic Standard for Geologic Map Symbolization. U.S. Geological Survey, Reston, VA, Federal Geographic Data Committee Document Number FGDC-STD-013-2006.

Federal Geographic Data Committee, 2008. National Vegetation Classification Standard, Version 2. U.S. Geological Survey, Reston, VA, Federal Geographic Data Committee Document Number FGDC-STD-005-2008 (Version 2).

Federal Geographic Data Committee, 2009. Wetlands Mapping Standard. U.S. Geological Survey, Reston, VA, Federal Geographic Data Committee Document Number FGDC-STD-015-2009.

Fisher, P.F., 1999. Models of uncertainty in spatial data. In: Longley, P.A., Goodchild, M.F., Maguire, D.J., Rhind, D.W. (Eds.), Geographic Information Systems, Volume 1, Principles, Techniques, Applications, and Management. John Wiley & Sons, Inc., New York, pp. 191–205.

Fisher, P.F., Tate, N.J., 2006. Causes and consequences of error in digital elevation models. Prog. Phys. Geogr. 30 (4), 467–489.

Fürst, J., 2002. Application of Geographical Information Systems in Operational Hydrology, Report to WMO RA VI. University of Agricultural Sciences, Vienna.

Geoscience Australia, 2012. Section 2 - National Topographic Map Series (NTMS) and General Reference Map Specifications. Commonwealth of Australia, Geoscience Australia, Canberra. <http://www.ga.gov.au/map-specs/topographic/v6/section2.html> (accessed 04.01.19).

Gergel, S.E., Stange, Y., Coops, N.C., Johansen, K., Kirby, K.R., 2007. What is the value of a good map? Example using high spatial resolution imagery to aid riparian restoration. Ecosystems 10 (5), 688–702.

Google LLC, 2019. Google Maps. Google LLC, Mountain View, CA, <https://maps.google.com/> (accessed 26.05.19).

Haining, R., Arbia, G., 1993. Error propagation through map operations. Technometrics 35 (3), 293–305.

Hengl, T., Nikolić, M., MacMillan, R.A., 2013. Mapping efficiency and information content. Int. J. Appl. Earth Observ. Geoinf. 22, 127–138.

IDON Technologies, Inc., 2016. Standards User's Guide for Geographic Information. Natural Resources Canada, Ottawa.

International Orienteering Federation, 2017. ISOM 2017 International Specifications for Orienteering Maps. International Orienteering Federation, Karlstad.

ISO Technical Committee 211, 2009. ISO/TC 211 Geographic Information/Geomatics. Standards Norway, ISO/TC 211 Secretariat, Lysaker. ISO/TC 211 N 2705.

Joint Technical Committee IT-004, 2013. Australian/New Zealand Standard: Geographic Information - Quality Assurance of Data Supply. Standards Australia Limited/Standards New Zealand. AS/NZS ISO 19158:2013.

Krause, K., 2011. A small contribution in the fight against immappancy. <https://commons.wikimedia.org/wiki/File:True_size_of_Africa.jpg> (accessed 30.05.19).

Langford, W.T., Gergel, S.E., Dietterich, T.G., Cohen, W., 2006. Map misclassification can cause large errors in landscape pattern indices: examples from habitat fragmentation. Ecosystems 9 (3), 474–488.

Lisitskii, D.V., 2016. Cartography in the era of informatization: new problems and possibilities. Geogr. Nat. Resour. 37 (4), 296–301.

Liu, E., Shi, W., 2016. A measure of average error variance of line features. Cartogr. Geogr. Inf. Sci. 43 (4), 321–327.

Ma, T., Chen, Y., Hua, Y., Chen, Z., Chen, X., Lin, C., et al., 2017. DEM generalization with profile simplification in four directions. Earth Sci. Inf. 10 (1), 29–39.

Metcalf, A.C., 2017. Who cares who made the map? La Carta del Cantino and its anonymous maker. e-Perimetron 12 (1), 1–23.

Nield, D., 2018. Google and Apple Maps Have Plenty of Errors. Here's How to Fix Them. Popular Science November 1, 2018 <https://www.popsci.com/edit-google-apple-maps> (accessed 27.05.19).

Open Geospatial Consortium, 2015. Geographic information - Well-Known Text Representation of Coordinate Reference Systems. OGC document 12-063r5, Version 1.0. <http://docs.opengeospatial.org/is/12-063r5/12-063r5.html> (accessed 06.01.19).

Open Geospatial Consortium, 2016. OGC® Land and Infrastructure Conceptual Model Standard (LandInfra). OGC document 15-111r1, Version 1.0. <http://docs.opengeospatial.org/is/15-111r1/15-111r1.html> (accessed 28.05.19).

Pascual, M.S., 2011. GIS Data: A Look at Accuracy, Precision, and Types of Errors. GIS Lounge November 6, 2011 <https://www.gislounge.com/gis-data-a-look-at-accuracy-precision-and-types-of-errors/> (accessed 27.05.19).

Rodriguez, F., Plaia, G.P., 2008. Mapping Errors. Coordinates March 2008 <https://mycoordinates.org/mapping-errors/> (accessed 27.05.19).

Rossiter, D.G., Zeng, R., Zhang, G.-L., 2017. Accounting for taxonomic distance in accuracy assessment of soil class predictions. Geoderma 292, 118–127.

Schaffer, G., Peer, M., Levin, N., 2016. Quantifying the completeness of and correspondence between two historical maps: a case study from nineteenth-century Palestine. Cartogr. Geogr. Inf. Sci. 43 (2), 154–175.

Schroeder, T.A., Wulder, M.A., Healy, S.P., Moisen, G.G., 2011. Mapping wildfire and clearcut harvest distur-
bances in boreal forests with Landsat time series data. Remote Sensing Environ. 115, 1421–1433.

Smith, C.S., 2008. Top Causes of Errors in Online Mapping Systems. Search Engine Land April 7, 2008 <https://
searchengineland.com/top-causes-of-errors-in-online-mapping-systems-13715> (accessed 27.05.19).

U.S. Department of Agriculture, Natural Resources Conservation Service, 2017. Web Soil Survey. U.S. Department
of Agriculture, Natural Resources Conservation Service, Washington, DC. <https://websoilsurvey.sc.egov.
usda.gov/App/HomePage.htm> (accessed 29.12.18).

U.S. Department of Defense, Defense Mapping Agency, 1990. Department of Defense Standard Practice:
Mapping, Charting, and Geodesy Accuracy. U.S. Department of Defense, Defense Mapping Agency, Fairfax,
VA. MIL-STD-600001.

U.S. Department of Justice, 2016. A Map of the United States Federal District Court Coverage. U.S. Department of
Justice, Washington, D.C. <https://www.uscourts.gov/about-federal-courts/court-role-and-structure>
(accessed 30.05.19).

University of Florida GeoPlan Center, 2018. Florida Parcel Data Statewide - 2017 (Uncompressed ESRI 10.4.1
FGDB, 2.2 GB). University of Florida GeoPlan Center, Gainesville, FL.

Utah Office of Administrative Rules, 2019. Utah Administrative Code, R655-3-5, Maps and Mapping Standards
for Reports of Conveyance. Office of Administrative Rules, Salt Lake City, UT. <https://rules.utah.gov/
publicat/code/r655/r655-003.htm#T5> (accessed 28.05.19).

Wood, D., 2015. This is not about old maps. Cartographica 50 (1), 14–17.

9

Maps in Popular Culture

Popular culture encompasses various forms of expression, such as music, dance, fashion, art, design, cinema, and television, that are used by, and that otherwise characterize a society (Cook and Taylor, 2013). Maps, as we will see, are pervasive objects for complementing and supporting these forms of expression. People can develop an emotional connection with maps, and in conjunction with their imaginative capacities, people can also develop and foster a fascination with maps as they are used in the different forms of popular culture (Brett and Laddusaw, 2017). The use of maps in popular culture is similar to, but much broader than, the field of *cultural mapping*. Cultural mapping is narrower in scope and purpose, and has the intent of helping people express a sense of self, place, and belonging (Cook and Taylor, 2013). For example, one may argue that the map being made in Fig. 9.1 is in fact a cultural map that represents the location of people's favorite markets. However, in a broader sense, it might also be considered a map of popular culture in that it represents a form of expression, albeit using very basic tools (pins and a printed map). Movies, television shows, games, and other forms of media designed to entertain or influence human behavior often use maps as objects to establish context. Some of these

FIGURE 9.1   Pinning locations of favorite farmer's markets during a U.S. Department of Agriculture Open House on the National Mall in Washington, D.C., January 18, 2013. *Source: U.S. Department of Agriculture through Publicdomainfiles.com.*

maps may have no basis at all in reality, and therefore may be outside the realm of cultural mapping.

Reflection 39
Think of a map that was used recently in television news show. Consider the quality of the map and the use of the map by the cast of the show. Describe the theme of the map you selected, how much area of the Earth it represented, and the features that were provided. Was the map of value in supporting the oral commentary provided by the news people?

In contrast to the blunt and pernicious ways maps have been used to characterize a society, our treatment of the use of maps as objects to support popular culture is much lighter in this book. At times, popular culture maps illustrate real landscapes portrayed within a fictional setting, or they more purposefully illustrate fictional landscapes (Brett and Laddusaw, 2017). At one extreme end of the spectrum, maps of the environment inhabited by *Pac-Man* or *Tolkien's Middle-earth* certainly represent fictional landscapes or maps of imaginary places. Toward the other end of the spectrum, the dynamic, simple map used to describe the route driven by the Griswolds in the movie *National Lampoon's Vacation* would represent the use of a real map to support a fictional adventure. At one time in the early 1980s, the most common popular culture maps (other than road maps) were those found on television, or in magazines, newspapers, or advertisements (Tyner, 1982). Today, thematic maps that illustrate real landscapes are still pervasively embedded in these media, but are also associated with many Internet websites and video games.

Maps as Expressions of Human Condition

As graphical devices, maps are often designed and used to attract the attention of the general public by portraying phenomena and conditions in a manner that intrigues interest. They are often created to address the tastes and perspectives of large masses of ordinary people, many of whom may not be interested in adherence to cartographic principles. For example, some maps used in popular culture distort the sizes of geographical land areas to emphasize relative values of resources of the demographics of the human population. These *cartograms*, as we noted in Chapter 2, Map Types, represent abstract maps of phenomena or circumstance. Other maps simply express how, as society, we may currently think about or utilize resources across a landscape. Through the use of creative color schemes these thematic maps often reflect the ways in which society behaves or thinks, and the ability of society to produce or consume resources. In developing these products, a map developer may omit common map elements so that map users can focus on the main message of the map, and not on cartographic minutiae.

Inspection 40
Examine the map of France (Fig. 9.2) that illustrates the additional capacity of each region to convert sunlight into electricity in 2013. As a potential user of this map, what information seems to be lacking that would help you interpret or otherwise comprehend the message that was meant to be communicated?

As we have seen over the last few years, political correspondents communicating through television and the Internet often use thematic maps to illustrate potential (or real) outcomes of political elections. Media coverage of recent national elections in the United States emphasized the colors blue and red to illustrate the political party receiving the majority of votes within various levels of political geographical units. For example, in one case the counties across the United States were colored blue when the majority of votes were cast for the Democrat Party, and were colored red when the majority of votes were cast for the Republican Party (Fig. 9.3). Thematic maps have also been developed to communicate geographical differences in other human preferences. One recent map of the United States indicates the preference for religious affiliation by state (Fig. 9.4). Whether the data are correct is another matter to be considered, but the map itself is interesting in that regional differences in religious affiliation might be concluded. Thus, maps such as these can support a forum for discussing how regional populaces differ.

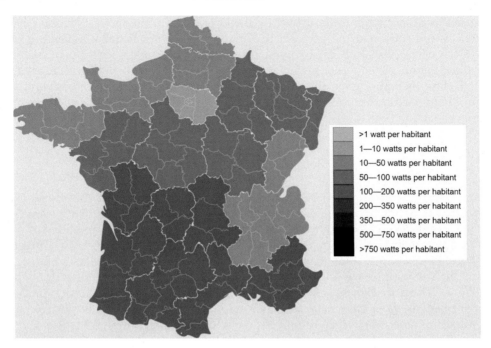

FIGURE 9.2   Photovoltaic installations, in watts per person, for various regions of France in 2013. *Source: Rfassbind through Wikimedia Commons.*

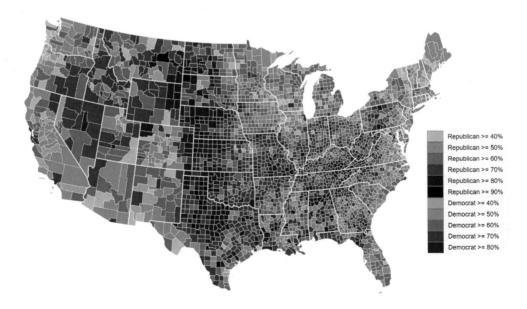

FIGURE 9.3   Map of 2016 US presidential election results, by county for the conterminous portion of the country. *Source: Magog the Ogre through Wikimedia Commons.*

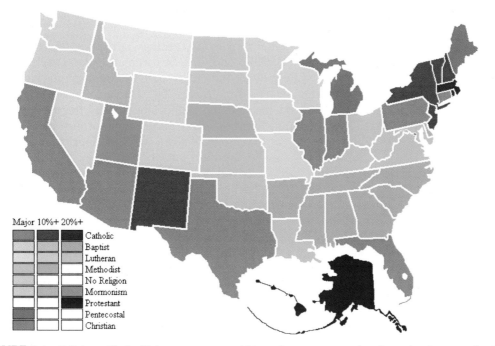

FIGURE 9.4   Religion with the highest percentage within each state, compared to the national average for that religion. *Source: Crayton through Wikimedia Commons.*

Diversion 46

Human use of the landscape is pervasive. Therefore, create a map using your preferred mapping program that presents the tree-covered areas as a percent of the land area of countries in North America, Europe, or some other area of interest to you. The data on forest area are provided by the Food and Agriculture Organization of the United Nations (2017), and are available on this book's website (mapping-book.uga.edu). Using a color scheme you feel appropriate, give each country a color value that is representative of the relative amount of tree-covered area it contains.

People who make maps, whether they consider themselves true cartographers or mapping neophytes, can attempt to exert influence and power through the creative display of real or fictional features (Lin, 2015). The ultimate expression of power is when maps are used to express political agendas or national identity, and in some sense these types of maps can be considered tools of propaganda (Yan, 2007). Yet maps can also be persuasive in the sense that they attempt to convince a map user that they truly represent the state of the world, perhaps to the detriment of other points of view (Tyner, 1982). While scientific maps are developed under the notion that facts and truth are paramount, with persuasive maps facts and truth may be irrelevant; they are often designed to affect the map user's feelings and emotions (Muehlenhaus, 2014a).

Maps in Cinema

Maps can be used as objects in cinema (movies) to facilitate intriguing dialog and suspense, and in essence to shape reality. While one form of communication (the map) may seem to be embedded in another (the movie), they both have explanatory, narrative, and argumentative purposes (Muehlenhaus, 2014b). In the 2002 Universal Pictures film *The Bourne Identity*, Alexander Conklin and his research technicians tracked the movements of Jason Bourne and Marie Kreutz through France using a printed map of Europe as a theatrical prop, pinpointing the places where Marie had lived in the 6 years prior to the events that were unfolding in the movie. In the background, attached to the wall of their office, rested a computer screen showing a rather trivial digital map of the world (white outlines of continental boundaries against a black background) that appears to have no relevance to the story other than to suggest that finding Bourne required a worldwide perspective. Earlier in the film, Bourne also used a printed road map of Paris to quickly understand the options for furiously eluding police during a scene that involved an extended car chase. Imagine having to do the same today using a GPS device. Dramatic use of a map as a prop can also be seen at the very end of the 20th century Fox film *Cast Away*, released in 2000. In this movie, after 4 years stranded on a deserted island, Chuck Nolan (played by Tom Hanks) returns home and searches for meaning and acceptance. In the final scene, he spreads a road map across the hood of his vehicle, and contemplates his options.

Reflection 40
With the exception of the movies mentioned in this chapter, in what movie was the last map you noticed? In what format was it offered (digital, print)? What geographical extent did it seem to cover? How was it used by the characters in the movie?

Maps used in movies may also relate to places that exist nowhere in the real world. The maps of these imaginary places often include roads, landmarks, and topography, yet arise solely from the mind of one person or a small group of people. Perhaps the following maps from recent movies are familiar: a map of Gotham City in association with recent Christopher Nolan *Batman* movies, a map of *The Realm of Middle Earth* to provide context to the *Lord of the Rings* trilogy, and a map of *Isla Nublar island*, the setting for several *Jurassic Park* movies. These maps are essentially cartographic works of art, expressions of space and place that may assist society in understanding the landscape within which the fictional movies are centered.

Maps in Television

Maps are important objects for influencing and reporting issues of popular culture within the television industry. Television news organizations rely extensively on maps in the presentation of weather-related information (Fig. 9.5). Current weather conditions and future projections of conditions are standard maps that one can view today during any

FIGURE 9.5   A television weather map from the American Forces Network Weather Center. *Source: Plueger (2017).*

news program. When severe weather or natural disaster events are imminent, or are actually occurring, various other dynamic types of weather maps may be developed to illustrate information of importance to society. For example, these maps may be developed very quickly to illustrate the recent and projected paths of a tornado, or the location of an earthquake. During the political election season, maps can often be found within television news reports displaying in opposing colors the potential outcomes by geographic region. These types of thematic maps provide people with projections of elections based on public opinion polling results. Journalists use them to communicate messages regarding the political dimension of our lives; they interpret public opinion from them often inferring how location affects public opinion (Chen and Perrella, 2016).

In television shows such as HGTV's *Living in Alaska*, often one will be presented with a digital map of the city or town featured in the show, and a north arrow to provide viewers with a sense of orientation. Similarly, Discovery Channel's *Gold Rush* and History Channel's *The Curse of Oak Island* often provide digital maps to add context to each episode's story. For example, on the *Rock Solid* episode of *The Curse of Oak Island*, a massive star map of the Taurus constellation was overlain upon a digital map of the island to suggest a pattern related to the presumed treasure buried there. Further, paper maps of the island are often referenced when discussions involve a drilling pattern or when historical search efforts are described. On the Travel Channel's *Josh Gate's Destination Truth*, an episode titled *Pterodactyl in Papua New Guinea and Chile's Monster* featured a hand-drawn map of cave locations in Papua New Guinea.

Diversion 47
This is a difficult exercise. Spend 1 hour watching television. Change the channels as often as you want. However, pay close attention to the appearance of a map.
Develop the following information for one full hour of time: when you see a map, describe its format (printed, digital), geographical scope (local, worldwide, etc.), and how it was used.

Dynamic digital maps, or maps that involve moving map features, are widely used by television shows today. The dynamic maps used by television news programs to illustrate current weather conditions are a good example of these. Another example is the maps indicating the current position and direction of travel of the taxicab in the show *Cash Cab*. These larger scale, digital maps allow one to develop a mental reference of the cab's position relative to major streets and avenues within the geography of New York City (Manhattan Island). The maps also highlight the path that the cab will be taking, and thus they provide a reference to the route and direction of travel as the host quizzes the contestants all the way to their destination.

Maps in Music

Maps are used at various times in song as metaphors for issues confronting people. Often, they are explicitly mentioned in the titles of songs. For example, the country song *Draw Me a Map* (Bentley and Randall, 2010), Dierks Bentley asks his love interest to draw him a map for the course or route that will lead back from heartache. The notions of maps guiding people through relationships can also be found in other contemporary songs. For example, in their adult alternative song *Maps* (Levin et al., 2014), Maroon 5 asserts that they are following *the map that leads to you* in their search for reconciliation with their partner. And in the adult contemporary song *Compass and Map* (Thicke, 2011), Robin Thicke suggests that in the stressful moments of relationship problems there may be no map one can follow to achieve a successful resolution.

Reflection 41
Does the mention of a map or a location (state and city) within a song help you visualize the context of the story? Can you think of an example?

Maps have also been used as explicit objects within the dialog among one or more people. For example, in the adult alternative pop/rock song *Get Out the Map* (Saliers, 1997), the Indigo Girls describe using a map to randomly decide where to travel (*get out the map and lay your finger anywhere down*). Finally, concepts of maps can be referred to in philosophical terms to describe the human condition; for example, R.E.M. refers to direction and color in their enigmatic alternative/indie rock song *Maps and Legends* (Berry et al., 1985). In other circumstances, the use of maps is simply mentioned in the lyrics of songs. For example, Alisan Porter, who was a contestant and the eventual winner of season 10 of *The Voice*, recorded the song *Down That Road* (Juber et al., 2016), where she sings

Let your heart draw a map in the stars . . .

This intertwining of music and maps, while perhaps only meant by songwriters to provide locational reference, can also result in a mental map exercise that helps educate society about cultural geography (Shobe and Banis, 2010).

Maps in Newsprint

In the early 20th century, when newspapers were the only mass media, artistically sketched maps accompanied stories of tragedies (e.g., the sinking of the Titanic, the San Francisco Earthquake, or the battles during the World Wars), wonders (polar explorations), and other undertakings (development of cities), and were meant to both influence public opinion and entertain (Cosgrove and della Dora, 2005). More recently, in the aftermath of the 2016 US presidential election, *The New York Times* printed maps that were intended to correlate voting outcomes and television show viewing habits (Larabee, 2017).

Inspection 41
Acquire the latest printed version of a newspaper from a city where you are currently situated. Browse each page and locate any maps that may be provided. How many maps did you locate? What was the geographical extent of each map? For what purpose was each map presented in the newspaper?

Maps can supplement many aspects of information contained in newspapers, they can expand upon weather forecasts, provide context for stories, and indicate the locations of merchants who have advertised. For example, in the Saturday, January 12, 2019 edition of *The Atlanta Journal-Constitution*, three maps were found:

- A small-scale (~1:40,000,000 scale) color weather map indicating average temperatures, cold and warm fronts, and potential precipitation events.
- A small-scale panchromatic map of the Atlanta area (~1:1,400,000 scale) indicating counties, major roads, and towns for real estate purposes.
- A small-scale (~1:1,000,000 scale) color map indicating the location of some select automobile dealerships.

In addition to the update of the weather map, the Sunday, January 13, 2019 edition of *The Atlanta Journal-Constitution* contained these three maps:

- A large-scale panchromatic map (~1:17,600 scale) illustrating the location of City Hall and the State Capital, for those who may be interested in traveling to downtown Atlanta to listen to hearings conducted during the legislative session.
- A medium-scale panchromatic map (~1:45,000 scale) illustrating the location of a wine shop with respect to nearby roads.
- A small-scale color map (~1:73,000,000 scale) illustrating locations in and around the Caribbean Sea and Atlantic Ocean where certain resorts might be found.

Maps in Magazines and Books

Magazines such as *Time, Newsweek, The Economist*, and even *Mad Magazine* and others often provide stylized, colorful maps in support of the stories offered in each issue. *National Geographic Magazine* has, for just over 100 years, published reference maps, cultural maps, and maps of other themes in the printed version of their magazine. At times, the magazine also included large, foldout maps and other supplemental maps to not only enhance their publishing business, but also to educate readers on geography. For example, a 1:5,000,000 scale map describing the geography and topography of Mexico appeared in 1911 (Fig. 9.6). More recently, in support of a *National Geographic Magazine* story on *The Other Tibet* (Teague, 2009), a map titled *Exploiting a Rich Frontier* (Mason and Hunseker, 2009) illustrated the paths of oil and gas pipelines, the locations of refineries, and the areas of coals reserves in the Xinjiang Uygur Autonomous Region, an area nearly two times the size of Texas. Although maps can be found in magazines as foldouts or as full-page supplements to the stories provided, they may also be found in magazine advertisements. For example, in the January 2019 issue of *National Geographic*, two maps could be found:

- A globe (~1:140,000,000 scale) illustrating North America, in an advertisement for the Cleveland Clinic.
- A piece of a world map in an advertisement for *National Geographic*, suggesting that the map will be offered for free as an incentive to purchase a subscription to the magazine.

Inspection 42

Acquire a recently printed news or popular science magazine of your choosing. Browse each page and locate any maps that may be provided. How many maps did you locate? What was the geographical extent of each map? For what purpose was each map presented in the newspaper?

With advances in computing power, magazines now offer online content, and in some cases, very vivid representations of maps that could be as engaging as their printed counterparts. For example, *National Geographic* offers a story on how Mars may have evolved over the past 3+ billion years through the use of global animated maps. The interactive media allows readers to scroll around the planet, viewing locations of crash sites of Earth-based space missions, as well as significant topographic features such as Olympus Mons, the highest point on the planet (Jacobs et al., 2016).

Like magazines, books may also provide maps as objects of popular culture that can help illustrate concepts and provide context for the story that is presented. With the exception of textbooks like this one, where maps are expected, other academic texts, such as *Introduction to Forestry and Natural Resources* (Grebner et al., 2013) or *Forest Plans of North America* (Siry et al., 2015), provide numerous maps regarding the location of countries and forest resources, as well as examples of forest management maps. These of course are considered nonfiction books. Fiction books may also include maps that describe real or imaginary places to support the story. For example, in Robert Louis Stevenson's *Treasure Island*, one may find a fictional map of the island that includes rudimentary depictions of topography and stream systems, as well as orientation and bathymetry (Fig. 9.7).

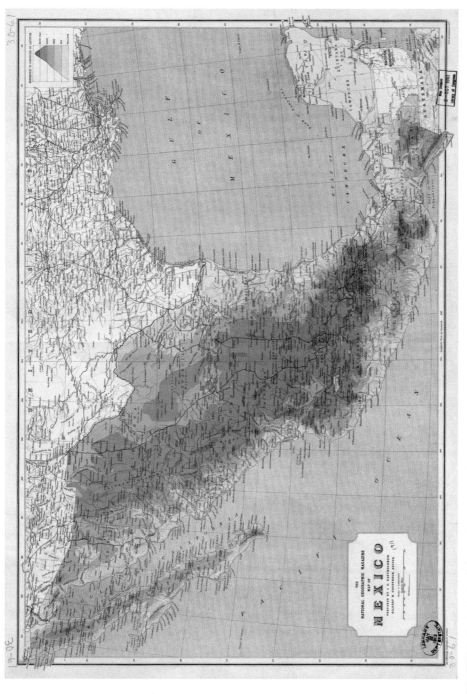

FIGURE 9.6   A *National Geographic Magazine* foldout map from 1911, featuring Mexico. *Source: Bartholomew et al. (1911).*

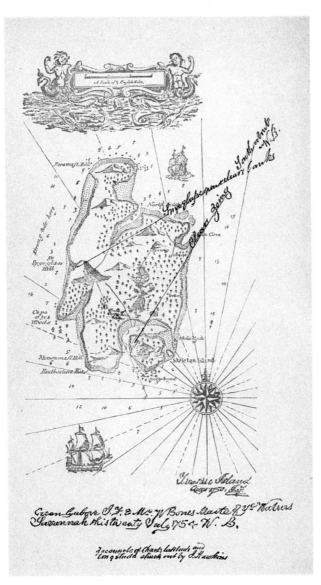

FIGURE 9.7 Map of *Treasure Island*. *Source: Stevensen (1883).*

Maps in Advertising

Maps that are found in advertisements should be considered persuasive maps, as they attempt to convince the reader of the advertisement of a true representation of a location, condition, or situation (Tyner, 1982). In the background of television commercials, a form of advertising meant to influence the commercial and social behavior, one might find well-placed maps. For example, it seems that many television commercials for cellular phone service contain a digital map illustrating the areas that their service covers. Postcards were

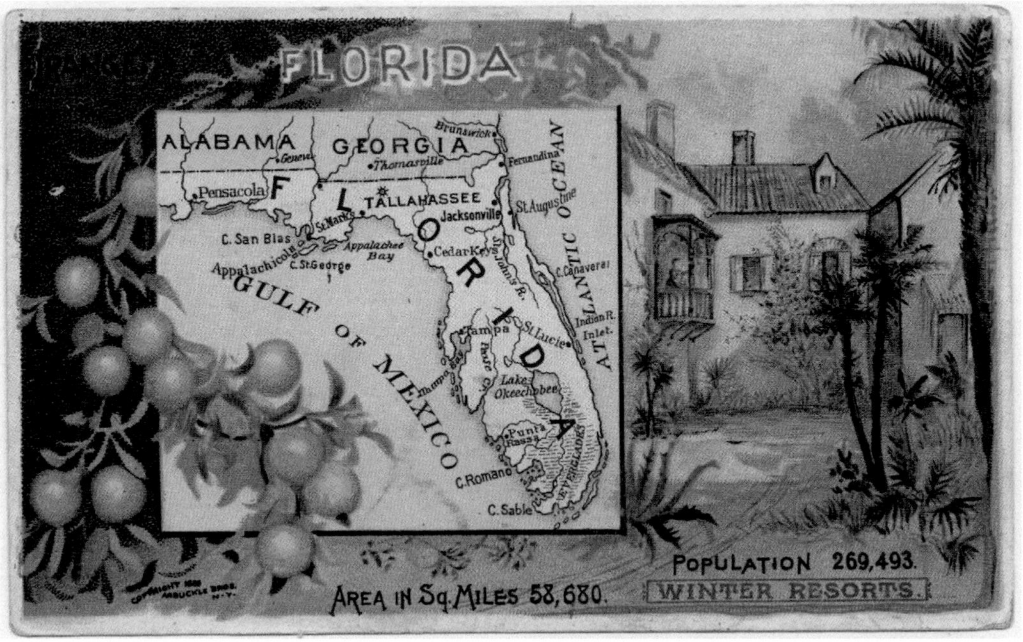

FIGURE 9.8   Florida postcard from 1899. *Source: Arbuckle Brothers Coffee, through the New York Public Library.*

also once widely used to advertise travel destinations. In addition to scenic pictures, post-cards employed (and still do) maps of places. In Fig. 9.8 one can see a map of the entire state of Florida, as it was popularly known in 1899, highlighting *winter resorts*.

Inspection 43
Examine closely the postcard presented in Fig. 9.8. If you were to create a postcard map of winter resorts in Florida for the year 2019, what would you change?

The logo of an organization might also be viewed as an advertisement of the brand. For example, a globe illustrating North and South America can be found in the center of the logo for the United States Marine Corps, suggesting its potential reach in protecting those who cannot otherwise protect themselves. Texas Instruments, a large American technology company, includes an outline of the State of Texas into their logo. Wikipedia utilizes a globe, stylized as a puzzle, to represent the brand. Even smaller companies, such as Stranco Products (www.strancoproducts.com), a Midwestern United States supplier of shrinkable tubing and electrical insulation products, utilize a map of the world in their company's logo.

Diversion 48
With the exception of organizational logos mentioned in this chapter, locate an organizational logo that integrates a map of a geographic entity (city, county, state,

province, country, etc.) in the design. Describe the prominence of the geographical portion of the image, and how it might be used to reflect upon the services provided by the organization.

As we mentioned earlier in the newsprint and magazine sections of this chapter, maps can also be embedded into printed advertisements. In many cases, these maps might vary considerably in scale, at one time illustrating a local neighborhood (small collection of streets) within which a business might be found, and at another time illustrating an entire country. The purpose, of course, is to illustrate *Here we are!* In other instances, maps embedded into advertisements may simply be used to assert importance. A company with several physical offices throughout a city, state, or region may use a map in its advertisement to indicate, in general, those locations. The purpose in this case may be to boast about their size or scope, or to illustrate that *We are never too far away from you!*

Maps on Digital Devices

Through cellular phones, tablets, and personal computers, society can quickly view interesting aspects of the world, facilitating the development of new types of maps and geographic knowledge (Dalton, 2013). Weather maps are some of the most frequently used maps delivered on digital devices, since we now have the technology to transmit this information electronically to computers or cellular phones. Daily weather forecasts are important guides that help inform the activities of our lives, but digital maps of more important weather events (Fig. 9.9) are of great value to society as well. Often, road maps, traffic maps, and weather maps are delivered across digital devices to allow one to understand their current situation (*Where am I? Where am I going? When will the rain end?*). Routing services are offered today through organizations such as *Mapquest* (Verizon Communications, Inc., 2018) and *Google Maps* (Google LLC, 2019). Assuming one has the appropriate digital device and that their service provider (e.g., AT&T, Verizon, etc.) adequately facilitates access to the routing services, a person may be able to determine very quickly the shortest path from one place to another. Furthermore, a person may be able to view the relative amount of vehicular traffic along these paths, and given alternative routes to select. This information may support decisions to diverge from the suggested path. Students attending our universities also use their digital devices to find their way to classes, particularly classes offered in unfamiliar buildings. Maps of the internal layout of these buildings are perhaps something mapping service providers will offer in the future, as locating the building may only be half of the fun in locating the correct classroom.

Translation 33
Imagine that you and your partner are lost in Rome, Italy. Your partner complains that if only you had a decent printed map of the city, you would be able to understand your location. Fortunately, you remember that your cellular phone contains a mapping application. In a few sentences, explain to your partner how this mapping application might be considered a *decent map alternative* to a printed map of the city.

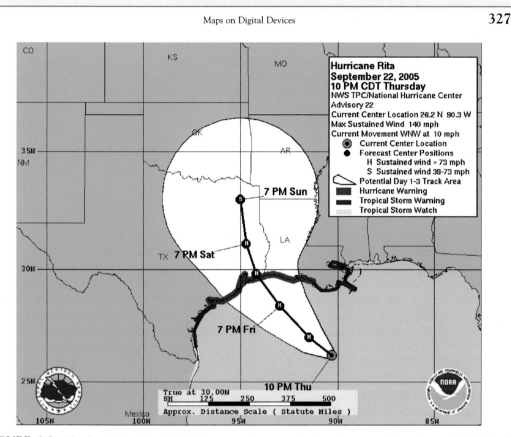

FIGURE 9.9 The September 22, 2005 3-day cone of uncertainty for Hurricane Rita. *Source: U.S. National Oceanic and Atmospheric Administration through Wikimedia Commons.*

Perhaps a step or two removed from the technical auspices of traditional cartography, the sharing of knowledge and subsequent mapping of experiences by ordinary people through Internet-based technologies have helped to form the concept of *public authoring* of maps (Angus et al., 2009). Of course, anyone with cartographic software, or even a pencil and paper for that matter, can author a map. However, these types of maps are developed by a collection of people, perhaps within some cartographic realm and using some data management protocols. No one person may be attributed as the author; however, a *community* of people would be. A community in this sense concerns people engaged in digital web-based interactions, and can involve Internet forums, news boards, social network sites, chatrooms, and other forms of interaction where people can exchange information on a topic without ever meeting in person (Häfele, 2009). These digital collaborations are prevalent today through crowd-sourced open mapping concepts. Public authoring of maps, and the general delivery of maps through digital devices, can foster the building of communities, public participation in mapping, advances in social knowledge, and (for better or worse) events such as organized resistance against governments and corporations.

The ability to create a map on a digital device makes map development relatively easy, and the perspectives map developers present with these maps can be distributed around the Earth virtually for free (Muehlenhaus, 2014a). Often within political or social contexts, how map users interact with online mapping tools, and what they produce with these, has been called *neogeography* (Elwood and Mitchell, 2013). The platforms for engaging people in the symbolization of features and events through online mapping processes can promote insight into various social phenomena that are otherwise unknown.

Maps in Video Gaming

The 1980 Namco/Midway video game *Pac-Man* is an iconic example of incorporating a hypothetical map into a video game. The Pac-Man travels around the map, and ideally consumes the Pac-Dots (gaining points), avoids the Ghosts (losing one of its four lives for every ghost encountered), and otherwise accumulates as many points as possible before all four lives are lost. One might view the Pac-Man landscape as roads or travel corridors within which movement can proceed in either direction. Unlike Pac-Man, modern video games are able to load different landscape data quickly and smoothly, shifting from global to local views and allowing gamers to navigate through hyperlocal, image-rich visualizations of a game's environment (Dalton, 2013).

Reflection 42
If you have ever played a video game, think about a map that may have accompanied a game or that may have been embedded within a game. If you have never played a video game, find someone who has, and ask them to think about a map that may have accompanied that game or a map that was embedded within the game. Which game was selected? How was the map used to facilitate the action or different levels within the context of the game?

The video game *Sid Meier's Civilization* was first developed in 1991 and a new version is scheduled for release in 2019. This is a map-based video game where the players exercise strategies to develop their civilization from prehistoric to futuristic societies. Maps provide the location of cities, towns, and various resources such as metal ore, coal, and oil that can be used to advance the civilization. As the cities gain size, more land falls within their boundaries. Players can attack one another to seize each other's cities, territories, and other resources, as one method for victory in this video game is conquest—to capture all of the existing cities or to capture territory that contains the resources needed for expansion. Thus, the game is based on the location resources contained on a map. *SimCity*, *Forge of Empires*, and other similar games allow video gamers to build cities, using maps as the base, locating buildings and roads across a landscape in an effort to earn fictional resources or currencies. Some of these games allow the user to both attack and defend their civilization (digitally, of course) from other users who have nearby civilizations, in their efforts to gain resources. Some of these games also allow users to act as a group in these efforts.

Concluding Remarks

Maps are pervasive tools in popular culture for assisting in the communication of messages (real or otherwise) meant to influence society. Whether the intent is to advance human society through explanation of geographic differences in phenomena or more simply to promote commerce or entertainment, our use of maps helps ground our perspectives on matters. As we have shown in this chapter, various forms of communication embed or include maps in attempts to further societal understanding. Certainly these examples were not inclusive of every form of communication human society has developed; for example, real or imagined maps have inspired the development of poems (Haft, 2014) and other literature (Laskow, 2018). In Holloway (2018) one can find a number of stone and plate lithograph *artist-editioned* maps (limited editions), accompanied by original poetry on geography, maps, nature, and emotion that binds the collection together. An excerpt of the poetry is provided below.

> ... In what sense can a map, the domesticated map, ever know the dreams of the heron in flight, in stillness. ... *Holloway (2018)*

While many maps in popular culture may be used simply for entertainment purposes, an uncritical acceptance of the map as a true model of a real system can be dangerous (Tyner, 1982). Popular culture maps can be devised for many reasons, even fictitious reasons. The amount of truth embedded in these types of maps needs to be considered carefully if important daily decisions are based on what they provide. Yet maps permeate, inform, and reflect popular culture through a variety of media, and some argue that much more can still be achieved in mapping aspects of popular culture (Larabee, 2017).

Reflection 43
We have mentioned a number of types of maps used in popular culture, admittedly from a North American perspective. Many maps used in popular culture adhere in a limited fashion to the mapping conventions (map components, color, etc.) we have noted earlier in this book. In thinking about your recent life experiences, describe a map developed for the purposes described in this chapter, and cite the source of the map as fully as possible. In other words, provide not simply an Internet address (or book, or magazine, etc.), but the date that the map was developed, the developer of the map, and the organization from which the map was distributed (if any). In what ways does the map contain a limited amount of traditional map conventions described in this book?

The selection of map elements, projections, coordinate systems, and even annotation should be made carefully in the development of a map of popular culture. Without knowing, one might easily create a product that offends a segment of society. Or, perhaps this was the intent of the map developer. In any event, whether a person likes a map or not, once a map is widely distributed it will undergo scrutiny. Therefore, it is important for a map developer to ponder the design of a popular culture map prior to distributing it, to mitigate or prevent false accusations about the intent of the map.

Maps, without question, will continue to be used as objects to support popular culture. People may include them in books or magazine articles to provide context for the story being told. Maps may be embellished in an attempt to inform or sway opinion of commercial products or world events. Maps can also be manipulated to falsely represent conditions, and fool people into changing their behavior. However, for more pleasant reasons, maps may also be used in video games, television, movies, music, and other media simply to entertain. Maps can be engaging, if constructed well. Color, symbology, themes, projections, and other map development considerations can be combined to help create a product to be admired, and perhaps a work of art.

References

Angus, A., Lane, G., Woods, O., Martin, K., 2009. Social tapestries. In: Cartwright, W., Gartner, G., Lehn, A. (Eds.), Cartography and Art. Lecture Notes in Geoinformation and Geography. Springer-Verlag, Berlin, pp. 353–357.

Bartholomew, J.G., Grosvenor, G.H., U.S. National Geographic Society, Edinburgh Geographical Institute, 1911. The National Geographic Magazine Map of Mexico. National Geographic Society, Washington, D.C. <https://www.loc.gov/item/2010593157/> (accessed 12.01.19).

Bentley, D., Randall, J., 2010. Draw me a map. Up on the Ridge. Capital Records Nashville, Nashville, TN.

Berry, B., Buck, P., Mills, M., Stipe, M., 1985. Maps and legends. Fables of the Reconstruction. Capital Records, Inc., Los Angeles, CA.

Brett, J., Laddusaw, S., 2017. Touring fantasyland: the "Maps of Imaginary Places" collection and exhibit at Cushing Memorial Library and Archives. J. Map Geogr. Libr. 13 (3), 280–299.

Chen, Y., Perrella, A.M.L., 2016. Interactive map to illustrate seat distributions of political party support levels: a web GIS application. Cartographica 51 (3), 147–158.

Cook, I., Taylor, K., 2013. A Contemporary Guide to Cultural Mapping, An ASEAN-Australia Perspective. Association of Southeast Nations, Jakarta.

Cosgrove, D.E., della Dora, V., 2005. Mapping global war: Los Angeles, the Pacific, and Charles Owens's pictorial cartography. Ann. Assoc. Am. Geogr. 95 (2), 373–390.

Dalton, C.M., 2013. Sovereigns, spooks, and hackers: an early history of Google geo services and map mashups. Cartographica 48 (4), 261–274.

Elwood, S., Mitchell, K., 2013. Another politics is possible: neogeographies, visual special tactics, and political formation. Cartographica 48 (4), 275–292.

Food and Agriculture Organization of the United Nations, 2017. FAOSTAT, Environment Land Cover All Data. Food and Agriculture Organization of the United Nations, Rome. <http://www.fao.org/faostat/en/#home> (accessed 26.05.19).

Google LLC, 2019. Google Maps. Google LLC, Mountain View, CA. <https://maps.google.com/> (accessed 26.05.19).

Grebner, D.L., Bettinger, P., Siry, J.P., 2013. Introduction to Forestry and Natural Resources. Academic Press, New York.

Häfele, T.A., 2009. Community Karlsplatz, Vienna video screen projection. In: Cartwright, W., Gartner, G., Lehn, A. (Eds.), Cartography and Art. Lecture Notes in Geoinformation and Geography. Springer-Verlag, Berlin, pp. 359–362.

Haft, A.J., 2014. The mocking mermaid: maps and mapping in Kenneth Slessor's poetic sequence The Atlas, Part four. Cartogr. Perspect. 79, 22–53.

Holloway, S.R., 2018. That map you love, that saved your life. Cartogr. Perspect. 90, 86–103.

Jacobs, B.T., Chwastyk, M.W., Jaggard, V., Treat, J., 2016. Rewind the red planet. National Geographic. <https://www.nationalgeographic.com/science/2016/11/exploring-mars-map-panorama-pictures/> (accessed 12.01.19).

Juber, I., Porter, A., Rise, E., 2016. Down That Road. Sony/ATV Songs, LLC, New York.

Larabee, A., 2017. Editorial: Mapping television and politics in popular culture. J. Pop. Cult. 50 (1), 7–8.

Laskow, S., 2018. How writers map their imaginary worlds. Atlas Obscura. October 22, 2018. <https://www.atlasobscura.com/articles/writers-maps> (accessed 23.03.19).

Levin, B., Levine, A., Malik, A., Tedder, R., Zancanella, N., 2014. Maps. V. Interscope Records, Santa Monica, CA.

Lin, W., 2015. Tracing the map in the age of Web 2.0. Cartographica 50 (1), 41−44.

Mason, V.W., Hunseker, M.B., 2009. Exploiting a rich frontier. Natl. Geogr. Mag. 216 (6). <https://docs.uyghura-merican.org/content_images/xerite/uygur-map-1400.jpg> (accessed 12.01.19).

Muehlenhaus, I., 2014a. Going viral: the look of online persuasive maps. Cartographica 49 (1), 18−34.

Muehlenhaus, I., 2014b. Looking at the big picture: adapting film theory to examine map form, meaning, and aesthetic. Cartogr. Perspect. 77, 46−66.

Plueger, J., 2017. 557th WW Airmen Lead in Global Forecasting for AFN. U.S. Department of Defense, Air Force, Offutt Air Force Base, Omaha, NE. <https://www.557weatherwing.af.mil/News/Photos/igphoto/2001750939/> (accessed 12.01.19).

Saliers, E., 1997. Get out the map. Shaming of the Sun. Sony Music Entertainment, Inc., New York.

Shobe, H., Banis, D., 2010. Music regions and mental maps: teaching cultural geography. J. Geogr. 109 (2), 87−96.

Siry, J.P., Bettinger, P., Merry, K., Grebner, D.L., Boston, K., Cieszewski, C. (Eds.), 2015. Forest Plans of North America. Academic Press, New York.

Stevenson, R.L., 1883. Treasure Island. Cassell & Company, Limited, London.

Teague, M., 2009. The other Tibet. Natl. Geogr. Mag. 216 (6). <https://www.nationalgeographic.com/magazine/2009/12/uygurs/> (accessed 12.01.19).

Thicke, R., 2011. Compass or map. In: Love After War. Star Trak, LLC, Virginia Beach, VA.

Tyner, J.A., 1982. Persuasive cartography. J. Geogr. 81 (4), 140−144.

Verizon Communications, Inc., 2018. MapQuest. Verizon Communications, Inc., Denver, CO. <https://www.mapquest.com/> (accessed 29.12.18).

Yan, S.-c, 2007. Mapping knowledge and power: cartographic representations of empire in Victorian Britain. EurAmerica 37 (1), 1−34.

Index